Engineering Design and Rapid Prototyping

To my wife Sonia and sons Arshya and Ariya

Ali Kamrani

To my parents, wife, and children Nada, Haidy, and Amr

Emad Abouel Nasr

Preface

Engineering design process consists of a set of activities arranged in a specific order with the clearly identified inputs and outputs. The process of engineering design is an iterative that supports the decision-making activities involved. Each activity in the process takes an input and transforms it into an output of some value defined by design specifications and objectives. The output of the process is either a product, a process, or a service. The objective of the process is to satisfy customer requirements and management objectives. This process is considered efficient when the output of the process satisfies general customers and defined requirements, meets management objectives and customer deadlines, and all these with less costs and resources. Establishing objectives and criteria, synthesis, analysis, construction, testing, and evaluation are considered as fundamental elements within the design process. Steps in the Engineering design process are illustrated in Fig. 1.

These steps and further described below:

1. **Identify the Need/Problem:** This step is not always realized by engineers. Problems are typically identified by the market and customers and then passed on to the engineering group to search and develop solutions.

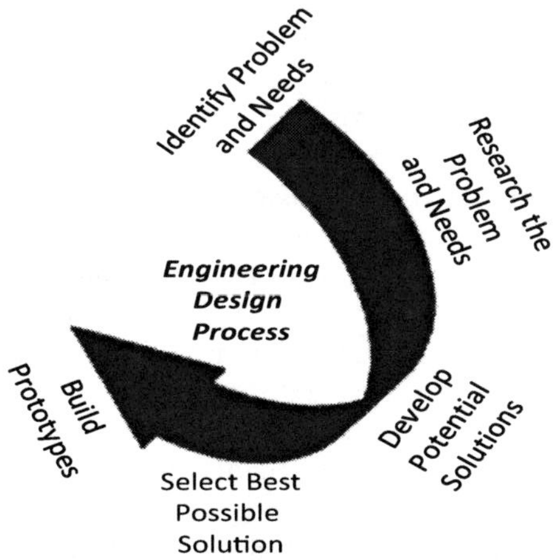

Fig. 1 Engineering design process

2. **Research the Need/Problem:** This step is of importance since a solution to the defined problem or portion of the problem may have already existed. Through proper research, time and money can be easily saved.
3. **Develop Possible Solutions:** In this step, the engineering group will propose different solution alternatives that will solve the problem. These possible solutions are made while taking into consideration the information found in the previous steps.
4. **Select Best Possible Solution:** The engineering group will then use the defined criteria and other methodologies to select the best possible solution. The parameters on which the engineering group determine if it is the best possible solution may vary depending on the defined constrains and/or the design criteria established at the beginning of the engineering design process.
5. **Construct a Prototype:** In this step, the engineering group proceeds to construct a prototype of the selected solution. Once the prototype is built, testing is performed to validate the proposed design solution. If the design does not meet the required performance, the engineering design process is repeated until a satisfactory solution is implemented.

New global economies and global markets changed business practices and focused on the customer as the major player in the economy. In order to compete in this fast-paced global market, organizations need to produce products that can be easily configured to offer distinctive capabilities compared to the competition. Furthermore, organizations need to develop and implement new engineering methods (e.g., modularity), and apply advanced techniques in design (e.g., Feature-Based Design) and technologies (e.g., Rapid Prototyping) to react rapidly to required changes in products and market trends and to shorten the product development cycle, which will enable them to gain more economic competitiveness. This requires that the tasks needed to develop products be made in parallel, starting at the early stages of product development. By developing such techniques, organizations will be able rapidly to design changed or new products, to change parts of a product, or to change manufacturing facilities to a new version of a product.

The concept of modularity can provide the necessary foundation for organizations to design products that can respond rapidly to market needs, and allow the changes in product design to happen in a cost-effective manner. Modularity can be applied to the engineering design processes to build modular products. An important aspect of modular products is the creation of a basic core unit to which different components (modules) can be fitted, thus enabling a variety of versions of the same module to be produced. The core should have sufficient capacity to cope with all expected variations in performance and usage. Components used in a modular product must have features that enable them to be coupled together to form a new product. Figure 2 illustrates the scope of modular design methodology.

Computer-aided design (CAD) and manufacturing (CAM) systems are based on modeling geometric data. The main advantage of CAD/CAM systems is the ability to visualize product design, and support design analysis and manufacturing activates. CAD/CAM systems need the standardization that gives them the ability

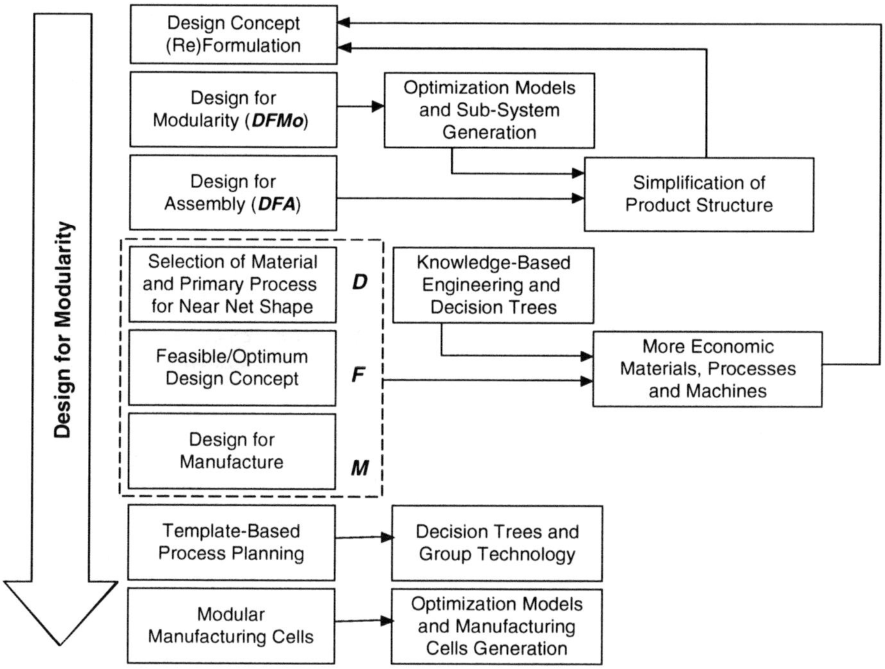

Fig. 2 Design for modularity life cycle

to communicate to each other. Different CAD or geometric modeling packages store the information related to the design in their own databases, and the structures of these databases are different from each other. As a result, no common or standard structure has been developed that can be used by all CAD/CAM packages. Therefore, a feature-based approach using IGES standard will provide the required standardization to achieve the integration between CAD and CAM. Figure 3 illustrates the stages that CAD/CAM and other advanced tools and technologies are used to support, and integrated and intelligent engineering design life cycle.

Rapid Prototyping (RP) is a technique for direct conversion of three—dimensional CAD data into a physical prototype. RP allows for automatic construction of physical models and has been used to significantly reduce the time for the product development cycle and to improve the final quality of the designed product. Before the application of RP, computer numerically controlled (CNC) equipments were used to create prototypes either directly or indirectly using CAD data. In RP process, thin-horizontalcross-sections are used to transform materials into physical prototypes. Steps in RP process cycle are illustrated in Fig. 4.

In the RP process, CAD data are interpreted into the Stereolithography (stl) data format. The stl is the standard data format used by all RP machines. By using "stl", the surface of the solid is approximated using triangular facets, with a normal vector pointing away from the surface in the solid. Within the last few years, corpora-

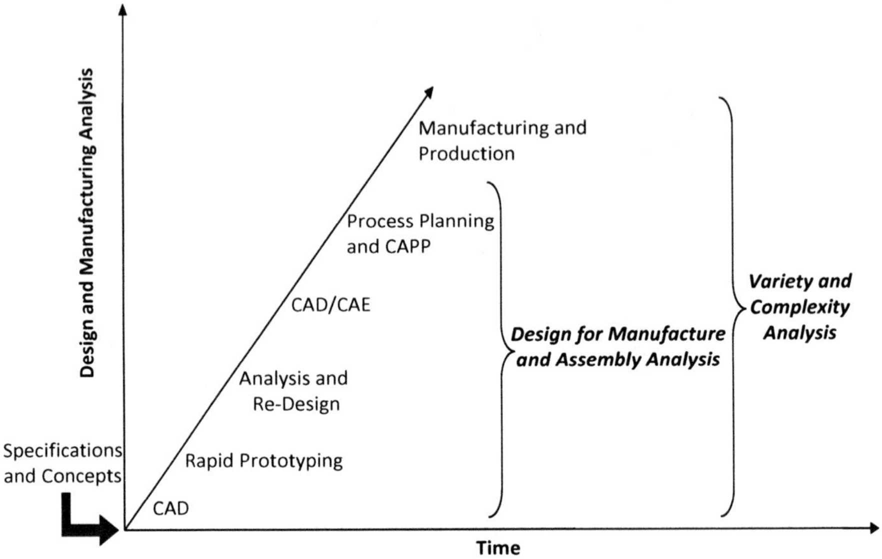

Fig. 3 CAD/CAM application for EDP

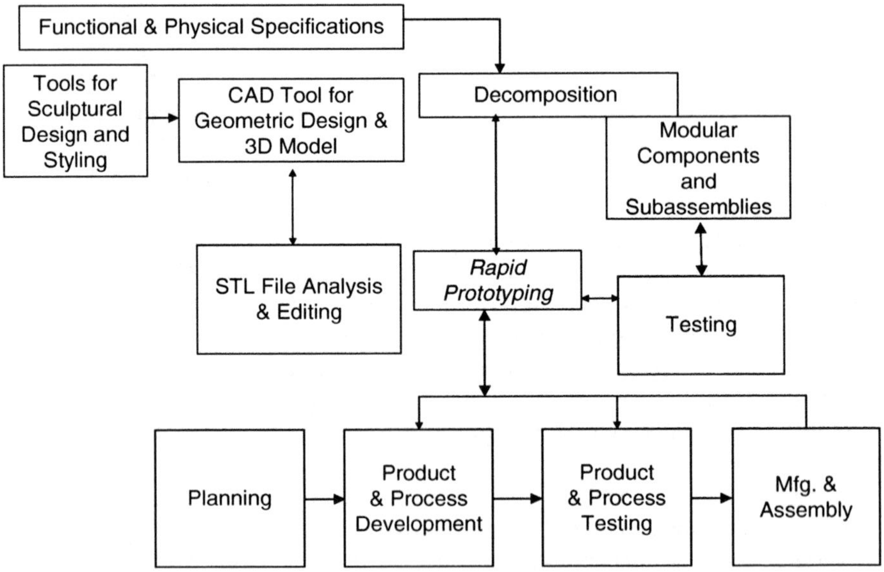

Fig. 4 Generic RP process cycle

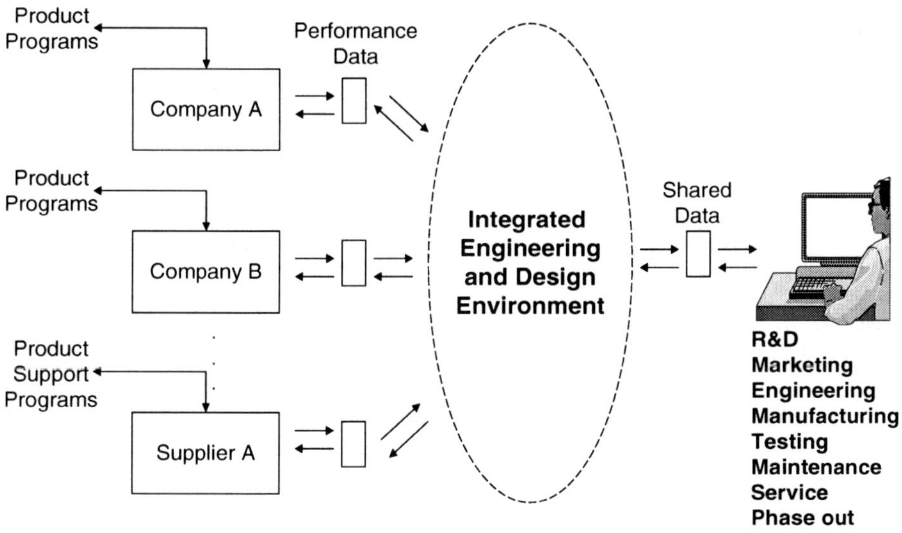

Fig. 5 Scope of an extended enterprise

tion has engaged in studies to integrate their distributed design processes, ranging from marketing to support. Recent government, academic, and industrial sector initiatives have sought advance technologies for developing and managing integrated product development environment. Many companies have established a distributed and concurrent design environment for their development purposes. Figure 5 illustrates the scope of today's extended enterprise.

The early phases of design life cycle are usually top-down. It is at the components level that the design process begins to integrate into sub-assemblies and assemblies till the design becomes a complete entity. The objective of the integration is not only to have compatible units, but also by integration all of the requirements as identified by the need statement are realized. This obvious goal has resulted in a very high cost of redesign and analysis due to poor analysis of design and manufacturing integration. By considering proper integration from the beginning, the problems with the final integration of activities will be significantly reduced.

This book offers insights into the methods and techniques that allow for implementing engineering designs by incorporating advanced methodologies and technologies in an integrated approach. This book is a collection of the latest methods and technologies, and it is structured in such a way that it could be used for a variety of advanced design and manufacturing courses. Using advanced CAD/CAM software tool (e.g., CATIA, and IDEAS) is encouraged to be used for courses that plan to use this book.

The book consists of three sections: (1) Product Development and Managements, (2) CAD/CAM and Features-Based Technologies, and (3) Rapid Design and Manufacturing. Below is a description of each chapter's content:

Chapter 1: Engineering Design and Innovations. This chapter provides the concept of information integration at every stage of product development, collaboration technology, which is needed for cooperative work. As the assistant of the design and development of new products, integrated design technology plays a very important role. The framework described in this chapter confirms design assumptions and predicts product performance in the early stages of the design process. This will result in a faster product development cycle, with lower associated costs, which can be achieved by eliminating the need to constantly build, test, and redesign.

Chapter 2: Product Development Process. This chapter discusses the product development life cycle, which can be defined as a sequence of all the required activities that a company must perform to develop, manufacture, and sell a product. These activities include marketing, research, engineering design, quality assurance, manufacturing, and a whole chain of suppliers and vendors. This chapter provides the concept of product development life cycle, the importance of product development during the product development life cycle, the major phases of product development process, benchmarking, the systematic procedure of generating concepts, and finally, a complete case study for understanding the presented concepts.

Chapter 3: Modular Design. This chapter provides the concept of modular design, which is a design technique that can be used to develop complex products using similar components. Modular design can be viewed as the process of producing units that perform discrete functions, and then connecting the units together to provide a variety of functions. This chapter provides the concept of modular design, modularity, product modularity representation, the modular systems development process, the product development process using the design structure matrix, and finally, a complete description for DSM building.

Chapter 4: Design for Modularity. This chapter presents a three-phase methodology, which is proposed for the development of complex products using the modularity concept. The proposed methodology matches the criteria set by the design for functionality, assembly, and manufacture. A detailed discussion of needs analysis, quality function deployment, product requirement analysis, product/concept analysis, product physical analysis, product functional analysis, product/concept integration, identification of the Impact of system-level specifications on general functional requirements, similarity index, and optimization-based solution methodology for grouping components into modules is presented in this chapter. Finally, genetic algorithm-based solution methodology and algorithm-based solution methodology for grouping components into modules are described for understanding the presented concepts.

Chapter 5: DFMo case study - Four-Gear speed reducer design. This chapter presents a design for modularity approach which is tested and validated using a test product. The selected test product is of moderate complexity to ensure that effort is focused on applying and validating the proposed approach rather than on attempting to understand a complex product. Maintaining moderate complexity will also show the potential for using the proposed approach in designing complex products or systems.

Chapter 6: Design for Manufacture and Assembly. This chapter presents the concepts of design for manufacturing and assembly. Product design is the first step in manufacturing and is where the critical decisions are made that will affect the final form and cost of the product. Design for manufacturing and assembly concentrates on simplifying designs while also evaluating assembly improvements to further enhance the overall design for manufacturability and quality. DFMA is a product development process and improvement methodology that provides a systematic process to achieve improved product design, robustness, and cost reductions through simplifications of the overall design. Finally, many case studies are shown for the purpose of understanding the concepts presented in this chapter.

Chapter 7: Computer-Based Design and Manufacturing. This chapter provides an overview of computer-aided design and manufacturing (CAD/CAM), the most important reasons of using CAD systems in the manufacturing environment, computer-integrated manufacturing (CIM), the implementation of the automation in the production organization, the role of CAD/CAM systems in the manufacturing facility, the CAM cycle in a feature-based design environment, and the different types of features.

Chapter 8: Feature Representations. This chapter presents discussions related to feature representation methodologies. It includes the definition of features, wire-frame modeling, surface modeling, boundary representation (B-rep), constructive solid geometry (CGS), and definition of interacting features. These methods are used to facilitate the feature recognition process.

Chapter 9: Feature Extraction Techniques. This chapter presents a brief review of the previous work on the related topics of feature representation and recognitions. The first section describes previous research efforts in the area of feature representation. Previous research in the area of feature recognition is described in the second section. In third section, a methodology for feature analysis and extraction of prismatic parts for CAM applications is developed and presented. This approach aims to achieve the integration between CAD and CAM.

Chapter 10: Engineering Materials - An Overview. This chapter provides an overview of material used for manufacturing. One of the most important aspects of material science involves the study of material's structure. Introducing a new product or changing an existing one involves many decisions. Although these decisions may seem to be independent, they will have a significant impact on the overall product life cycle, which will in turn influence cost, performance, and service.

Chapter 11: Geometric Dimensioning and Tolerancing. In this chapter, an introductory discussion is provided on the topics of geometric dimension and tolerancing, and its impact on selecting the right manufacturing processes. Also, improvements in measurements provide a need to better understand that variation is unavoidable in manufacturing. However, acceptable levels of variation will result in a good assembly.

It is important to understand the limits of this variation. After the development of GD&T, drawings became the main tool for communication among design, manufacturing, testing, etc.

Chapter 12: Rapid Prototyping. This chapter provides an overview of the rapid systems: stereolithography (SLA), solid ground curing (SGC), laminated object manufacturing (LOM), selective laser sintering (SLS), direct shell production casting (DSPC), and fused deposition modeling (FDM).

Chapter 13: DATA Mining Methodology and Techniques. The chapter provides the basic concepts of data mining. Data mining deals with the discovery of hidden knowledge, unexpected patterns, and new rules from large databases. It is regarded as the key element of a much more elaborate process called knowledge discovery in databases, or KDD, which is closely linked to data warehousing. Data mining can bring significant gains to organizations, for example, through better-targeted marketing and enhanced internal performance.

Chapter 14: Group Technology and Application. The chapter provides the concepts of group technology and cellular manufacturing. Grouping objects into groups based on the object features has been done using group technology (GT) approaches. Similar components can be grouped into design families, and new designs can be created by modifying an existing component design from the same family. The philosophy of group technology is an important concept in the design of advanced integrated manufacturing systems. Group technology is not an answer to all manufacturing problems, but it is a good management technique to standardize efforts and eliminate duplication. A well-designed classification and coding system may result in several benefits for the manufacturing plant.

Ali K. Kamrani, Ph.D., P.E.
Emad Abouel Nasr, Ph.D.

Acknowledgments

We thank our students that contributed to this book as well as our Publisher, Springer for giving us the opportunity to publish our work.

Contents

Part I
Product Development and Managements

Chapter 1
Engineering Design and Innovations

Abstract This chapter provides the concept of the information integration at every stage of product development, collaboration technology which is needed for cooperative work. As the assistant of the design and development of new products, integrated design technology plays a very important role. The framework described in this chapter confirms design assumptions and predicts product performance in the early stages of the design process. This will result in a faster product development cycle, with lower associated costs which can be achieved by eliminating the need to constantly build, test, and redesign.

Keywords Engineering • Design • Innovations • Product development

1.1 Introduction

During the past decade, global manufacturing competition has increased significantly. This increased competition needs faster product introduction, team collocation, and electronic data sharing. In such a dynamic marketplace, companies are under intense pressure to introduce new products to keep up with their competitors. Consequently, the manufacturing industry in the United States has been undergoing some fundamental changes, including a move to low-cost, high quality systems and a shift in focus from large business customers to diffused commodity market for all size and type of customers. The challenges include shortened product life cycle, high quality product, highly diversified and global markets with unexpected changes of technologies and customer needs. For such a dynamic market, product development requires the involvement of customers and business partners during the development process. As a result, the manufacturers are trying to get most of the work done by the vendors so as to minimize the time to market.

One of the most important steps to address these issues is the development of the appropriate business strategies to meet the challenges of present and future markets. In doing so, manufacturing organization has not only to understand the customer demands but also to develop mechanisms to instantly respond to the changing

A.K. Kamrani and E.A. Nasr, *Engineering Design and Rapid Prototyping*,
DOI 10.1007/978-0-387-95863-7_1, © Springer Science+Business Media, LLC 2010

customer demands. The next step is to determine the right kind of resources to support these strategies. This may include the selection of personnel, technology, and their interaction. The most important task here would be to synchronize these resources to achieve the corporate objectives. Corporate objectives such as growth in market share, profitability, and other financial measures essentially originate from the understanding of the marketplace. The marketing department is responsible for understanding the customer requirements and the corresponding business plans. The contingency of achieving the objectives depends mostly with design and manufacturing departments, which explore various products and processes available to satisfy the objectives.

1.2 Technological Innovation

Increasing opportunities in international markets have led to significant new competitive pressures for many industries. Recently, these pressures have brought about changes to organizational structures at the level of product design and have altered the philosophies of engineering more toward innovation and technological innovation. Innovation makes a significant difference to how organizations operate in today's global market. A simplified model for managing innovation consists of generating new ideas, selecting the best one, and finally implementing them. Generating new ideas could be from inspiration, listening to expert users, frontier research, or by just by combining existing ideas and technologies into something new. Of course, innovation is risky. Managers must strike a balance between the associated risks and the potential of success. Selecting the best idea is about a strategic choice and managers should ask questions such as follows:

1. Does this idea fit our business strategies and goals?
2. Is it being developed on something that the company would have the knowledge and the expertise?
3. What are the potential risks and consequences?
4. Does the company have the skills and resources to take it forward?

If an organization does not change what they offer and how they offer it, it could risk their survival and to be overtaken by others that do. Implementing the innovation as part of an organizational culture is a complex process that requires resources. Organizations must invest time, money, and people to find out whether or not it will work for them. A systematic process should facilitate how organization manages the innovation process. This process requires research and development, market studies and simulation, competitor analysis, risk model development, testing, etc. to gradually improve their understanding of the innovation and its potential benefits. Innovation can take many forms but it is summarized into four categories: These are known as follows:

1. *Product innovation*: Products and services that an organization offers.
2. *Process innovation*: Methods by which products and services are developed and delivered.

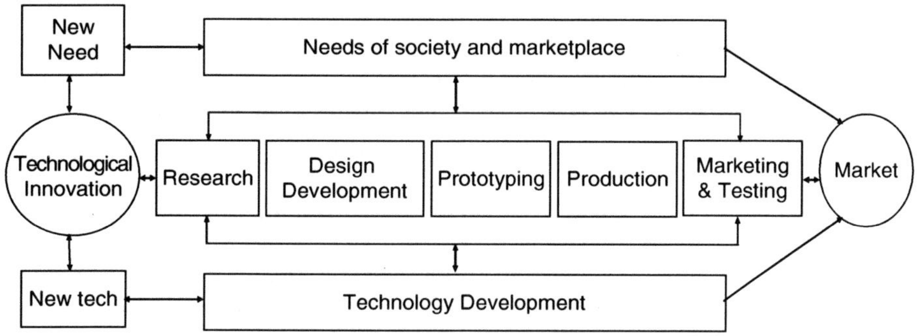

Fig. 1.1 Integrated technology pushed and market-pull process

3. *Position innovation*: The environment in which the products and services are introduced.
4. *Paradigm innovation*: The philosophy and the mental models that define how the organization does business.

Recent government, academic, and private sector initiatives have sought to advance technologies for developing, as well as managing, innovative product development environments. Many companies have established a concurrent design process for their product development and have likely recognized a need for evaluating engineering risk factors for monitoring and measuring the effectiveness of the process. A product development process includes a set of activities arranged in a specific order with the clearly identified inputs and outputs. Each activity in a process takes an input and transforms it into an output of some value to a customer. The output of the process is either a product or a service. The objective of the process is to satisfy requirements and management objectives. The process is considered efficient when output (product) satisfies general customer and product requirements, meets management objectives and customer deadlines, and all these with less costs and resources. Figure 1.1 illustrates a generic model for an integrated process for innovative design and development.

It is well documented that incremental product innovation is managed by the cooperation of marketing and technological knowledge. A product is an embodiment of both market and technological knowledge. In case of emerging technology, technology seemed to serve as a major driving force to introduce the relevant new product and it is also recognized that market knowledge can stimulate successful new product development and assist accumulation of technological knowledge. It is therefore concluded that technology knowledge and market knowledge are able to serve as key factors that enhance successful new and innovative product development.

1.3 Market Trend

Two of the most important elements in today's changing environment are increased product sophistication and variation. To remain competitive, manufacturers must minimize total costs while being quick to develop and market new products.

This involves integrating many diverse functional areas of an organization into a process of creating a better design when viewed across the entire product life cycle. Today's products are complex, composed of several mechanical and electronic components. Successful product development requires the input of people from several functional areas including engineering, finance, marketing, and manufacturing. The focus on customer-driven market, coupled with increased competition, requires fast updating of designs, flexibility in manufacturing systems, and responsiveness in production schedules.

A major problem in achieving effective and timely product designs and innovations has been the long design and product development process. The long design development process can be attributed to the need for many design iterations and refinement. Many times, work is accomplished in informal collaboration, where the emphasis is on exploration of the ideas, compared to formal collaboration, which mostly consists of confirming designs that are brought to the meetings. However, traditional collaboration is geographically limited. Colleagues are not easily able to collaborate and exchange their ideas when situated at different locations. In an organization for which the design teams are geographically dispersed, it is necessary to provide a medium to link their data or resources. Nowadays with the advent of World Wide Web (WWW) applications, it is possible to transfer the data or link to the design team members.

Collaboration is one of the central requirements for today's engineering. The shift from traditional design and manufacturing paradigm to a new, virtual, and agile model is generally observed. The traditional model is the one with very limited information sharing, static organizational structure, and almost no cooperation among the competitors. Engineering paradigm like participatory design, concurrent design, and TQM all focus on teamwork. Participatory design supports cooperation between users and system designers. *Concurrent engineering* especially focuses on cooperation between design and production. *TQM* requires cooperation between all departments of an enterprise. *Collaborative engineering* (CE) is an innovative method for product development that integrates the widely distributed engineers for virtual collaboration. The reasons for widely geographically dispersed teams are various, such as locality of certain resources and competence, or perhaps different production costs. Computer modeling is used in the whole engineering design process resulting in virtual prototypes. The high edge technology is required to assure real time, interactive engineering process. This includes high performance workstations with advanced visualization and modeling software, high speed networks for data transfer, compatible data exchange mediums, and appropriate standards including those for product data representation.

A rapid development of Internet-based technologies with steadily increasing easiness in accessing any kind of information through WWW profoundly changes the engineering practice. The WWW has been conceived as a means of collaboration and as such proves to be extremely useful. Already, much of the collaboration work is based on the Internet. New models are being created with tele-working and mobile working. The evolving design published on the WWW can be used to negotiate design modifications and refinements. Future design teams could utilize this information to aid in design of similar products. The unique synergy of technologies developed is responsible for the profound change in information infrastructures becoming available to enterprises. A group of collaborating individuals working toward a common goal has a need for accessing some shared workspace through a common interface. The Internet with WWW provides an interface and infrastructure for worldwide access to data. Interaction on the WWW can be realized through the Virtual Reality Modeling Language (VRML). It is an emerging and developing standard for representing three-dimensional models and scenes on the WWW.

The challenge facing American and European management today is to ensure that their organizations remain competitive in a rapidly changing and volatile marketplace. CE is considered to be important for any engineering company wishing to survive in the present market. The growing complexity of products and the design process has elevated the importance of efficient product design and development. The pressure for reducing the cycle time of product development, engineering, and manufacturing has made it necessary to improve the integration of these functions and to manage the interfaces.

Two of the most important elements in today's changing environment are increased product sophistication and variation. To remain competitive, minimizing the total cost and the development cycle time are the key issues. The most effective step toward achieving these goals could be the integration and collaboration of the available resources. This involves integrating many diverse functional areas of an organization into a process of creating a better design when viewed across the entire product life cycle. One of the consequences of the demand for this integration of resources has been the necessity for teams of engineers, offer from several areas and geographical locations, to work together over networks, supported by information and computer services. The aim here is the effective integration of these tools and resources in order to improve the product development process. There is a need for a generic product development framework to support the collaborative design process.

Large design problems often require specialized knowledge from different fields. Several designers, each with specific core competencies, interact in the design process and will need to alter the design as per respective demands. During the conceptual phase of the design life cycle, emphasis is mainly given on establishing the requirements and their proper interaction. New technologies and tools along with advancement in Information Technology (IT) are helping these organizations in several ways. However, there is no established generic implementation model for a wide range of industries. Software vendors may provide "Custom" software package for an individual firm. Different industries have different product development strategies, which demands a generic framework that will help them collaborate efficiently irrespective of their product, organizational structure, and/or geographical location.

1.4 Design Process

Design refers to those activities involved in creating the styling, look, and feel of the product, deciding on the product's mechanical architecture, selecting materials and processes, and engineering the various components necessary to make the product work. Development refers collectively to the entire process of identifying a market opportunity, creating a product to appeal to the identified market, and finally testing, modifying, and refining the product until it is ready for production. Great products are not simply designed, but instead they evolve over time through countless hours of research, analysis, design studies, engineering, and prototyping efforts, and finally, testing, modifying, and retesting until the design has been perfected. Regardless of organizational differences, a good product is the result of methodical development effort with well-defined distribution of work among the design team members and their interaction.

Concept development is a crucial stage in new product development process. The basis for the concept development includes analysis of potential market needs, competitive products reviews, and well-defined product specifications. Based on these results a concept is selected for an economic analysis, and the project outline. In the final stage of concept development, the team prepares a detailed development plan, which includes a list of activities, the necessary resources and expenses, and a development schedule.

A system can be defined as a collection of components in which individual components are constrained by connecting interrelationships, such that the system as a whole fulfills some specific functions in response to varying demands. System-level design, or the task of designing the architecture of the product, is one of the most important stages in effective development process. In the concept development phase, the design team focuses on the core product idea, and the prospective design is largely based on overviews rather than in-depth design and engineering. Once the development plan is approved, marketing may begin to develop ideas for additional product options and add-ons. Designers and engineers develop the product architecture in detail, and manufacturing determines which components should be made and which should be purchased, and identifies the necessary vendors.

The product architecture defines the product in the primary functional systems, subsystems, and how they interact so as to work as a unit. The architecture of the product, i.e., the way it is decomposed into systems and subsystems, and their integration, impacts a number of important attributes such as standardization of components, modularity, options for future changes, and ease of manufacture, etc. With outsourced components, the supplier may contribute much of the associated design and engineering. The effective system-level design is the basis for success of overall development and the quality of the product.

Detail design is the stage wherein the necessary engineering is done for every component of the product. During this phase, each part is identified and engineered. Tolerances, materials, and finishes are defined, and the design is documented with drawings or computer files. Increasingly, manufacturers and developers are turning to three-dimensional solid modeling. The detailed specification for individual components is very important during this phase. In many cases, the design is refined if there are any problems at downstream activities.

During the testing and refinement stage, a number of prototypes are built and tested. Even though they are not made from production components, prototypes emulate production products as closely as possible. These prototypes are necessary to determine whether the performance of the product matches the specifications, and to uncover design shortfalls and gain in-the-field experience with the product in use. Later, prototypes are built from the first production components received from suppliers. Three-dimensional computer models form the core of today's rapid prototyping and rapid manufacturing technologies. Once the detailed specification of components is prepared, prototype components can be rapidly built on computerized machines such as CNC mills, fused deposition modeling devices, or stereo lithography systems. Once satisfactory results are obtained from the prototype tests, the product specifications are finalized for the production. The manufacturing personnel are responsible for the process planning, scheduling, and finally the production.

1.5 Traditional Product Development Process

In the traditional (phased) development process, the research and development department along with the marketing department develops several concepts. The feasible concept is then forwarded to the engineering (design development) department, where it is further developed after it has undergone several iterations, refinement, and rework. Once the design is finalized it is sent to the manufacturing department to identify different manufacturing processes and resources required. The manufacturing personnel decide which components to be made in-house and which to outsource. While doing so, manufacturing department may suggest some changes to refine the current design. This leads to increased development cost and time. Once the changes are made and the design is modified it is finally sent to the production department. The production department is responsible for laying the process plans and productions schedules. The distinct expertise from several fields forces separation of these functional departments. The development process takes a relatively long time due to such sequential nature. Also, there are high chances of misunderstanding and the lack of communication among the functional areas. The integrated product development (IPD) can effectively resolve these issues.

1.6 Integrated Product Development

Very few products are developed by an individual working alone. It is unlikely that an individual will have all the necessary skills in marketing, industrial design, mechanical and electronic engineering, manufacturing processes and materials, toolmaking, packaging design, graphic art, and project management, etc. Product development process generally involves the design teams. The design teams are the groups formed based on the area of expertise. The team leader draws on talent from a variety of disciplines, often from both outside and inside of the organization. As a general rule, the cost of a development effort is a factor of the number of people involved and the time required fostering the initial concept into a fully refined product.

The impulse for a new product normally comes from a perceived market opportunity or from the development of a new technology. Consequently, new products are broadly categorized as either market-pull products or technology-push products. With a market-pull product, the marketing department of the company first determines that sales could be increased if a new product were designed to appeal to a particular segment of its customers. Engineering is then been asked to determine the technical feasibility of the new product idea. This interaction is reversed with a technology-push product. When a technical breakthrough opens the way for a new product, marketing attempts to determine the idea's prospects in the marketplace. In many cases, the technology itself may not actually point to a particular product, but instead, to new capabilities and benefits that could be packaged in a variety of ways to create a number of different products. With either scenario, manufacturing is responsible for estimating the cost of building the prospective new product, and their estimations are used to project a selling price and estimate the potential profit for the company. If the decision has been taken to outsource some of the components in the final product, the vendors come into direct consideration. The vendors become a part of the design team, as they will be contributing toward the final product. Hence, it is very important to consider the vendors' involvement in the design process beginning from the initial stages of the design and development of the product. Some of the basic principles and guidelines for an IPD are as follows:

1. *Understand Customer Needs and Manage Requirements:* Better customer relationships, frequent communication, and feedback systems lead to better understanding of the customer's/user's needs. Customer involvement increases the probability of the product meeting those needs and being successful in the market. Methodologies such as quality function deployment aid in defining customer needs and translating those needs into specific product, process, and quality requirements. Once customer requirements are defined, one should track and tightly manage those requirements and minimize creeping elegance that will stretch out development.
2. *Plan and Manage Product Development:* Integrate product development with the business strategy and business plans. Determine the impact of time to market on product development and consider time and quality as a source of competitive advantage. Communicate these plans to concerned development personnel. Undertake fewer development projects at any point in time to allow a greater focus on resources and shorten development time.
3. *Use Product Development Teams:* Early involvement of marketing/program management, manufacturing, material, test, quality, and product support personnel in product development provides a multifunctional perspective and facilitate the parallel design of product and process, reducing design iterations and production problems. Collocation improves communication and coordination among team members. Empowered, self-directed teams have greater ownership and are more committed to development objectives, improving the chance of a successful design on schedule and budget.

4. *Integrate Process Design:* The design of manufacturing and product support processes must be integrated with the design of products in order to optimize the performance, availability, and life cycle cost of the product. Understand existing and planned process capabilities and constraints. Develop and apply producibility and supportability guidelines and utilize formal tools such as Design for Manufacture/Assembly (DFMA) and maintainability analysis tools to improve product designs. Integrate design and business rule checking into automated design processes.
5. *Involve Suppliers and Subcontractors Early:* Suppliers know their product technology, product application, and process constraints best. Utilize this expertise during product development and optimize product designs to the capabilities of the "virtual factory" which includes these suppliers. Reduce your supplier base to focus more attention on a long-term relationship and incentivize early involvement. Maintain schedule stability and be open to improvement suggestions from suppliers to create better products at lower costs.
6. *Develop Robust Designs:* Quality engineering and reliability techniques such as design of experiments provide an efficient way to understand the role and interaction of product and process parameters with a performance or quality characteristic leading to robust designs and enhanced reliability. Learn from the past experience to avoid repeating past mistakes and conduct formal design reviews to assure all design issues and risks have been appropriately addressed.
7. *Integrate CAD/CAM & CAE Tools:* Integrated CAE/CAD tools working with a common digital product model facilitate capture, analysis, and refinement of product and process design data in a timelier manner. Integration of CAM tools facilitates process design/definition with fewer errors, greater accuracy, and a reduction in lead time to production. A single repository of product data minimizes data handling, redundancy, and errors as product data is used by various functions. Feature-based solids modeling, parametric modeling, electronic design frameworks, and ultimately standards such as STEP and EDIF facilitate the downstream interpretation, analysis, and use of this product data. These tools, when intelligently and cost effectively applied, can lead to a streamlined development process and project organization.
8. *Simulate Product Performance and Manufacturing Processes Electronically:* Solids modeling with variation analysis and interference checking allows for electronic mock-ups. Analysis and simulation tools such as FEA, thermal analysis, NC verification, and software simulation can be used to develop and refine both product and process design inexpensively. These tools should be used early in the development process to develop a more mature design and to reduce the number of time-consuming design/build/test iterations for mock-ups and developmental prototypes.
9. *Create an Efficient Development Approach:* Form compact product development teams with highly experienced and motivated members. Minimize the number of development teams and the coordination required with skillful architecture and a minimum of interfaces.

10. *Improve the Design Process Continuously:* Reengineer the design process and eliminate non-value-added activities. Constantly question why things are done and how they could be done better. Make quality the basis for decisions. Continued integration of technical tools, design activities, and formal methodologies will improve the design process. Use benchmarking as an objective basis for comparing the organization and its products to other companies and their products and identifying opportunities for improvement. Invest in training and develop personnel to improve productivity.

1.7 Teams

Product Design Teams (PDTs) are formed to support different stages of the process, including production and services as shown in Fig. 1.2. This early involvement will result in a complete understanding of all the requirements and a consensus approach to the design of both the product and its manufacturing and support processes. Product development teams promote open discussion and innovative thinking resulting in superior products, more efficient processes, and, ultimately, a more satisfied customer. The focus of the team will be to satisfy the external customer's product and support requirements as well as the internal customer (functional department) requirements related to factors such as producibility, cost, supportability, testability, etc. Although PDTs require more resources early in the development cycle, the result will not only be superior designs, but reduced resources over the

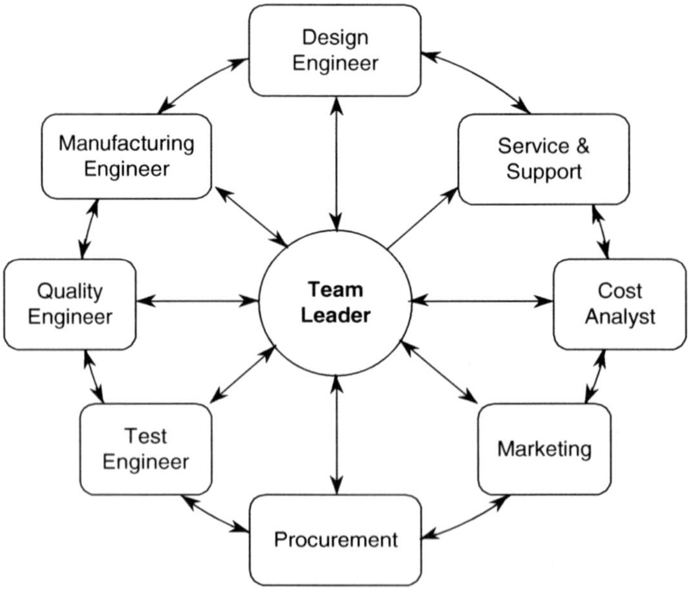

Fig. 1.2 Scope of PDT

life cycle of development, production, and support through reduced design iterations. The team approach will lead to greater commitment to the design and will result in a smoother transition to production.

1.8 Effectiveness of PDT

A key factor in the effectiveness of the PDT is the opportunity for regular interaction among team members and working together as a true team. If a team only meets periodically much as a committee would, interaction, working relationships, collaboration, and effectiveness are limited. The majority of potential communication among team members regarding a product development effort is informal. The physical proximity of an "expert" in another discipline will trigger asking a question and seeking or sharing information. Collaboration not only facilitates this communication but improves the nature of working relationships and leads to a more streamlined development process. This improved informal communication and coordination accelerates development activities and truly enables a parallel mode of development.

The important stage of collaboration is breaking down barriers between departments that result in the "over the wall" design process. As personnel have an opportunity to interact and develop relationships, issues, and questions can more easily be resolved. Personnel learn of others' expertise, and there is a greater opportunity to take advantage of "hidden knowledge." As the organization moves toward implementing PDTs, the closer proximity of the functional departments makes it easier for the team members to work together and coordinate activities. As individuals are working on a day-to-day basis in close proximity, they have an opportunity to develop a close working relationship that improves overall team dynamics. This enhances the frequency and quality of communication. There is greater opportunity for feedback and discussion. Team members can respond more rapidly to issues and initiate process tasks more quickly.

Product design requires team efforts and it cannot be initiated and completed by an individual single-handedly. Currently industry recognizes the need for overlapping of tasks to reduce the design cycle time. Any product development activity starts with the development of a product model. This product model is utilized for tolerance analysis manufacturing feasibility, etc. In a collaborative environment, the product model representation should capture all the geometric and technological information, which would be required for future analysis. The visual representation is usually developed in CAD packages and the technological data such as surface roughness, material, etc. should be closely associated with it. This will help the designers to easily access and modify the information, if necessary.

In a conventional, sequential product design process, design engineers usually design a part and put their design related information on a CAD system or as an engineering drawing and send it to a manufacturing department. It is then manually read by the process planner who recommends the changes to the product if it is not manufacturable. This sequential approach can lead to long product development cycles.

For reduction in product development cycles, it is necessary to capture complete product model information that can be utilized by product design modules for further processing. Also, the product models should be accessible to geographically dispersed product developers (i.e., designers and process planners) as well as remotely located product design modules. This necessitates *Collaborative Integrated Framework* that can meet the need for this increased agility in production organizations.

1.9 Collaborative Engineering

Collaborative product design and development have been viewed by researchers and industry practitioners as the key for reducing cycle times and improving product quality. The practitioners as well as researchers in the field of engineering, manufacturing, and production management are working on different collaborative techniques and methodologies. Different groups have focused on different sets of requirements. Some researchers have focused on the changes that are required in the design of products to reduce production costs, setup requirements, and lead times. Another body of work has focused on design of variant products and on improving manufacturing flexibility as a basis for producing a larger mix of products. Other researchers have focused on methods of improving creativity in the product development process. A useful point of this changing trend in the effort to understand changes that are necessary in design procedures to cope with the new competitive requirements is to examine the nature of design processes and the coordination problems that arise in design.

The success of the collaborative product design depends on the effectiveness of the collaboration and agreements among the participating disciplines. These product development teams are taking on a global character that involves geographically dispersed participants. For instance, in the automotive industry the design of the car for a global market may involve teams dispersed in time and space in Europe, South America, and North America. The collaborative design teams that are dispersed must be coupled with a generic system for effective sharing of information, designs, data, etc. Development in the computer-supported collaborative work (CSCW) systems can potentially enhance the productivity and effectiveness of such teams. Cooperative team-based activities are changing the way people collaborate and share information across an enterprise. Competition has forced enterprises to improve product development and decision-making processes and to explore concurrent engineering, collaborative design, IPD, and total quality control.

The CE is the systematic approach to the integrated, concurrent design of products and related processes, including manufacturing and support. This approach considers all the elements of the product life cycle from conception through disposal, including quality, cost, schedule, and customer requirements. The objective of CE is to reduce the system/product development cycle time through a better integration of resources, activities, and processes.

The basic principle of CE is the integration of methodologies, processors, human beings, tools, and methods to support product development. CE is a multidisciplinary technique that includes aspects from knowledge-based systems, optimization algorithms, database management systems, and CAD/CAM. CE involves the interaction of diverse group of individuals who may be scattered over a wide geographic range. To enable effective and complete communication among them, there are certain technological concepts that must also become organized into concurrent layers. Distributed information sharing and collaborative work are important techniques to provide a basis for the CE. The CE takes advantage of shared information and allows simultaneous focus on different phases of the product development life cycle.

1.10 Collaborative Development Process

In a traditional paradigm, engineering occurs in a sequential manner. Design must occur largely before manufacturing; testing must occur after manufacturing has started but before full-scale production is undertaken. However, CE is based on the fact that there is no necessary condition forcing that expertise to occur only at the required stage. It is often the case that all kinds of expertise are absolutely essential at almost every stage of product development. There is a need to provide variety of expertise concurrently in each stage of a sequential process. These expertises are grouped in different teams. Each team is responsible for its respective contribution throughout the process. One should not interpret the term *expert* as a consultant. These are the active members throughout the development process. With these experts, we can begin to parallelize tasks. Doing engineering tasks in parallel tends to shorten lead times, which improves the overall efficiency of the product development process.

The team of experts exists during each stage of the product development process. It is also present during the design stages of that process. So now, the kinds of tasks that need to be carried out during the design phases must account for the needs of the CE process. In order for a CE approach to be effective, a strong level of communication between the developers and end users is required. In the context of CE, a customer is both internal and external to the development process. Each member of the CE staff is an internal customer for intermediate products during development. CE advocates an integrated, parallel approach to design. By paying attention to all aspects of the design at each phase, errors are detected prior to being implemented in the product. The integrated design process must include a strong information sharing system, an iterative process of redesigns and modifications, trade-off analysis for design optimization, and documentation of all parts of the design. Integration of computerized systems largely enhances the benefits of CE with automatic knowledge capture during the development and lifetime management of a product, and automatic exchange of that knowledge among different computer systems.

A major problem in achieving effective and timely product designs and product innovations has been the long design and product development process. The long design development process can be attributed to the need for many design iterations and problems are typically solved with meetings and many phone calls. Many times, work is accomplished in informal collaboration, where the emphasis is on exploration of the ideas, compared to formal collaboration, which mostly consists of confirming designs that are brought to the meetings. However, traditional collaboration is geographically limited. Colleagues are not easily able to collaborate and exchange their ideas if they are situated in different locations.

The collaborative marketplace has been evolving over the last 15 years, delivering technologies that enable coordination and information sharing, virtual meetings, and more recently virtual collocation. The promise of these technologies is to improve organizational ability to collaborate, coordinate, and share information in order to facilitate inter- and intra-organizational teamwork. While these new collaborative technologies promise to reduce cost and time of information exchange, they have implications on collaborative design processes. The need to bring together the process of collaboration and computing has led to the development of various collaborative systems. A major purpose of using collaborative systems is to have meaningful interactions with other people. Such interaction richness can be achieved when barriers of space, time, and media/document formats are overcome when interacting with others. This includes the ability to talk, see, write, and draw in both synchronous and asynchronous manner, access to relevant information, archiving of interactions for future review, and debate of issues on a global basis. The participants are geographically dispersed and each member presents their own viewpoint, but understanding of the process as a complete product is understood. Communication between different members of a design team is a notoriously difficult problem, especially at the early stages of the design process. A recurrent problem in many design domains is communication between members of the design team involved in the different stages of concept creation, embodiment and detail design, fabrication, and production. It is frequently the case that misunderstandings arise when incorrect and/or incomplete specifications are passed from one team to another, and that errors or inconsistencies in the design are discovered at late stages of the process. Sometimes production or technical staff resolves the design problems. But further re-corrections may be too late or too costly to attempt. A response to this problem has been to develop broader, more integrated teams, and the introduction of "Collaborative Design Process." The communication problem in design teams has also been addressed by the development of computer-based models of the elements being designed. Such models allow simultaneous access by different members of the team, and the element design evolves through collaborative input and evaluation from all the team.

1.11 A Template for Collaborative Design and Implementation

Advances in IT have enabled designers to effectively communicate and collaborate during the product design and development. This allows engineers to obtain and exchange a wide range of resources and information, since companies are capable of publishing their product information on the Internet. Some Internet-based companies are providing design information for manufacturers and act as information brokers between vendors and customers. This has resulted in the development of a new product design and development paradigm based on the network-oriented design environment. This integrated environment allows for design and development of models for complex problems and provides the necessary tool to assist in evaluation, optimization, and selection of alternatives. Furthermore, a design problem constructed from modeling components made available over Internet might facilitate the collaborative development of analytical system models in addition to the exchange of design information. Modeling of product design problems is a way to help the designer predict the effects of decisions or choices [1]. A well-defined model is used to predict product properties and to evaluate how well the solution alternatives meet stated design objectives [2, 3]. Key to the analysis of any problem is by the identification of what functions are performed and the relationships between them. A CE development process includes a set of activities and functions arranged in a specific order with clearly defined inputs and its transformation process into the outputs. Figure 1.3 illustrates conceptual architecture of the CE design environment.

Key to the analysis of any problem is the identification of what functions are performed and the relationships between them. A process is considered efficient when output of the process satisfies the general requirements, meet management

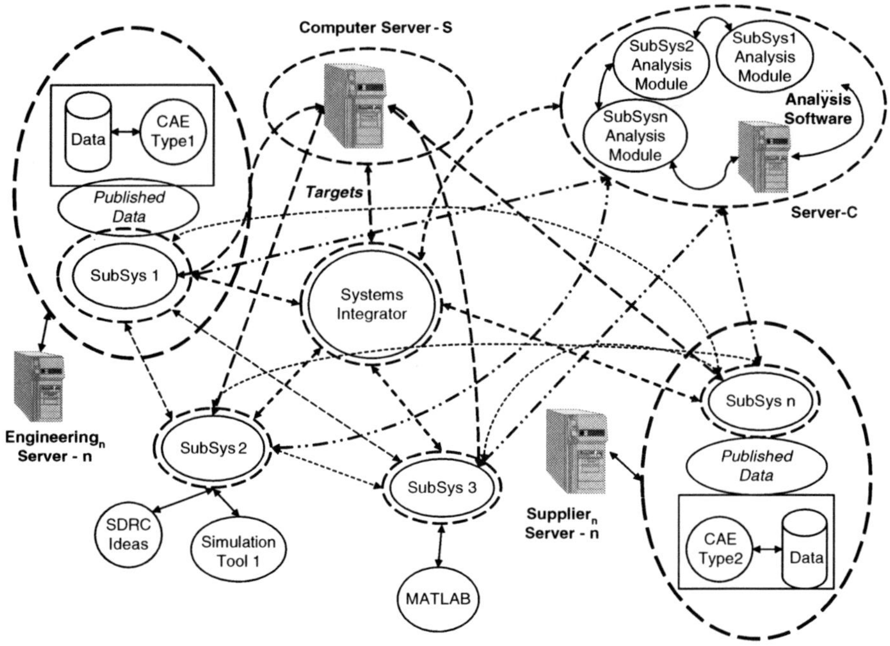

Fig. 1.3 Scope of collaborative engineering design and analysis

objectives, and deadline. Some of the features that are necessary during process modeling include [4]:

- *Detail* (detail which make the process useful)
- *Completeness* (information flow, process management, and material management)
- *Separation of concerns* (process description which addresses the specific needs and concerns of one type of user at a time)
- *Conceptual clarity* (aspects of the process should be clearly described)

In order to develop the necessary templates the Interrelationship Digraphs and Activity Modeling techniques were used. IDEF0 and IDEF3 are the most applied function modeling methods for analyzing and communicating the functional and process perspectives of a system.

IDEF0 modeling method establishes the scope of analysis for a particular functional analysis. IDEF0 is also a communication tool, which allows for decision-making based on developed graphical models. IDEF0 assists the designer in identifying the functions and their relationship. IDEF0 in summary provides

- Activities including their relationships
- Identification of core activities
- Selection of activities for reengineering task

IDEF0 graphical modeling methodology provides the capabilities of cell modeling graphics, hierarchical decomposition, and disciplined teamwork. Using IDEF0, relationships and interactions between activities are easily identified and then modeled [5, 6].

IDEF3 captures "how" a process within an organization is done. It describes the process and provides detailed descriptions at any desired level through decomposition analysis. IDEF3 allows an analyst to capture the AS-IS process and it is used to develop a standard template for operations. It is further used for capturing and analyzing the TO-BE scenarios. Table 1.1 lists the major characteristics for IDEF0 and IDEF3 modeling.

Figure 1.4 illustrates the top level of the IDEF0 hierarchy developed for a Collaborative Product Development (CPD) system.

The A0 level is further decomposed to identify specific activities for the implementation of the CPD environment. The next step is the development of IDEF3, the process model which includes the logics and the timing necessary for the implementation of the CPD environment. Figure 1.5 illustrates the first level of

Table 1.1 Comparing IDEF0 and IDEF3

IDEF0	IDEF3
What is?	How to?
No timing incorporated	Time incorporated
No logic incorporated	Cause-and-effect logic
Target activities for improvement	To improve targeted processes
Functional dependencies	Precedence and cause-and-effect based

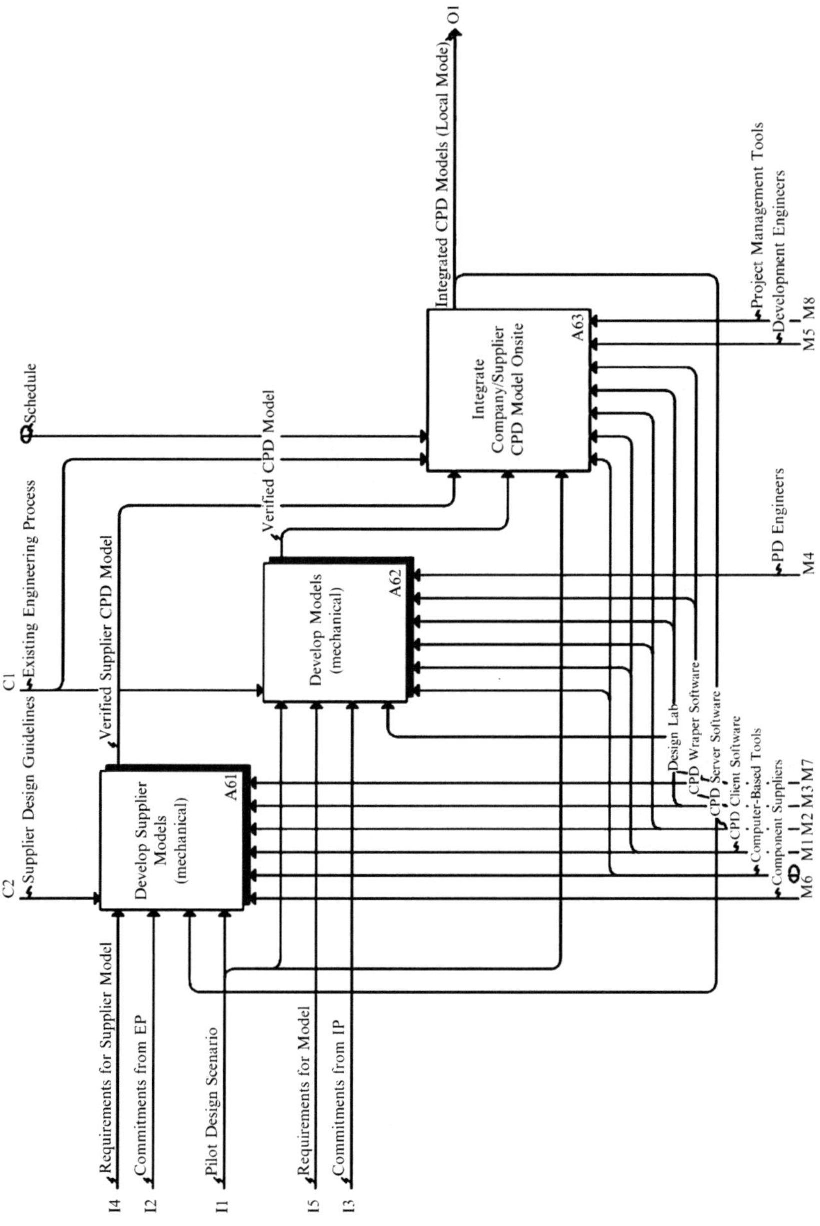

Fig. 1.4 IDEF0 representation of sample CPD hierarchy

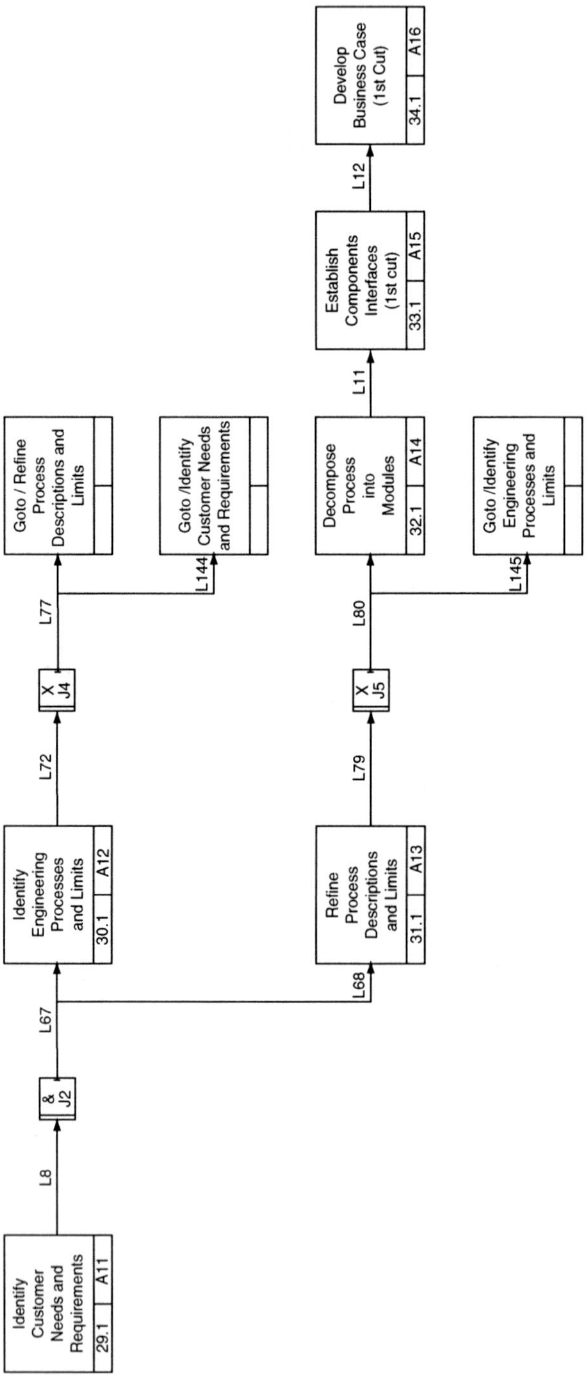

Fig. 1.5 IDEF3 representation of sample CPD hierarchy

the process model. The process models are then simulated in order to plan the required workforce for CPD implementation.

1.12 Summary

Today's manufacturers encounter various difficulties involved in the product development process and that must be overcome for international competitiveness. The obstacle includes shortened product life cycle, high quality product, highly diversified global markets, and unexpected changes of technologies and customer needs. If there are delays in the development, you run the risk of losing revenue to your competition. Also the companies are heading toward vendor-based manufacturing, i.e., manufacturers are trying to get most of the work done by the vendors so as to minimize the time to market. Hence it is essential to utilize a computer-aided system in designing, manufacturing, testing, and distributing the products to minimize the time to market.

For the integration of information at every stage of product development, collaboration technology is needed for cooperative work. As the assistant of the design and development of new products, integrated design technology plays a very important role. The framework described in this chapter confirms design assumptions and predicts product performance in the early stages of the design process. This results in a faster product development cycle – with lower associated costs – achieved by eliminating the need to constantly build, test, and redesign.

1.13 Review Questions

1.1. Define the technological innovation.
1.2. What are the four categories for the engineering innovation?
1.3. Discuss product design and development process.
1.4. Explain in detail how the IPD can be involved in the CE environment.
1.5. What are the basic principles and guidelines for IPD?
1.6. Define collaborative engineering.
1.7. What is the importance of collaborative engineering?
1.8. Describe briefly the collaborative engineering approach.
1.9. How can IPD phase be integrated through collaborative engineering?
1.10. What is the role of product development teams in the collaborative engineering environment?
1.11. How can the product development teams be efficient in the collaborative engineering environment?
1.12. Discuss the product design in collaborative engineering environment.

1.13. The collaborative marketplace has been evolving over the last 15 years, delivering technologies that enable coordination and information sharing, virtual meetings, and more recently virtual collocation. Discuss this statement and show how collaborative marketplace should be effective in the collaborative engineering environment.
1.14. List the advantages of the collaborative system addressed in this chapter.

References

1. Kamrani AK, Sferro P (2002) Design interaction and integration across the supply chain. 6th international conference on engineering design and automation, pp 86–91
2. Pahng F, Senin N, Wallace D (1998) Distributed object-based modeling and evaluation of design problems. Comput Aided Des 30(6):411–423
3. Wallace DR, Abrahamson S, Senin N, Sferro P (2000) Integrated design in a service marketplace. Comput Aided Des 32(2):97–107
4. Mandel K (1991) Graphical process description – views and diagrams. In: Mize JH (ed) Guide to systems integration, revised edn. Industrial Engineering and Management Press, Norcross, GA, pp 283–296
5. AI0Win (1996) User's manual for AI0W™. KBSI, College Station, TX
6. ProSim (1996) User's manual for ProSim™. KBSI, College Station, TX

Chapter 2
Product Development Process

Abstract This chapter discusses the product development life cycle, which can be defined as a sequence of all the required activities that a company must perform to develop, manufacture, and sell a product. These activities include marketing, research, engineering design, quality assurance, manufacturing, and a whole chain of suppliers and vendors. This chapter provides the concept of product development life cycle, the importance of product development during the product development life cycle, the major phases of product development process, benchmarking, the systematic procedure of generating concepts, and finally, a complete case study for understanding the presented concepts.

Keywords Product development • Life cycle • Benchmarking • Generating concepts

2.1 Introduction

The product development life cycle is a sequence of all the required activities that a company must perform to develop, manufacture, and sell a product. These activities include marketing, research, engineering design, quality assurance, manufacturing, and a whole chain of suppliers and vendors. The process also comprises all strategic planning, capital investments, management decisions, and tasks necessary to create a new product.

An important part of product development is the *engineering design process*, which can be defined as the process of devising a system, component, or process to meet desired needs [1]. Engineering design consists of several sequential and/or parallel activities that begin with identifying a need and conclude with a ready-to-manufacture product (prototype). The prototype is considered to be the first product completed in the production process. It is produced using all manufacturing processes and test procedures called for by the design drawings and specifications.

A.K. Kamrani and E.A. Nasr, *Engineering Design and Rapid Prototyping*,
DOI 10.1007/978-0-387-95863-7_2,

2.2 The Evolution of Product Development

Product development is evolving from a sequential process carried out primarily by engineers to an integrated process incorporating a cross-functional team. Similar steps are followed in either case, but they are accomplished concurrently and with higher speed in the integrated process environment.

Four logical groups of activities can be identified in product development [2]:

1. Identifying an opportunity or demand for a new product.
2. Creating the technical specifications for the new-product idea.
3. Developing the manufacturing process to produce the new product.
4. Fabricating the new product.

In the first group, markets or potential markets are analyzed to generate customer needs, meaning that the customer will eventually generate the requirements for the desired product features and functions. Market information is usually compiled by marketing specialists, who translate it into a set of product features or product descriptions that are intended to satisfy a certain target customer base. This process also includes analyzing other products that meet the target needs, offered by competitors, to find their points of both strength and weakness so that efforts can be made to overcome weaknesses and improve desired features. Selling-price ranges are also estimated at this point by analyzing the pricing of similar products. This, in addition to a value of desired profit margin, will set the criteria for the economic feasibility of the new product. These data are translated into cost and quality specifications [3].

The next step is to formulate the product into a concept based on the product feature set identified by marketing in the previous step, i.e., a first vision of how the product will look and perform is created. Then, the technical specifications of the product are developed. Using this initial conceptual vision, the design process proceeds to design and test the product until a preliminary design is completed. Then, a prototype can be created and tested to make sure that the product is functioning as it should. The prototype is considered the first finished product in the sense that it must be produced using all the manufacturing processes that the actual products will go through. Prototype testing may reveal a need for design modification; thus, the design will be refined and a new prototype produced. This will continue until no more modifications are required. The next step is to finalize the product documentation, and then the manufacturing process development may be initiated [4].

Manufacturing processes must be created so that the product can be produced in the production facility. Purchasing new equipment and training workers may be required if new technology is to be used. Tools, fixtures, and the sequence of steps in the manufacturing processes must all be developed to allow rapid, high-quality, cost-effective production. Also, it may be necessary to rearrange the production facility to adapt to the new manufacturing processes.

After the product design and development of manufacturing processes are compeleted, the business of producing and shipping the product begins. Raw materials can be purchased, and the production facility can go into operation. During first production periods some problems may arise as a result of technical production problems, which will lead to design modification to resolve these new problems and reach the expected production rate with the intended quality [5].

2.3 Sequential Product Development

In the traditional development environment, each of the four logical groups occurs sequentially (Fig. 2.1). Research precedes the development of the new product concept, then concepts are developed by the research and development department through an iterative process until an agreed-upon concept is found. Then, a formal description of the concept is sent to the engineering department (design department), where a sequence of design work, review, and rework of design takes place as the concept is being developed. When the design is completely finalized it is "released" to manufacturing to define the manufacturing processes. An important step that is included in the manufacturing work is to determine which components will be made and which will be purchased. When the manufacturing department finishes its study and a make/buy decision is reached, other departments such as those responsible for production planning and procuring materials can start to act. Finally, materials must be ordered, necessary production equipment installed, workers trained, and the product produced and shipped [6].

The division of labor among distinct and separate departments prescribes this sequential nature of the steps. By the time a product is produced, each department would have performed its role in the long sequence of events, leading to the production of this new product. For the most part, each department has completed its work within its own functional area, consulting other departments only to obtain information needed or to review the results of a task in the sequence. The development process takes a relatively long period of time because of the nature of the sequential operations [7]. Also, technical problems can occur as a result of the lack of communication between functional borders. Technical problems can cause rework, scrap, and customer complaints in addition to design changes if discovered later.

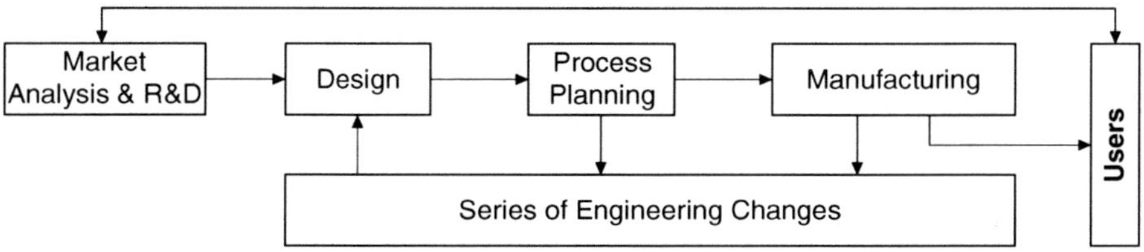

Fig. 2.1 Sequential product development

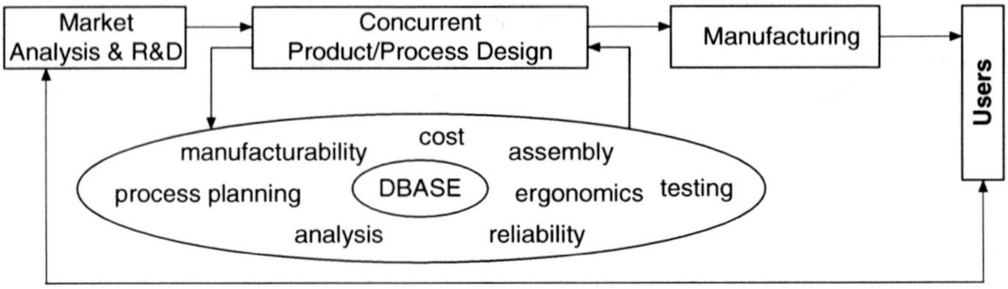

Fig. 2.2 Simultaneous/integrated product development

2.4 Simultaneous/Integrated Product Development

The main disadvantage of the sequential method is the weakness of links between the functional departments that should be cooperating to develop a new product. To overcome this weakness, it is necessary to change the steps of development into a more simultaneous and less sequential process [8], as illustrated in Fig. 2.2.

The change of the development steps from sequential into simultaneous can be facilitated by the use of the *concurrent engineering* (CE) philosophy [1–5, 8–15]. CE can be defined as an integrated and systematic approach to the design of products and their related processes, including manufacturing, testing, and services. CE improves quality, reduces costs, compresses cycle times, increases flexibility, and raises productivity as well as efficiency.

CE can be implemented in an integrated product development environment in which concept development proceeds simultaneously with research into possible technologies. Engineers design components of the product that can be completed as information and technology become available. Previous designs that fit the new application are reused or modified, reducing engineering time. Simulation and prototyping occur simultaneously within engineering design activity. As design work progresses, development begins on the manufacturing process. All major functional areas participate in the design effort, and a cross-functional team must be formed [16].

2.5 Generic Product Development Process

A generic product development process can be constructed starting with needs recognition and ending with the marketing of a finished product [17]. The major phases are illustrated in Fig. 2.3.

- ***Needs Recognition***

The goal of this stage is to explore and investigate customer requirements and needs in an attempt to discover potential "opportunities." Acquiring and using information

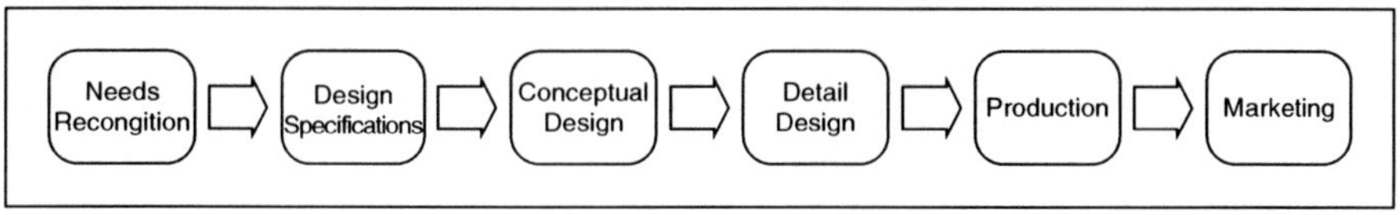

Fig. 2.3 Product development process

is emphasized in this stage and can be accomplished through conducting market research and using available external data relative to the need under study and consumer behavior toward similar needs.

- ***Establishing Design Specifications***

Once a need is realized, the next step is to interpret these needs into technical terms and specifications capable of describing the desired functional characteristics of the product under study.

- ***Conceptual Design***

Several design alternatives are generated and evaluated for their functionality and cost effectiveness. Solutions or concepts that meet the design specifications are generated in the form of ideas or alternatives. A number of design alternatives are generated with no detailed analysis of any alternative. At the end of this phase, the most acceptable concept is selected for further development and analysis.

- ***Detail Design***

In this phase, specifications are refined and trade-offs are made. The selected concept is finalized according to the refined specification. A final cost analysis is performed and a prototype model is produced as the final step in the development process.

- ***Production***

Manufacturing processes capable of producing the parts according to the specified requirements are identified in this phase. Manufacturing sequence and manufacturing costs are also assessed.

- ***Marketing***

Product promotion and distribution to the target markets occur in this phase. Packaging and storage requirements need to be addressed by the development team to ensure the product's safe delivery. Following is a detailed discussion of each phase.

2.5.1 Needs Recognition

Product development begins with identifying needs. The design process can be identified based on an idea for a solution to an existing or identified need or from an idea for a product process for which it is thought a need can be generated [14].

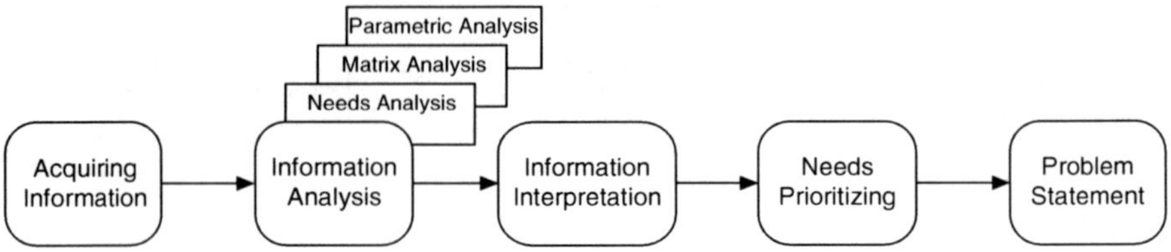

Fig. 2.4 Needs recognition

The product idea needed must look promising given the current market situation, technology available, company needs, and economic outlook.

It is important to analyze the needs clearly and in detail before launching the product development process. Needs analysis should be aimed at collecting information about the requirements that must be fulfilled by the product and about the existing constraints and their importance. Therefore, a requirement list can be formulated, which will form the basis for and guide the subsequent phases. Finding and analyzing needs can be performed systematically as illustrated in Fig. 2.4.

- ***Acquiring Information***

Different types of information are needed to recognize a need or market opportunity. This information includes all the necessary information about similar products (competitive products) obtained from published reference books, handbooks, and manufacturers catalogs. It is essential to obtain information about registered designs, trademarks, patents, and copyrights. This information will be analyzed to establish a competition analysis through a benchmarking study.

- ***Information Analysis***

At this stage all the information collected should be analyzed to gain greater insight about the proposed product or opportunity. The result of this analysis will be the preparation of a "needs" list that represents a comprehensive statement structured to state just what should be designed to satisfy the user need. Three main techniques for information analysis can be used [10, 14]: parametric analysis, needs analysis, and matrix analysis.

- ***Parametric Analysis***

Parametric analysis is a form of desk research that can be used as a tool for both marketing and engineering. It is used to perform a competition analysis by determining the product place in the market relative to the competition. Also, parametric analysis is used to gain insight into the structure and interrelationship between parameters inherent in the product under consideration by identifying the relationships between parameters for the particular product area under consideration. This is done by cross-plotting such parameters to see whether a relationship exists between them.

Figure 2.5 shows an example of a parametric plot. It is clear from the figure that parameter A decreases as parameter B increases. Such plots are useful for identifying desirable parameters and comparing different products with respect to some desired parameters.

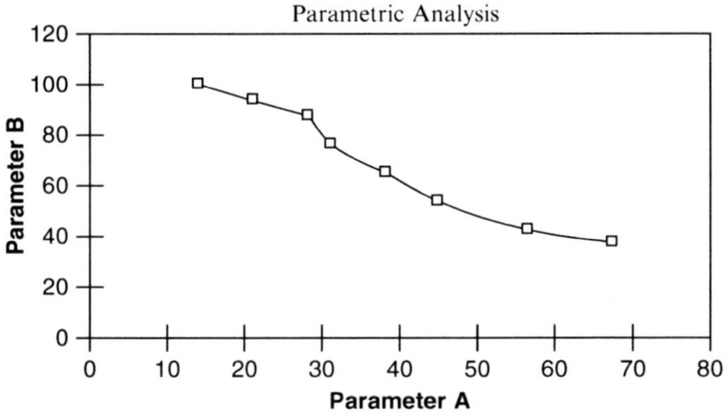

Fig. 2.5 Parametric analysis plot

	Modules									
Feature	M_1	M_2	M_3	M_4	-	-	-	M_m	Graphic Representation of percentage	%
F_1	x	x			-	-	-	X		85
F_2		x		X	-	-	-			45
F_3	x				-	-	-	x		10
F_4	x	x			-	-	-	x		25
F_5			x	X	-	-	-			15
F_6		x		X	-	-	-			65
-					-	-	-		-	-
-					-	-	-		-	-
-					-	-	-		-	-
-					-	-	-		-	-
F_n	x	x	x	X	-	-	-	x		100

Fig. 2.6 Matrix analysis

- ***Needs Analysis***

The true needs of the customer – "The Voice of the Customer" (VOC) – is the main concern of the needs analysis. Customer needs are established by conducting an extensive examination of the market data, product reports, structured interviews, and customer questionnaires.

- ***Matrix Analysis***

A matrix is drawn up with all the features of the competitors' comparable products on the vertical axis and the model type on the horizontal axis. The matrix is then completed to show which models incorporate which features; these are then summed simply and represented graphically on the right-hand side of the matrix.

Figure 2.6 shows an example of a matrix analysis. It can be concluded from the matrix that feature F_1 is incorporated in 85% of the models compared, which may indicate a special importance of this feature.

- ***Information Interpretation***

At this step, the information is translated into a detailed list of customer/market requirements that must be satisfied by a product. That is, the information gathered is interpreted into customer/market needs. A list of product specifications is prepared, which guides the product development process. Although this list is not a rigid item and can be changed and refined when necessary, it is important to maintain the basic structure of the specifications in order not to violate customer needs.

- ***Needs Prioritizing***

Customer/market needs specified earlier must be arranged in a hierarchy, beginning with the most general needs at the top level, termed primary needs. The primary needs will be further characterized by a set of more detailed secondary and tertiary needs at the bottom levels. The needs hierarchy may consist of several levels; the main point here is to start with a general need and progress toward detailed needs.

The needs hierarchy does not convey any importance of the needs, so theses should be established based on either engineering assessment of the needs or a customer survey. The establishment of need importance is critical in making a trade-off analysis and allocating design resources later on in the design process. The importance of needs is usually expressed using an ordinal scale in which the most important needs are placed at the top of the scale and the least important at the bottom.

- ***Problem Statement***

After the needs are identified and their importance established, a problem statement is prepared. The problem statement is an abstraction of what the product is supposed to do to meet its needs. This step is very important for the successive steps, since it will be treated as a "mission statement" for the design process.

2.5.2 Design Specifications

Establishing the design specifications is one of the most important and difficult elements in the overall design process. The design specifications both drive and control the design throughout the process. They are especially important during the early phases of the design effort because they serve as the principal guidelines for the project team at this point in the process. The specifications are so critical to the ultimate design capability and its cost that they must be established early in the process. They have to be established using sound judgments, with wide and in-depth coordination among key participants in the process and with test and analysis support when appropriate. The design specifications need to be as specific to a system and component level as possible. Although specifications are established to be permanent and inviolate, they should nevertheless be continually reviewed and revalidated during the design process, at least until the design is frozen, to ensure that they continue to reflect the goals and objectives of the project.

To establish the design specifications it is necessary to prepare *a list of metrics* that reflects the degree to which the product meets the predefined needs.

Competitive benchmarking [18] can be used to determine the relationship of the new product to the competitive products. Once the *target values are assigned* to the selected metrics and the *specifications are refined*, the next stage (conceptual design) can start.

The process of establishing design specifications can be further explained by the steps shown in Fig. 2.7.

- ***Prepare a List of Metrics***

Customer needs specified in the previous steps are translated into measurable characteristics that will reflect the degree to which the product satisfies the needs (metrics). The major assumption here is that the translation from customer needs to metrics is possible and each need can be represented by one (and only one) metric; thus, meeting the metrics will lead to customer satisfaction. Theoretically, this assumption is valid, but there are needs that cannot be measured or that are difficult to represent by a single metric. In this case, engineers can make the assumption that satisfying more than one metric will eventually lead to satisfying a certain need up to an acceptable degree.

A useful tool that can be used when preparing the metrics list is the needs-metrics matrix [13], in which the rows of the matrix will correspond to the customer needs and the columns correspond to the metrics. A generic needs-metrics matrix is illustrated in Fig. 2.8, where a mark in a cell in the matrix indicates that the need and

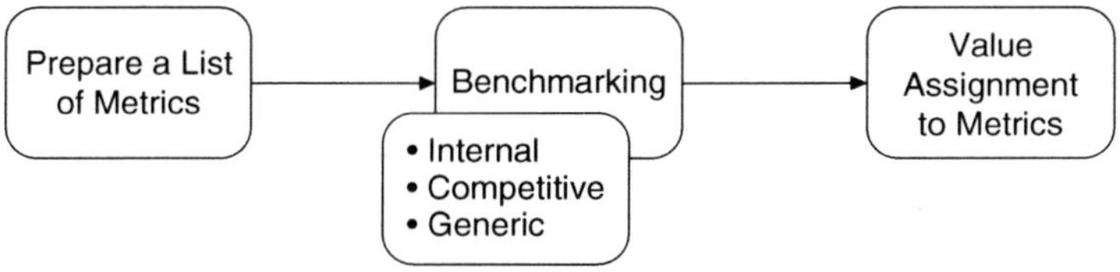

Fig. 2.7 Establishing design specifications

		Metrics						
		M_1	M_2	M_3	M_4	-	-	M_m
Needs	N_1	*		*				
	N_2		*					
	N_3			*				
	N_4							*
	N_5				*			
	N_6							
	N_7		*		*			
	N_8				*		*	
	-							*
	-				*			
	N_n						*	

Fig. 2.8 Needs-metrics matrix

the metric associated with the cell are related. The needs-metric matrix will represent the relationship between needs and metrics and ensure that all the customer needs are considered.

- ***Benchmarking***

Benchmarking is defined as the continual search for the implementation of practices that could provide a competitive edge [18]. Companies differ in the way they implement benchmarking, but it is usually adapted as a corporate strategy used to identify the industrial leaders, promote proven techniques and approaches, establish meaningful goals, perform business forecasting, and analyze the overall internal process. Benchmarking can be categorized into three major categories: (1) Internal benchmarking, (2) Competitive benchmarking, and (3) Generic benchmarking.

Internal Benchmarking: In this type of benchmarking, similar activities in various locations, departments, and units are evaluated to gain data accessibility.

Competitive Benchmarking: This type of benchmarking is concerned with the identification and evaluation of direct competitors to obtain data relevant to the product under investigation and to find comparable processes to gain a competitive edge [10].

Generic Benchmarking: The objective here is to evaluate the organizations and their functions that are considered to be the industry standard to achieve procedure standardization.

- ***Value Assignment to Metrics***

In this step, the design team synthesizes all the information acquired to set actual values for the metrics. Two values are usually assigned to each metric: one is the ideal value, which can be defined as the optimal value that the design team hopes to accomplish, and the other is the minimum acceptable value, which can be considered as the lower limit that can satisfy the needs. Usually, design will progress to achieve a metric value between the ideal and the lower limit; this is due to trade-offs performed throughout the design. In all cases, the design team should have the ideal value as their primary objective.

2.5.3 Conceptual Design

After the problem has been clarified and completely described, viable solutions are identified and the optimum approach is selected. Problem solutions or "concepts" are defined as an approximate description of the product or technology that meets the stated needs. The conceptual design stage is concerned mainly with the generation of solutions/concepts that satisfy the needs, and it selects a concept that is most suited for matching the predefined design specifications. Three major steps can be identified in the conceptual design phase.

2.5.4 Concept Generation

The concept generation, or generation of ideas, begins with a defined problem statement that includes customer/market needs and design specifications and ends with several product concepts from which the design team will select the most suitable one. A systematic procedure may be followed in generating concepts as shown in Fig. 2.9.

- ***Problem Formulation***

Prepare an abstraction of the problem to broaden it out and clarify it so it is easier to understand the important issues. It may be necessary at this point to break the problem down into several easier, understandable, and manageable sub-problems.

- ***Overall Function Analysis***

Analyze the overall function by describing what the product or system is supposed to do. It is important here to focus on the main functions and to describe functions in general terms as much as possible. Furthermore, action statements in the form of verb–noun should be used in representing the functions, i.e., "to transform materials" or "to transmit information."

- ***Sub-function Analysis***

The overall function is now broken down into several sub-functions necessary for the product or the system to operate. The sub-functions, when reassembled, should support and lead to the accomplishment of the overall function. Sub-functions may be thought of as specifications or requirements imposed on the overall function.

- ***Function Diagram***

A function diagram is a representation of the function structure, in which the function under study is represented by a block and the input and outputs are represented by arrows entering and leaving the block.

- ***Generation of Ideas and Solutions***

At this step, ideas and solutions for the sub-functions or the sub-problems are first generated and then combined together to form the overall function or to solve the overall problem. Ideas and solutions can be generated using different techniques. The development team can use "brainstorming," in which a group of participants generates, in a set period of time, many ideas that can be used to solve the problem. Brainstorming sessions usually aim at producing a large quantity of ideas irrespective of their quality. Also, the development team can search published literature,

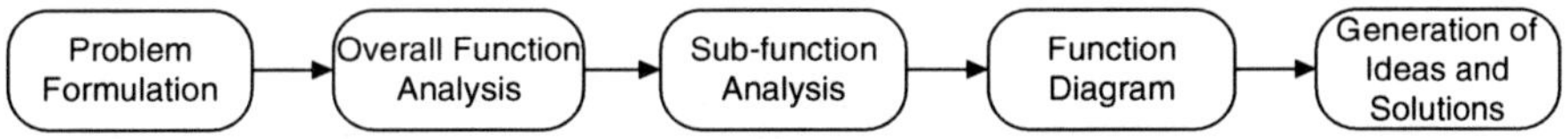

Fig. 2.9 Concept generation

patents, and catalogs for possible solutions. Interviewing users can trigger some ideas for solutions. A good practice in generating ideas is to try to reuse existing solutions or products in solving new problems.

2.5.5 Concept Selection

Concept selection is the process of evaluating and comparing alternative concepts with respect to the customer/market needs and design specifications, leading to the selection of the most suitable one or a set of concepts for further investigation and/or development. Concept selection can be performed according to the following guidelines in Fig. 2.10.

- ***Select Solution Principles***

Suitable solution principles that can satisfy the needs are selected individually or in combination with other solutions. Selected solutions should be able to perform the required function effectively and efficiently.

- ***Combine Solution Principles into Complete Design Concepts***

The selected principles in the previous step are arranged into a complete conceptual design that corresponds to the overall function, that is, concepts that correspond to sub-functions are arranged together to form a larger concept that can accomplish the overall function.

- ***Evaluate Concepts Technically***

Concepts are now evaluated with respect to the degree to which they meet the design specifications; concepts that do not meet the specification are eliminated at this point.

- ***Evaluate Concepts Economically***

Concepts are evaluated with respect to their cost, that is, concepts are investigated for their economic feasibility. Non-feasible concepts with unjustified high costs are eliminated.

- ***Select Final Concept***

A final decision must be made among technically and economically feasible concepts by utilizing a scoring technique that can incorporate needs and customer satisfaction.

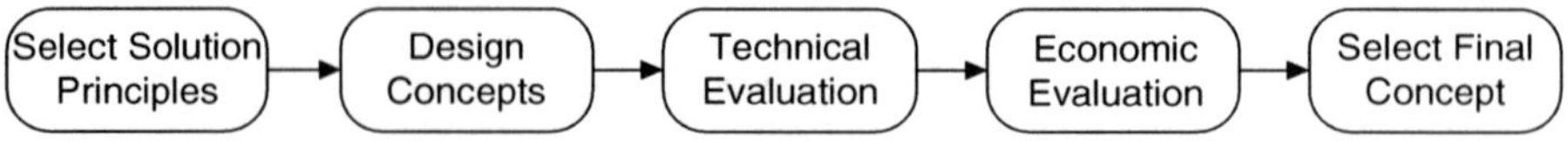

Fig. 2.10 Concept selection

2.5.6 Final Concept Preliminary Design

This step of the design process bridges the gap between the conceptual design phase and the detailed design phase of the design effort. The final concept is further defined during this step; the overall system configuration is defined; and a schematic diagram, definition drawing, or other engineering documentation is developed to provide early project configuration control. System-level – and, to the extent possible, component-level – design requirements should be established during this phase of the design process in a manner that corresponds to the design specifications previously defined.

2.5.7 Detail Design

Detail design (Fig. 2.11) is that part of the design in which, starting from a concept of a technical product, the design is developed in accordance with technical and economic criteria. At this phase, the design concept is resolved into its component parts, components are evaluated to validate previously established requirements, and the effect of the component requirements on the overall system requirements is evaluated. Also, all the arrangement, forms, dimensions, and surface properties of all the individual parts are finally laid down; the materials specified; production possibilities assessed; costs estimated; and all the drawings and other production documents produced. The intent of the detail design phase of the project is to develop a system of drawings and specifications that completely describes a proven and tested design so that it can be manufactured.

2.5.8 Component Final Design

The overall product concept generated in the previous step is now designed by designing its components. Components are designed to meet the product specifications identified earlier. The result of the component design should be a component specification list, which in most cases will be part of the overall product specifications. The component specifications, in general, will contain a list of all the necessary information required to procure or manufacture the component such as operating parameters, component dimensions, materials, etc. Component final design is

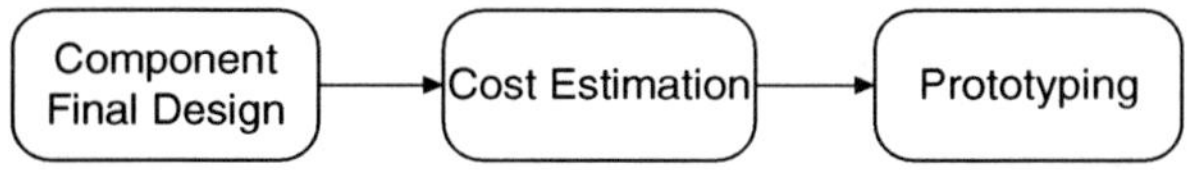

Fig. 2.11 Detail design

represented in several documents such as detail drawings, assembly drawings, and bills of materials.

2.5.9 Cost Estimation

The cost of producing or developing the selected concept/product is estimated. Justification of the trade-offs considered must also be included in this study.

2.5.10 Prototyping

A functional prototype model of the product is made at this point. Further investigation concerning the actual functionality and appropriateness of the product developed can be made on this model as a final step before starting production and introducing it to the market.

2.5.11 Production

Production process planning aims at constructing a production plan that uses the available machinery to produce products efficiently and effectively. Production planning begins by analyzing the detail design documentation, which includes information about the product's geometrical features, dimensions, tolerances, materials, and surface finish. This information is treated as targets that must be met. The process proceeds to identify the appropriate machinery capable of achieving the design targets. The sequence of operations is also identified.

Design for manufacture and assembly is an important concept used in the production phase to gain greater insight about how the product design interacts with the manufacturing system and uses this knowledge to design better-quality products that can be produced for lower cost and in less time [1, 2, 5, 8–12, 19, 20]. Design for manufacture and assembly will identify the product design alternatives that will facilitate the optimization of the manufacturing system as a whole.

Following are some general guidelines that can be used in design for manufacture and assembly:

- ***Simplify the Design***

The design should be simplified before production automation can be considered. Product design should be made with automated assembly in mind.

- ***Eliminate Operations that Require Skill***

Effort should be made to eliminate manufacturing tasks that require special skills.

- ***Minimize the Total Number of Part***

Parts used in a certain product should be those that are critical for the product to function appropriately. It is necessary to design parts that perform several functions. Reducing the number of parts will decrease the production cost significantly.

- ***Use a Modular Design***

It is easier to automate the production of a modular product.

- ***Use Standardized Parts***
- ***Use a Multifunctional Design***

Components should be designed to perform more than one function.

2.5.12 Marketing

Although design engineers are not involved directly in product promotion and distribution, information about problems that occur during the marketing and distribution of products should be integrated into the product design. Design engineers should design packaging to protect products from damage during transport and storage. The design engineer must specify any special shipping and storage requirements. Design engineers can also be involved in the promotional activity by interpreting customers' questions and criticism about the products and relating them to design specifications. This will enable the design engineer to modify the product deign and improve it to correspond to customer needs.

2.6 An Automatic Ice Maker Case Study (Based on the Work Done by Madriz and Saenzay, University of Houston, 2005)

2.6.1 Overview

The redesign of a product must first be fully evaluated on the basis of customer satisfaction, manufacturability, assembly design, and cost. This case study evaluates the design of an ice maker and breaks down the function of each individual component used by the ice maker to generate new design features. An activity diagram along with customer needs analysis by use of a questionnaire was used to begin understanding the functions. A cost breakdown was performed and evaluated in a modular system. Options included a number of new features and different ways to eject the ice. The ice maker was redesigned incorporating an interchangeable and rotatable mold including other features essential to the customers.

Most icemakers use an electric motor, an electrically operated water valve, and an electrical heating unit. To provide power to all these elements, the icemaker has to be hooked up to the electrical circuit powering the refrigerator. Also, the icemaker has to be hooked up to the plumbing line in the house, to provide fresh water for the ice cubes. The power line and the water-intake tube both run through a hole in the back of the freezer.

When everything is hooked up properly, the icemaker begins its cycle. The cycle is usually controlled by a simple electrical circuit and a series of switches. Closing this switch lets electrical current flow through a heating coil underneath the icemaker. As the coil heats up, it warms the bottom of the ice mold, loosening the ice cubes from the mold surface. The electrical circuit then activates the icemaker's motor. The motor spins a gear, which rotates another gear attached to a long plastic shaft. The shaft has a series of ejector blades extending out from it. As the blades revolve, they scoop the ice cubes up and out of the mold, pushing them to the front of the icemaker. Since the cubes are connected to one another, they move as a single unit. At the front of the icemaker, there are plastic notches in the housing that match up with the ejector blades. The blades pass through these notches, and the cubes are pushed out to a collection bin underneath the icemaker.

The revolving shaft has a notched plastic cam at its base. Just before the cubes are pushed out of the icemaker, the cam catches hold of the shut-off arm, lifting it up. After the cubes are ejected, the arm falls down again. When the arm reaches its lowest resting position, it throws a switch in the circuit, which activates the water valve to begin another cycle. If the arm cannot reach its lowest position, because there are stacked-up ice cubes in the way, the cycle is interrupted. Quality level such as materials' quality, production time, higher ice store capacity, and others are also considered in this study. In this case study, the objective is to redesign an Icemaker which has failed several times in the past. This product was chosen because of its failure to operate satisfactorily to costumers. The objectives of this project are as follows:

- Redesign for electrical component isolation from environment.
- Redesign for mechanical integrity.
- Redesign for size, make more compact to make more Freezer space available.
- Evaluated the option to incorporate a filter to clean and pure water.

2.6.2 Procedure

2.6.2.1 Reverse Engineering Phase

The reverse engineering phase begins by analyzing the current product to understand the market of the current product.

Activity Diagram

The customer lifecycle needs of the ice maker are analyzed in the Activity Diagram as one of the first steps. The Activity Diagram starts with the comparison and selection

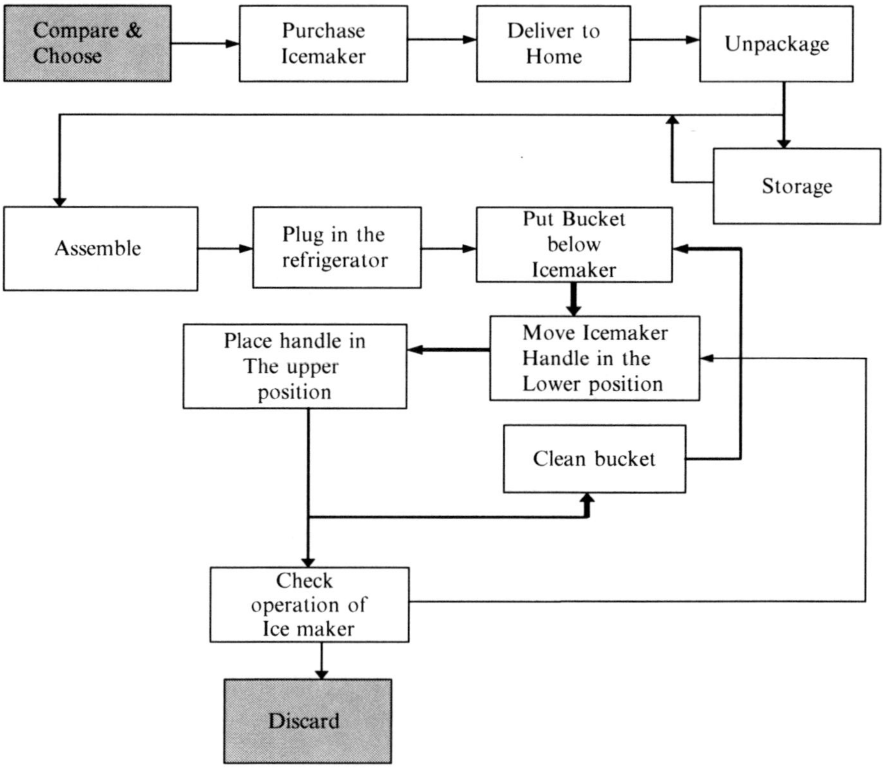

Fig. 2.12 Activity diagram

of the type of ice maker. Many times this selection is limited by the branch of refrigerator that the customer has at home. At home the ice maker needs to be installed inside the refrigerator and an initial turn on is required, and then the ice maker autumnally stops and starts new ice production. Figure 2.12 shows all the activities in the ice maker operation.

Kano Classification and Affinity Diagram

The Kano method of characterizing customer requirements is an effective tool. The results of the affinity diagram helps to understand the potential customer needs and its possible importance. Kano Methods complements the last analysis bringing the customer voice. A Kano questionnaire was developed including the main potential needs from the last analysis. A total of 15 participants were part of the study (students from the course) with age between 25 and 35 years. Kano Analysis is a systematic method to understand how a customer's sense of

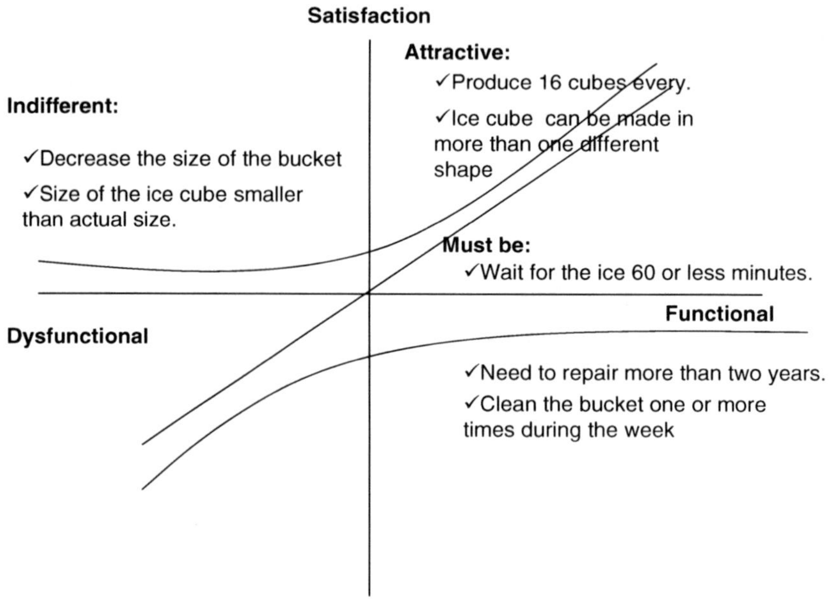

Fig. 2.13 Kano diagram of ice maker

satisfaction is affected when a product or service succeeds or fails to meet either spoken or unspoken customer expectations. This approach was developed by Dr. Noriaki Kano of Tokyo Rika University. Dr. Kano's model is usually shown in the form of the Kano Diagram or the 3-Arrow Diagram. Figure 2.13 shows a Kano Diagram. An Affinity Diagram is a tool that gathers large amounts of ideas, opinions, and issues and then organizes them into groupings based on their natural relationships. The potential customer needs collected inside the design team were grouped into seven categories: Time, Availability, Modularity, Capacity, Weather, Power and water source, and Cost. Brainstorm technique was applied for this purpose. The categories are shown in Fig. 2.14.

After classification, the customer needs were analyzed to make a second classification, but now according to the importance for the customer. The new categories were as shown in Fig. 2.15. The most important customer needs are related with the saving time and automatic start and stop.

2.6.2.2 Conceptual Design

The conceptual design phase consists of flow sheet development and the selection of the major process technologies. Many of the key design decisions made during

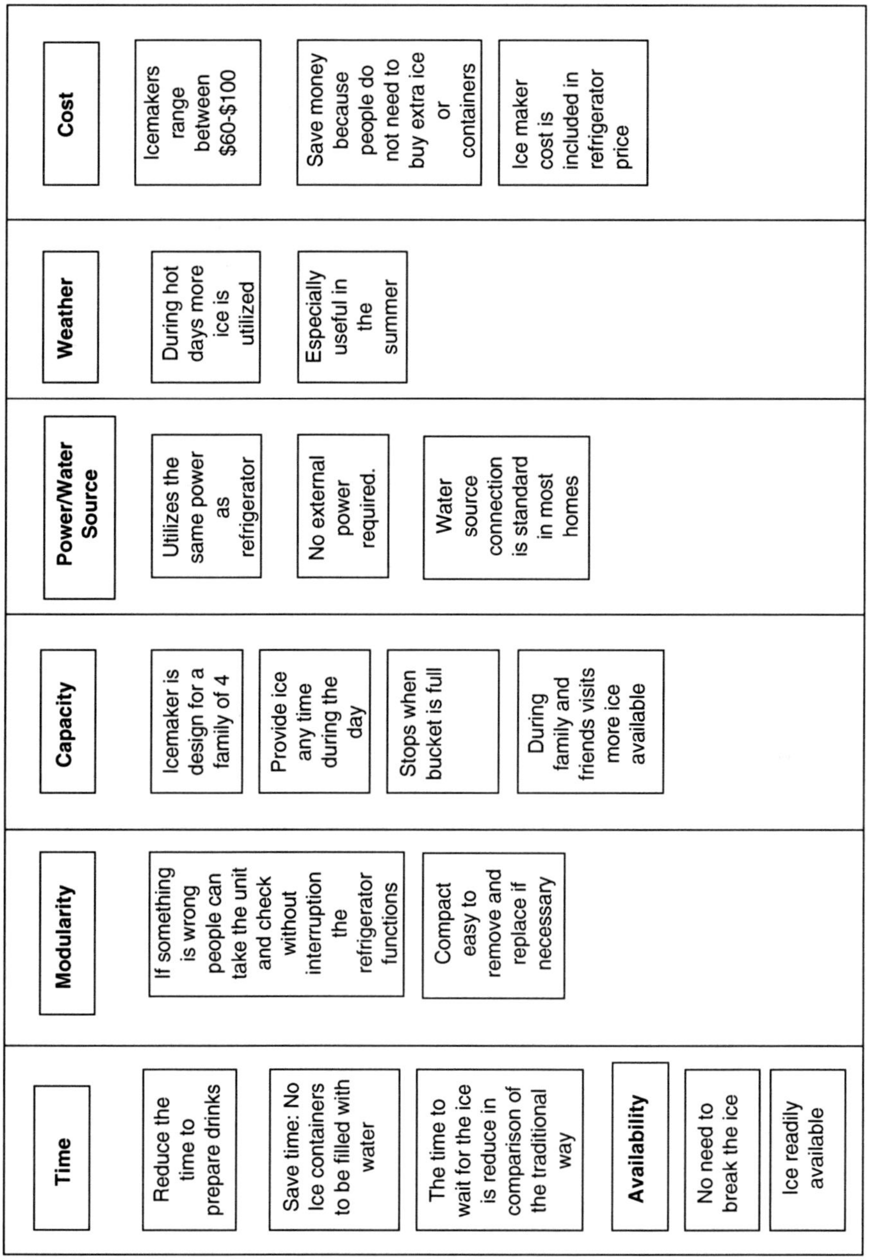

Fig. 2.14 Affinity diagram

Need	Good	Should	Must
Time			
Reduce the time to prepare drinks	√		
Save time: No Ice containers to be filled with water			√
The time to wait for the ice is reduce in comparison of the traditional way		√	
No need to break the ice		√	
Modularity			
If something is wrong people can take the unit and check without interruption the refrigerator functions		√	
Compact easy to remove and replace if necessary	√		
Capacity			
Icemaker is design for a family of 4.	√		
Provide ice any time during the day			√
Stops when bucket is full			√
During family and friends visits more ice available	√		
Power/Water Source			
Utilizes the same power as refrigerator	√		
No external power required.	√		
Water source connection is standard in most homes		√	
Weather			
During hot days more ice is being utilized.	√		
Especially useful in the summer	√		
Cost			
Icemakers range between $60-$100		√	
Save money because people do not need to buy extra ice or containers	√		
Ice maker cost is included in refrigerator price		√	

Fig. 2.15 Customer needs classification

this phase can significantly influence process wastage. Different techniques were applied in this phase; the following are some of those:

- Functional Modeling (Black box, Physical Domain, FAST and Functional requirement).
- Exploded View.
- Cost Analysis.
- Benchmarking.
- QFD.
- Design Structure Matrix.

Black Box Model

The ice maker's main function is represented in the simplest way by using the black box model.

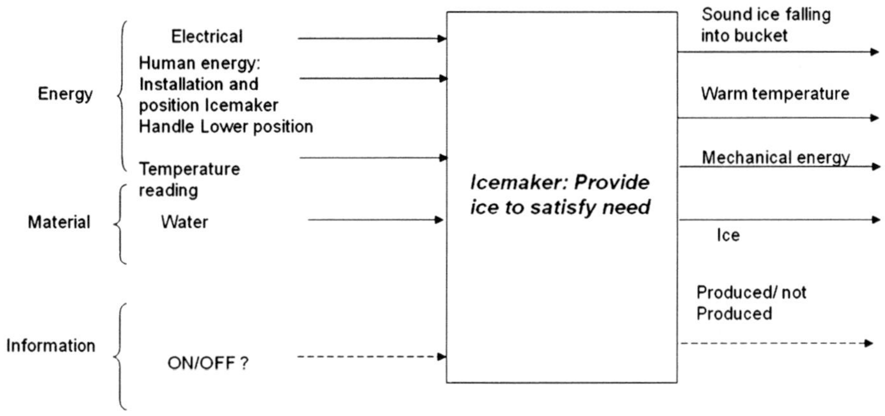

Fig. 2.16 Black box model for ice maker

The general input and output were identified for functional decomposition (Fig. 2.16). The main function of the ice maker is to provide ice to satisfy needs. To reach this function several inputs are required: Energy, material, and information. As an energy is identified the electrical energy that is provided for the refrigerator, the human energy which install and position the icemaker handle in the lower position in order to start the ice production. The only material is the water that is converted from liquid to solid. The information deals with the off or on state of the ice maker.

Aggregated Function Structure

The main function is now divided in the several sub-functions as shown in Fig. 2.17. Each sequence of sub-functions for the full set of needs are aggregated (combined) to represent the functions of the entire product. This step is accomplished by appropriately connecting flows between each sequence and adding sub-functions that interact or provide control states.

FAST Diagram

FAST stands for functional analysis systems technique. It is a methodology used to systematically organize and represent the functional relationships of a technical system. Function is expressed as verb–noun. In FAST, the functions of the system are identified and displayed graphically with respect to a "how?–why?" relationship. The FAST diagram presents the whole technical system with the inter-relationships of its various functions as shown in Fig. 2.18.

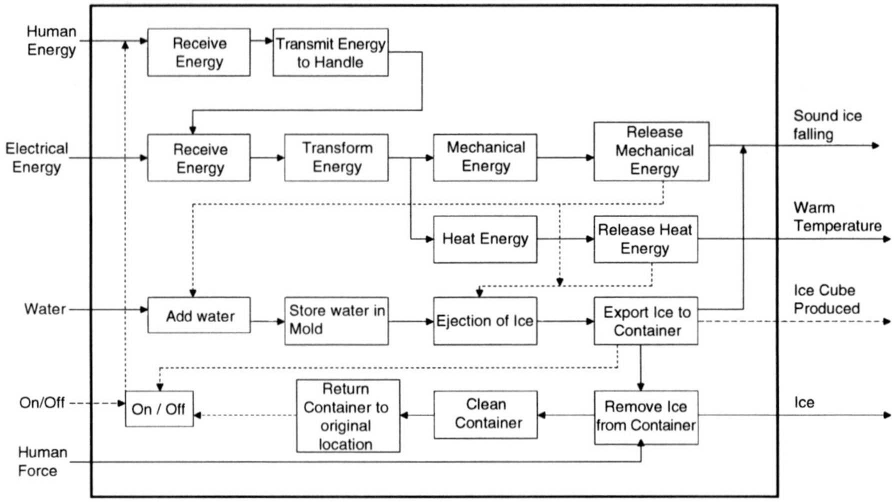

Fig. 2.17 Ice maker aggregated function structure

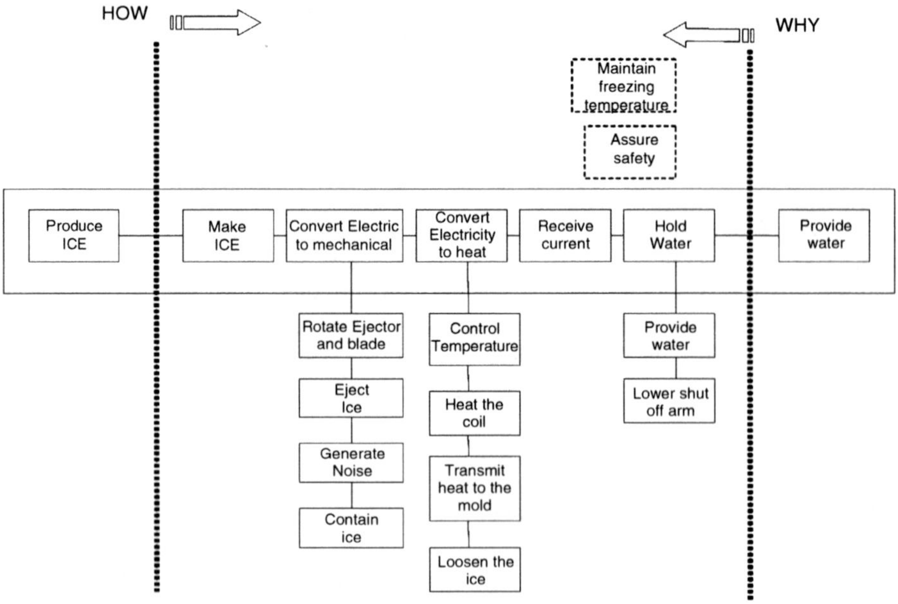

Fig. 2.18 FAST diagram

Function	Design Requirements	Physical Domain
Make ice	Means to convert water to ice (liquid to solid)	Ice Maker
Convert Electrical to mechanical Energy	Means to convert electrical energy to mechanical energy	Electrical motor
Receive current	Means o receive current	Coil housing
Hold water	Means to hold the water in the mold	Ice mold
Rotate ejector blade	Means to rotate ejector blade	Sprocket
Control temperature	Means to control the temperature level to heat the mold	Thermostat
Provide mold	Means to provide the mold	Ice mold
Eject ice	Push out the ice from the mold.	Ejector blade
Heat the coil	Means to heat the coil	Current
Lower shut off arm	Means to move the lever arm down	Shut off arm
Generate noise	Means to produce noise when the ice is produced	Ice
Transmit heat to mold	Means to heat the mold	Coil housing
Contain ICE	Means to hold the ice in the container	Bucket
Loosing the Ice	Means to separate the ice from the mold	Warm mold ice

Fig. 2.19 Functional requirements relations

According to the FAST analysis, the functional requirement was determined and the relationship between those and the design parameter as well the relationship with the physical domain. Figure 2.19 shows this relationship.

Exploded View Analysis

The product teardown process helps to understand how the product works and where and why the components have a specific location. Also, it is a help for a competitive benchmarking. The teardown of the ice maker took about 5 h. Each part was measured, identified its function, kind of material, and cost. A total of 24 components were identified in the analysis. The 24 components were grouped into 8 clusters as illustrated in Fig. 2.20.

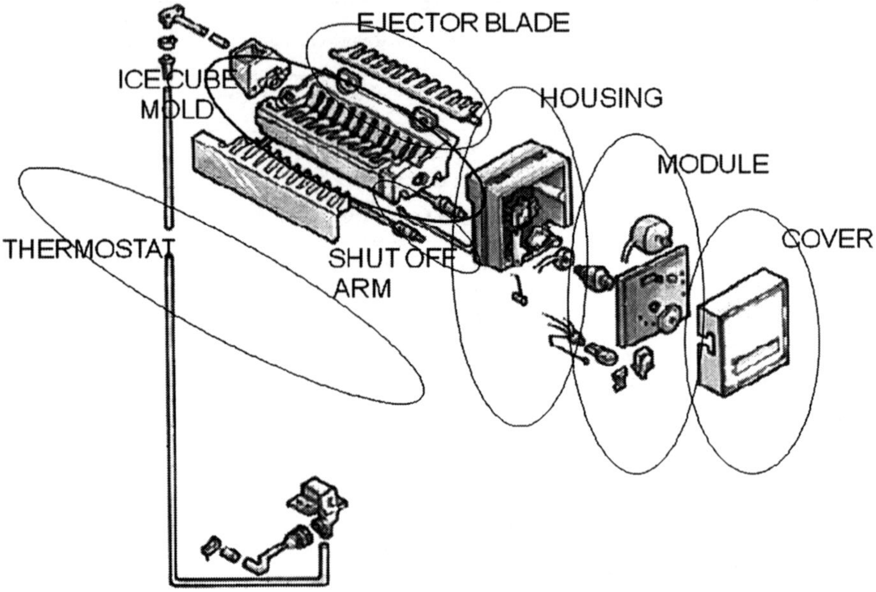

Fig. 2.20 Exploded view of the ice maker by cluster

The cost for each cluster was calculated as shown in Fig. 2.21. It is important to explain that the cost reported for each part is a unit price; therefore, the total cost for the ice maker is \$477.35. This cost is almost three times the current price of the ice maker in the market. Therefore, a correction factor of 0.255577668 was determined and used to reduce each part cost. The housing cluster represents the most expensive part with a cost of \$51.81, followed for the module cluster with a cost of \$20.05.

Competitive Benchmarking

Benchmarking is the process of determining who is the very best, who sets the standard, and what that standard is. In the ice maker benchmarking, three different ice maker brands were considered. According to the results, no big differences between the ice maker brands were founded.

Quality Function Deployment

A prerequisite to QFD is Market Research of the ice maker. This is the process of understanding what the customer wants, how important these benefits are, and how

Group part	Module/ Part #	Description/Name	Unit price	Base price	Group cost
Cover	17	Enclosure cap white box	$28.05	$7.17	$7.17
Ejector blade	21	Ice ejector guide	$16.85	$4.31	
	23	Ejector blade	$10.90	$2.79	$7.09
Housing	1	Electric motor	$51.05	$13.05	
	2	TCO	$27.50	$7.03	
	D_8	Hold switch	$19.85	$5.07	
	D_8	Water fill switch	$19.85	$5.07	
	D_8	Shut off switch	$19.85	$5.07	
	12	Rocker arm	$7.40	$1.89	
	13	Shout arm spring	$3.95	$1.01	
	16	Enclosure brown box	$33.50	$8.56	
	27	Wire	$19.35	$4.95	
	30	Thumb screw	$0.45	$0.12	$51.81
Ice cube mold	19	Heater coil	$26.75	$6.84	
	20	Ice Mold	$28.65	$7.32	
		Water receptor guide	$8.20	$2.10	$16.25
Module	10	Sprocket 48 t	$10.75	$2.75	
	14	Mounting plate	$35.90	$9.17	
	18	Connector with 4 leads	$24.30	$6.21	
	22	Coupling	$7.50	$1.92	$20.05
S_Off arm		Shut off arm	$4.95	$1.27	$1.27
Thermostat	5	Thermostat	$31.10	$7.95	$7.95
Bucket		Bucket	$40.70	$10.40	$10.40
		Total	**$477.35**	**$121.9**	**$121.9**

Fig. 2.21 Cost for cluster of the ice maker

well different providers of products that address these benefits are perceived to perform. Quality Function Deployment (QFD) is a systematic process for motivating a business to focus on its customers. Therefore, the customer needs were evaluated for each competitor and also the relationship of functional requirement and customer needs.

The analysis suggests that the redesign mainly focus on the functional requirement of holding water, provide mold, lower shut off arm, and contain ice.

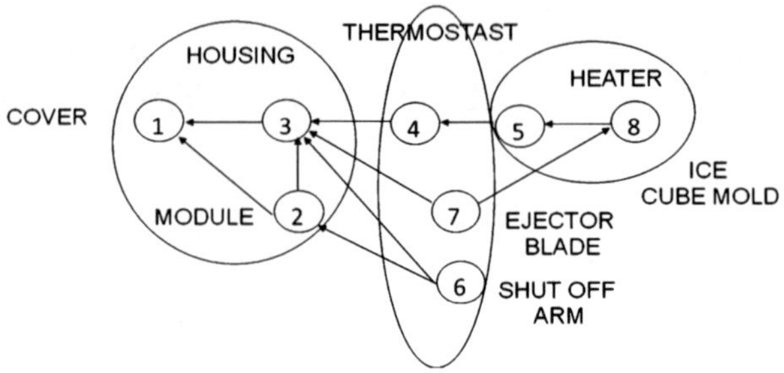

Fig. 2.22 Digraph with the cluster

Design Structure Matrix

The design structure matrix (DSM) provide insights about how to manage a complex system/project and highlights issues of information needs and requirements, task sequencing, and iterations. As a first step of the DSM, a Digraph with the eight clusters for the ice maker was developed. The Raw matrix of the relationship between clusters was analyzed and output suggests a specific order of the cluster. Figure 2.22 shows this result. Three groups can be identified as a result of the analysis. Gourp one includes housing, module, and covering. These have a very strong spatial relationship. The second group includes thermostat, ejector blade, and shut off arm. The last group includes heater and ice cube mold, which also have a strong spatial relationship.

2.6.2.3 Parametric Re-design

The parametric redesign phase includes Morphological Analysis, Alternatives, Design for Assembly, and Concept Selection.

Morphological Analysis

The morphological analysis for the ice maker was developed by considering alternatives that does not increase the cost for the customer. Figure 2.23 shows the results of the analysis. According to the Morphological matrix three options were configured. *The first option* consists of changing the metal mold to a plastic mold. The plastic mold is significantly less expensive than the metal mold. In addition, the function of ejecting the ice has been replaced with a different method of ejecting the ice, eliminating the ejector blade and the ejector guide. The new method consists of rotating the mold and utilizing gravity to eject the ice after the mold

Function	Current	Alternatives		
Convert Electrical energy to mechanical Energy	electric motor			
Hold water	metal mold	plastic mold	changeable plastic mold	rotable mold
Rotate ejector blade	mechanical			
Control temperature	termostast			
Eject ice	ejector blade	swing latch	rotate upside down	
Heat the coil	Current			
Lower shut off arm	metal arm	rotational metal arm	magnetic contact sensor	manual shut off
Transmit heat to mold	heat coil	heating sheet		
Contain ICE	bucket	smaller bucket	adjastable bucket	angled bucket
Loosing the Ice	heat			

Fig. 2.23 Morphological matrix for the ice maker

has been heated. The lower shut-off arm has been modified to a rotational arm. The existing problem with the lower shut off arm is that the ice begins to stack up on top of the arm, resulting in continues ice production.

The current heating method has also been changed from a heating coil to a heating sheet. Because plastic does not conduct heat as effective as metal, the heating coil would present heat transfer issue and uniform heating of the plastic mold. Therefore, the heating coil was changed to a heating sheet, which can be evenly distributed under the surface of the mold. The last change for option 1 is changing the current fix volume bucket to an adjustable one. From the Kano diagram, this alternative is an attractive alternative, whether the household has one or four members, all benefit from the use of an adjustable bucket. *The second option* also changes the metal mold to a plastic mold, but the plastic mold is interchangeable. The mold incorporates a designer ice cube shapes essential for special events. The method of ejecting the ice used a swing latch construction. The mold swung downward, enabling the ice to fall freely into the bucket. The lever arm is eliminated by the use of a magnetic contact. The bottom of the bucket is fitted with two magnetic contacts to indicate when the bucket is full and transmitting this information to the shut-off electronics. This option also incorporates the heating sheet, due to the plastic material of the mold. The bucket is smaller to offset the cost of magnetic contact and to lessen the design difficulties of a larger bucket. *The third option* maintains the metal mold but eliminates one ice cube section. This enables a compact construction and frees up more freezer space. The rotational mold, as in option one, eliminates the ejector blade as well as the guide. The ice drops into the bucket utilizing gravity after the mold has been heated. The heating coil is unchanged. The bottom of the bucket is angled to push the ice forward; as a result, the oldest ice gets used first. The results are shown in Figs. 2.24–2.29. Design for asselby analysis was performed to determine the part cost from assembly point of view. The DFA methods will be discussed later in the book. Fig. 2.30 shows the DFA analysis for base part. The cost of the ice maker for each option was determined as presented in Figs. 2.31–2.33.

Function	Current	Alternatives		
Convert Electrical energy to mechanical Energy	electric motor			
Hold water	metal mold	plastic mold	changeable plastic mold	smaller rotable metal mold
Rotate ejector blade	mechanical			
Control temperature	termostast			
Eject ice	ejector blade	swing latch	rotate upside down	
Heat the coil	Current			
Lower shut off arm	metal arm	rotational metal arm	magnetic contact sensor	manual shut off
Transmit heat to mold	heat coil	heating sheet		
Contain ICE	bucket	smaller bucket	adjastable bucket	angled bucket
Loosing the Ice	heat			

Fig. 2.24 Configuration of the option 1

Function	Current	Alternatives		
Convert Electrical energy to mechanical Energy	electric motor			
Hold water	metal mold	plastic mold	changeable mold	smaller rotable metal mold
Rotate ejector blade	mechanical			
Control temperature	termostast			
Eject ice	eljector blade	rotate upside down	swing latch	
Heat the coil	Current			
Lower shut off arm	metal arm	rotational metal arm	magnetic contact sensor	manual shut off
Transmit heat to mold	heat coil	heating sheet		
Contain ICE	bucket	smaller bucket	adjastable bucket	angled bucket
Loosing the Ice	heat			

Fig. 2.25 Configuration of the option 2

Function	Current	Alternatives		
Convert Electrical energy to mechanical Energy	electric motor			
Hold water	metal mold	plastic mold	changeable mold	smaller rotable metal mold
Rotate ejector blade	mechanical			
Control temperature	termostast			
Eject ice	eljector blade	rotate upside down	swing latch	
Heat the coil	Current			
Lower shut off arm	metal arm	rotational metal arm	magnetic contact sensor	manual shut off
Transmit heat to mold	heat coil	heating sheet		
Contain ICE	bucket	smaller bucket	adjastable bucket	angled bucket
Loosing the Ice	heat			

Fig. 2.26 Configuration of the option 3

2.6.2.4 Criteria for Concept Selection

The concept selection matrix shown below lists the product requirements or technical characteristics considered to select the best option. These include cost, size, shape for ice, time to make the ice, and the design for assembly value.

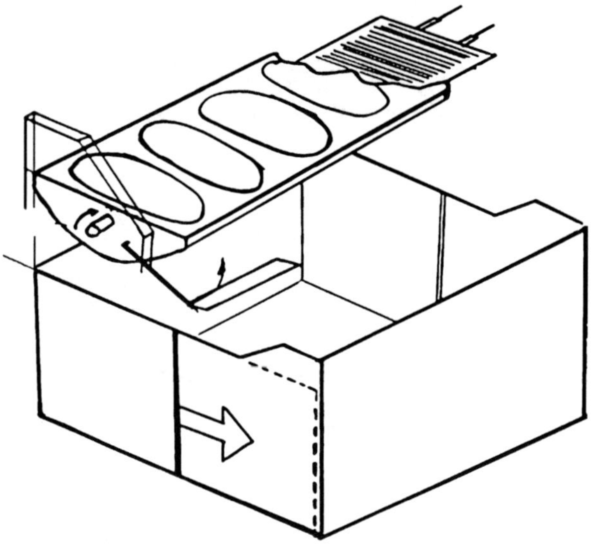

Fig. 2.27 Sketch of the option 1

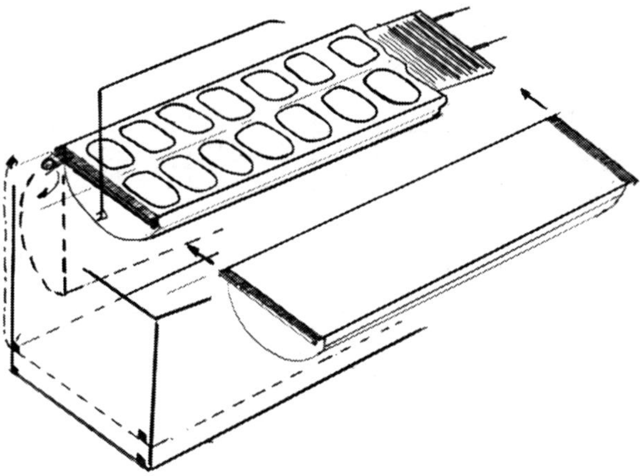

Fig. 2.28 Sketch of the option 2

Technical characteristics are derived from the QFD product planning matrix. Each alternative is rated on how well it meets the technical characteristics or requirements. The result is listed in Fig. 2.34.

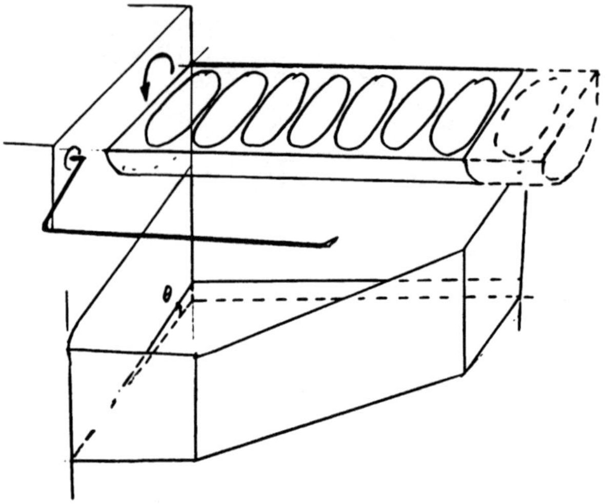

Fig. 2.29 Sketch of the option 3

DESIGN FOR MANUAL ASSEMBLY WORKSHEET											
1	2	3	4	5	6	7	8	9	10	11	NAME OF ASSEMBLY
Part I.D No.	number of times the operation is carried out consecutively	two-digit manual handling code	manual handing time per part	two-digit manual insertion code	manual insertion time per part	operation time, seconds (2) x [(4) + (6)]	Operation cost, cents 0.4 x (7)	Figures for estimation of theoretical minimum parts	α	β	
1	1	1,0	1.5	0,6	5.5	7	2.8	1	360	360	ice cube mold
2	1	2,1	2.1	3,1	5	7.1	2.84	1	360	360	Shutoff arm
3	1	0,2	1.88	3,0	2	3.88	1.552	0	0	180	ejector blade
4	1	2,0	1.8	3,2	4	5.8	2.32	1	360	360	Heater
5	1	1,2	2.25	3,0	2	4.25	1.7	1	0	360	Thermostat
6	1	2,0	1.8	1,3	6	7.8	3.12	1	360	360	Housing
7	1	2,0	1.8	1,3	6	7.8	3.12	1	360	360	Module
8	1	1,0	1.5	3,0	2	3.5	1.4	1	360	90	Cover
						47.13	18.852	7		0.445576	= Design efficiency = 3 x NM =
						TM	CM	NM			TM

Fig. 2.30 DFA analysis for original product

2.7 Summary

The product development life cycle can be defined as a sequence of all the required activities that a company must perform to develop, manufacture, and sell a product. These activities include marketing, research, engineering design,

Group part	Description/Name	Base price	Group cost
Cover	Enclosure cap white box	$7.17	$7.17
Ejector blade			$ 0.00
Housing	Electric motor	$13.05	
	TCO	$7.03	
	Hold switch	$5.07	
	Water fill switch	$5.07	
	Shut off switch	$5.07	
	Rocker arm	$1.89	
	Shout arm spring	$1.01	
	Enclosure brown box	$8.56	
	Wire	$4.95	
	Thumb screw	$0.12	$51.81
Ice cube mold	Silicone Rubber Blanket	$11.50	
	Ice Mold	$3.00	
	Water receptor guide	$2.10	$16.60
Module	Sprocket 48 t	$2.75	
	Mounting plate	$9.17	
	Connector with 4 leads	$6.21	
	Coupling	$1.92	$20.05
Shut off arm	Shut off arm	$1.27	$1.27
Thermostat	Thermostat	$7.95	$7.95
Bucket	Bucket	$14.04	$14.04
	Total	$118.88	$118.88

Fig. 2.31 Cost for the option 1

quality assurance, manufacturing, and a whole chain of suppliers, and vendors. This chapter provides the concept of product development life cycle, the importance of product development during the product development life cycle, the major phases of product development process, benchmarking, the systematic procedure of generating concepts, and finally a complete case study for understanding the presented concepts.

Group part	Description/Name	Base price	Group cost
Cover	Enclosure cap white box	$7.17	$7.17
Housing	Electric motor	$13.05	
	TCO	$7.03	
	Hold switch	$5.07	
	Water fill switch	$5.07	
	Shut off switch	$5.07	
	Rocker arm	$1.89	
	Shout arm spring	$1.01	
	Enclosure brown box	$8.56	
	Wire	$4.95	
	Thumb screw	$0.12	$51.81
Ice cube mold	Silicone Rubber Blanket	$11.50	
	Ice plastic Mold (3 option)	$5.00	
	Water receptor guide	$2.10	$18.60
Module	Sprocket 48 t	$2.75	
	Mounting plate	$9.17	
	Connector with 4 leads	$6.21	
	Coupling	$1.92	$20.05
Shut off arm	Magnetic contact sensor	$2.56	$2.56
Thermostat	Thermostat	$7.95	$7.95
Bucket	Bucket (3/4 of actual size)	$7.80	$7.80
	Total	$115.93	$115.93

Fig. 2.32 Cost for the option 2

2.8 Review Questions

2.1. Define the product development life cycle.

2.2. List the logical groups of activities that should be identified in product development.

2.3. Discuss in detail the importance of product development during the product development life cycle.

2.4. Why is the sequential product development process not efficient?

2.5. The sequential product development process has disadvantages, explain how the organization can improve this process.

2.6. What are the major phases of product development process? Explain

2.7. What is Benchmarking? Discuss in detail.

Group part	Description/Name	Base price	Group cost
Cover	Enclosure cap white box	$7.17	$7.17
Housing	Electric motor	$13.05	
	TCO	$7.03	
	Hold switch	$5.07	
	Water fill switch	$5.07	
	Shut off switch	$5.07	
	Rocker arm	$1.89	
	Shout arm spring	$1.01	
	Enclosure brown box	$8.56	
	Wire	$4.95	
	Thumb screw	$0.12	$51.81
Ice cube mold	Heater coil	$6.84	
	Ice Mold	$7.32	
	Water receptor guide	$2.10	$16.25
Module	Sprocket 48 t	$2.75	
	Mounting plate	$9.17	
	Connector with 4 leads	$6.21	
	Coupling	$1.92	$20.05
Shut off arm	Shut off arm	$1.27	$1.27
Thermostat	Thermostat	$7.95	$7.95
Bucket	Bucket	$10.40	$10.40
	Total	$114.90	$114.90

Fig. 2.33 Cost for the option 3

		Concept 1		Concept 2		Concept 3	
Criteria	**Weight**	**Rating**	**Weight score**	**Rating**	**Weight score**	**Rating**	**Weight score**
Cost	1/5	2	2/5	3	3/5	4	4/5
Ice maker Size	1/5	4	4/5	3	3/5	5	1
Shape of Ice	1/5	1	1/5	4	4/5	1	1/5
Time	1/5	1	1/5	2	2/5	3	3/5
Easy Assembly	1/5	4	4/5	3	3/5	1	1/5
TOTAL SCORE	1		2 2/5		**3**		2 4/5

Fig. 2.34 Concept selection

2.8. Describe the systematic procedure of generating concepts.
2.9. Discuss briefly the product development categories.
2.10. Select any product in the market and do all the required analysis that you learned in this chapter.

References

1. Ertas A, Jones JC (1996) The engineering design process, 2nd edn. Wiley, New York
2. Erhorn C, Stark J (1994) Competing by design: creating value and market advantage in new product development. Essex Junction, Vermont, Omneo
3. Rogers GG, Bottaci L (1997) Modular production systems: a new manufacturing paradigm. Intell Manuf 8(2):147–156
4. Sabbaghian N, Eppinger S, Murman E (1998) Product development process capture & display using web-based technologies, Proceedings of the IEEE international conference on systems, man, and cybernetics, San Diego, pp 2664–2669
5. Matzler K, Hinterhuber HH (1998) How to make product development projects more successful by integrating kano's model of customer satisfaction into quality function deployment Technovation 18(1):25–38
6. Kamrani AK, Salhieh SM (2002) Product design for modularity. Springer, Berlin
7. Crabb HC (1998) The Virtual Engineer: 21st Century Product Development (Society of Manufacturing Engineers), Hardcover
8. Jo HH, Parsaei HR, Sullivan WG (1993) Principle of concurrent engineering, concurrent engineering: contemporary ssues and modern design tools. In: Parsaei HR, Sullivan WG (eds) Chapman & Hall, London
9. Frank DP (1993) Re-engineering to achieve a concurrent engineering environment. J Design Manufac 3(1):75–89
10. Dan D, Dwenger K (1996) World-class new product development: benchmarking best practices of agile manufacturers. American Management Association, New York
11. Huang GQ (1996) Design for X: concurrent engineering imperatives, 1st edn. Chapman & Hall, London
12. Pimmler UT, Steven D (1994) Integration analysis of product decompositions, design theory and methodology – DTM'94, DE-68(1), ASME
13. Ulrich K, Karl T, Steven D, Eppinger S (1995) Product design and development. McGraw-Hill, New York
14. Roozenburg NFM (1995) Product design: fundamentals and methods. Wiley, New York
15. Smith GP, Donald G (1991) Developing products in half the time. Van Nostrand Reinhold, New York
16. Snead CS (1989) Group technology: foundation for competitive manufacturing. Van Nostrand Reinhold, New York
17. Pugh S (1991/1993) Total design: integrated methods for successful product engineering. Addison-Wesley, Wokingham, England; Reading, MA
18. Kamrani AK (1997) "Modular Design Methodology for Complex Parts", 6th IIE Annual. Engineering Research Conference, Miami Beach, Florida, USA, pp 391–396
19. Boothroyd G, Dewhurst P, Knight W (1994) Product design for manufacture and assembly. M. Dekker, New York
20. Cohen M, Eliashberg J, Ho T (1995) New product development: the performance and time-to-market tradeoff, Manag Sci 3(2):89–95

Chapter 3
Modular Design

Abstract This chapter provides the concept of modular design, which is a design technique that can be used to develop complex products using similar components. Modular design can be viewed as the process of producing units that perform discrete functions, and then connecting the units together to provide a variety of functions. Moreover, this chapter provides the concept of modular design, modularity, product modularity representation, the modular systems development process, the product development process using the design structure matrix, and finally, a complete description for DSM building.

Keywords Modular design • Modularity • Product modularity representation • Structure matrix

3.1 Introduction

Modular design is a design technique that can be used to develop complex products using similar components [1, 2]. Components used in a modular product must have features that enable them to be coupled together to form a complex product. Modular design can be viewed as the process of producing units that perform discrete functions, then connecting the units together to provide a variety of functions. Modular design emphasizes the minimization of interactions between components, which will enable components to be designed and produced independently. Each component designed for modularity is supposed to support one or more functions. When components are structured together to form a product, they will support a larger or general function. This shows the importance of analyzing the product function and decomposing it into sub-functions that can be satisfied by different functional modules. There are different definitions that have been established to define what modularity is. Most of them are based on the concept of module but there is not a unique consensus about it. Some differences lay on the reason why a company should do it, others on the drivers of modularity and the

A.K. Kamrani and E.A. Nasr, *Engineering Design and Rapid Prototyping*,
DOI 10.1007/978-0-387-95863-7_3,

ways how a product should be clustered. Many authors define it in different ways; here are some of those definitions [3, 4]:

Modules:

- Group of standard and interchangeable components [5].
- A structurally independent building block of a larger system with well-defined interfaces. A module has fairly loose connections to the rest of the system allowing an independent development of the module as long as the interconnections at the interfaces are well thought of.
- A complex group that allocates a function to the product and which could be changed and replaced in a loose way and be produced independently [6].
- A component or group of components that can be removed from the product non-destructively as a unit, which provides a unique basic function necessary for the product to operate as desired [7].
- A set of components and the architecture consists of all the components in the product plus relationships among the components [8].
- Modules have the following characteristics [9]:
  - They are co-operative subsystems that form products, manufacturing systems, and so on.
  - Functional interactions occur within rather than between modules.
  - They have one or more, well-defined functions that can be tested in isolation from the system and are a composite of components of the module.
  - They are independent and self-contained, and can be combined and configured with other modules to achieve overall function.

Modular Systems and Design:

- Modular systems are those whose interfaces are well defined and shared with only a few other systems [4].
- Modular design can group components into easily detachable modules such that they can be easily re-used or re-manufacture. Material compatibility has to be considered for recycling apart from ease of disassembly [10].

Modularity:

- Three fundamental elements to modularity [11, 12]:
  - Independence of a module's components from external components.
  - Similarity of components in a module with respect to their life-cycle process.
  - Absence of similarities to external components.
- Using structurally independent modules to form product architecture [13].
- It arises from the decomposition of a product into subassemblies and components. This division facilitates the standardization of components and increased product variety [14].
- A degree to which a product's architecture is composed of modules with minimal interactions between modules [7, 15].

- Most important characteristic of product architecture. The most modular architecture is one in which each functional element of the product is implemented by exactly one chunk (subassembly) and in which there are few interactions between chunks. Such a modular architecture allows a design change to be made to one subassembly without affecting the others [16].

From these definitions, it is obvious that there is some consensus about the meaning of modularity, especially around the concept of module and the idea of modularity as a grouping of components. However, there are still some differences among the characteristics of modularity.

3.2 Modularity Types

Modularity can be applied in the areas of product design, design problems, production systems, or all three. It is preferable to use modular design in all three types at the same time; this can be done by using a modular design process to design modular products and to produce them using a modular production system or modular manufacturing processes.

3.2.1 Modularity in Products

Modular products are products that fulfill various overall functions through the combination of distinct building blocks or modules [17], in the sense that the overall function performed by the product can be divided into sub-functions that can be implemented by different modules or components. An important aspect of modular products is the creation of a basic core unit to which different elements (modules) can be fitted, thus enabling a variety of versions of the same module to be produced. The core should have sufficient capacity to cope with all expected variations in performance and usage. A good example of modular products is the personal computer (PC). Any PC consists of several components or building blocks such as hard drive, RAM, CPU, CD-ROM, video card, and many other modules. Many modules can be modified or changed with little or no modification to the other modules. For example, a CPU can be sold with different combinations of hard drives, RAM, and other options. Through the use of such modular components, a company can choose from a variety of major components and form a product that can meet the customers' needs.

3.2.2 Modularity in Design Problems

Most design problems can be broken down into a set of easy-to-manage simpler sub-problems. Sometimes complex problems are reduced into easier sub-problems, where a small change in the solution of one sub-problem can lead to a change in other sub-problems' solutions. This means that the decomposition has resulted in functionally dependent sub-problems. Modularity focuses on decomposing the overall problem into functionally independent sub-problems, in which interaction or interdependence between sub-problems is minimized. Thus, a change in the solution of one problem may lead to a minor modification in other problems, or it may have no effect on other sub-problems.

3.3 Modular Systems Characteristics

3.3.1 Categories of Modules

Modular systems are built from independent units or modules. Two major categories of modules are identified. They are *function modules* and *production modules* [17]. Function modules are designed to accomplish technical functions independently or in combination with other modules. Production modules are designed based on production considerations alone and are independent of their function. Function modules can be classified based on the various types of functions reoccurring in a modular system that can be combined as sub-functions to implement the different overall function (Fig. 3.1). These functions are basic, auxiliary, special, adaptive, and customer-specific functions [18].

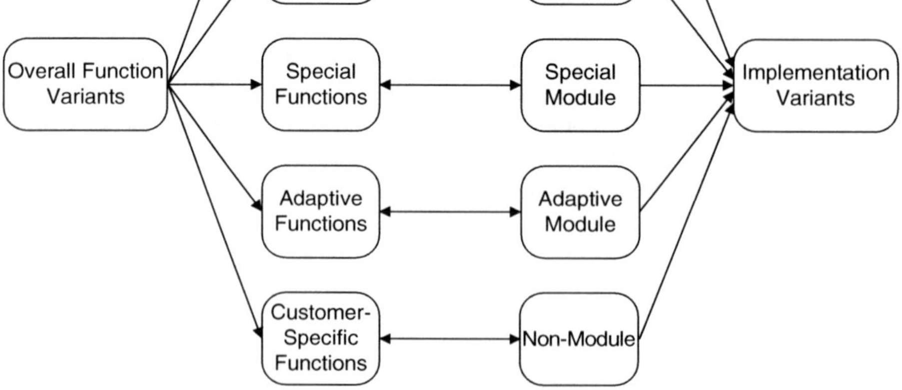

Fig. 3.1 Function and module types

3.3.1.1 Basic Functions

These are functions that can fulfill the overall function simply or in combination with other functions. Basic functions are not variable in principle and they are implemented in *basic modules*.

3.3.1.2 Auxiliary/Secondary Functions

These are implemented using *auxiliary modules* in accordance with basic modules.

3.3.1.3 Special Functions

These are task-specific sub-functions that may not appear in all overall function variants and are implemented by *special modules*.

3.3.1.4 Adaptive Functions

These are the functions that permit the adaptation of a part or a system to other products or systems. They are implemented by *adaptive modules* that allow for unpredictable circumstances.

3.3.1.5 Customer-Specific Functions

These are functions that are not provided by the modular system, and they are implemented by *non-modules* which must be designed individually. If they are used, the result is a mixed system that combines modules and non-modules.

3.3.2 Product Modularity Representation

Product modularity can be represented based on the types of combinations between the modules. Combinations between modules are analyzed based on the types of interactions between the different modules within a product. Four categories of modularity are defined in [19].

3.3.2.1 Component-Swapping Modularity

Different product variants belonging to the same product family are created by combining two or more alternative types of components with the same basic

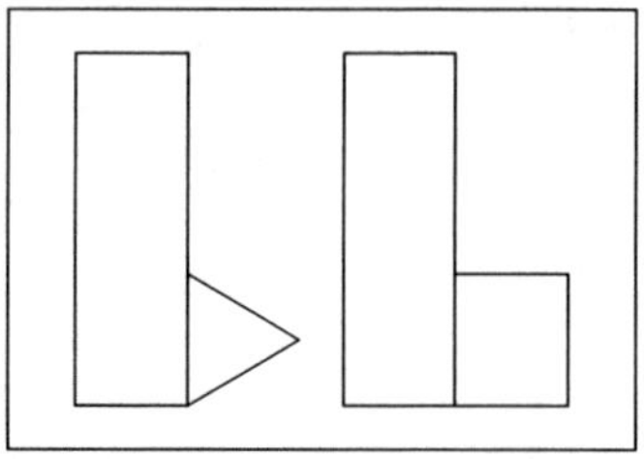

Fig. 3.2 Component-swapping modularity

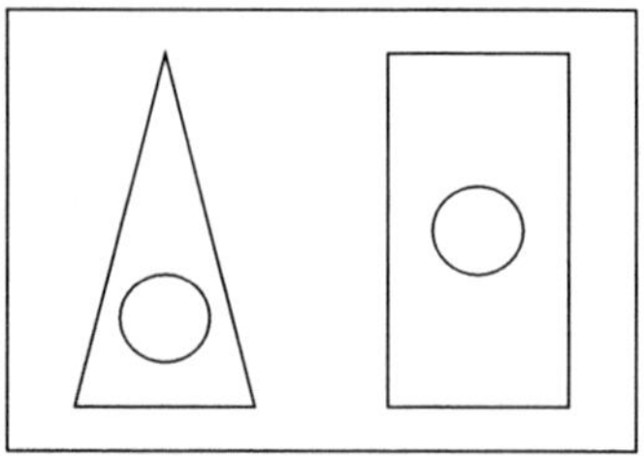

Fig. 3.3 Component-sharing modularity

component or product. Figure 3.2 illustrates the swapping modularity in which two alternative components (the small rectangular block and the triangular) are combined with the same basic component (the big block), forming product variants belonging to the same product family.

An example of component-swapping modularity in the computer industry is illustrated by matching different types of CD-ROMs, monitors, and keyboards with the same motherboard. This allows for different models of computers to be implemented.

3.3.2.2 Component-Sharing Modularity

In this category, different product variants belonging to different product families are created by combining different modules sharing the same basic component. Component-sharing is considered the complementary case to component-swapping. Component-sharing and component-swapping modularity are identical except that swapping involves the same basic product using different components and sharing involves different basic products using the same component. The difference between them lies in how the basic product and components are defined in a particular situation. Figure 3.3 shows two different basic components (block and triangular) sharing the same component (the circle). Component-sharing modularity in the computer industry is represented by the use of the same power cord, monitor, or microprocessor in different product (computer) families.

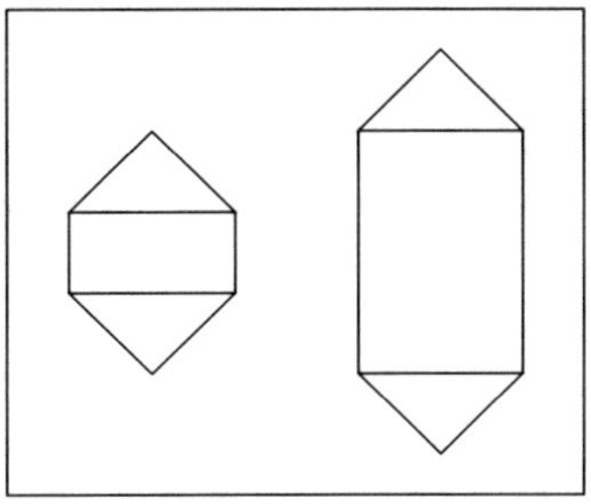

Fig. 3.4 Fabricate-to-fit modularity

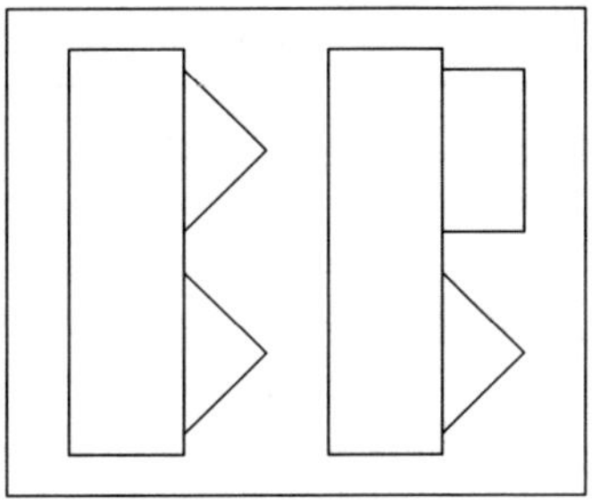

Fig. 3.5 Bus modularity

3.3.2.3 Fabricate-to-Fit Modularity

One or more standard components are used with one or more infinitely variable additional components. Variation is usually associated with physical dimensions that can be modified. Figure 3.4 illustrates a component with variable length (the block) that can be combined with two standard components (the triangular) forming product variants. A common example of this kind of modularity is cable assemblies in which two standard connectors can be used with an arbitrary length of cable.

3.3.2.4 Bus Modularity

This type of modularity occurs when a module can be matched with any number of basic components. Bus modularity allows the number and location of basic components in a product to vary. Bus modularity is illustrated in Fig. 3.5. An example of bus modularity is a computer where different input and output units, in addition to different types of mice, RAMs, and hard drives, can exist and vary in both their location and number.

3.4 Modular Systems Development

In general, modular systems can be developed by decomposing a system into its basic functional elements, mapping these elements into basic physical components, then integrating the basic components into a modular system capable of achieving the intended functions. This approach faces two important challenges: (1) Decomposition: Finding the most suitable set of sub-problems may be difficult. (2) Integration: Combining the separate subsystems into an overall solution may also be difficult. To fully comprehend the underlying foundations of modular systems development, decomposition categories are further discussed.

3.4.1 *Decomposition Categories*

System decomposition is expected to result in two benefits: (1) Simplification: Decomposing large systems into smaller ones will lead to a reduction in the size of the problem that needs to be solved, which will make it easier to manage. (2) Speed: Solving smaller problems concurrently (parallel solutions) will reduce the time needed to solve the overall problem. Decomposition methods can be categorized according to the area into which they are being applied: product decomposition, problem decomposition, and process decomposition.

3.4.1.1 Product Decomposition

Product decomposition can be performed at various stages of the design process and can be defined as the process of breaking the product down into physical elements from which a complete description of the product can be obtained. Two approaches are used in product decomposition, *product modularity* and *structural decomposition.*

Product Modularity

Product modularity is the identification of independent physical components that can be designed concurrently or replaced by pre-designed components that have similar functional and physical characteristics. Product modularity relies on the lack of dependency between the physical components. The computer industry provides an excellent example of modular products. The major components of a computer are shown in Fig. 3.6. Major components are manufactured by different supplier that allows the manufacturers of microprocessors to choose from a wide library of products.

Structural Decomposition

The system is decomposed into subsystems, and those are further decomposed into components leading to products, assemblies, sub-assemblies, and parts at the detailed design stage. The decomposition is represented in a hierarchy structure that captures the dependencies between subsystems. For example, the structural decomposition of a vehicle system is shown (Fig. 3.7), and the carriage unit is further decomposed into components (Fig. 3.8).

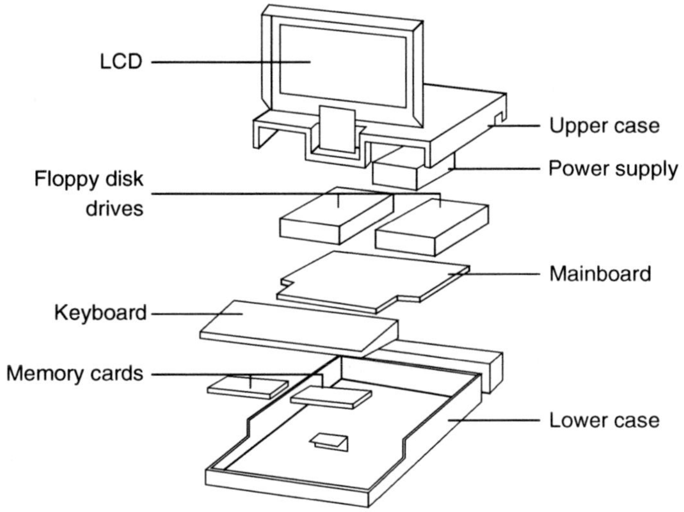

Fig. 3.6 PC assembly diagram

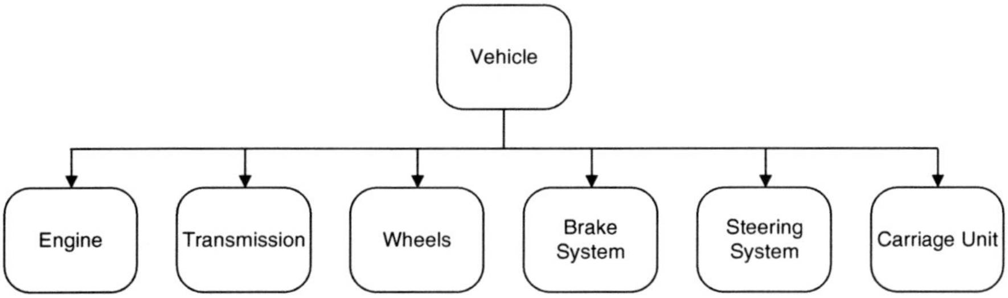

Fig. 3.7 Structural decomposition of a vehicle system

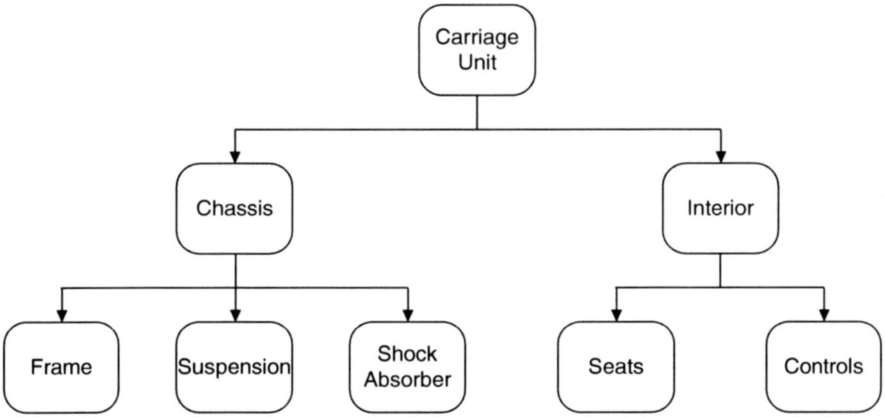

Fig. 3.8 Structural decomposition of a carriage unit

3.4.2 *Problem Decomposition*

For centuries, complex design problems were handled by breaking them into simpler, easy-to-handle sub-problems. Problem decomposition should continue until basic independent products or units are reached. The interaction between the basic products should be identified and introduced as constraints imposed by higher sub-problems. Problem decomposition is divided into *requirements decomposition*, *constraint-parameter decomposition*, and *decomposition-based design optimization*.

3.4.2.1 Requirements Decomposition

Requirements represent an abstraction of the design problem, starting with the overall requirement (general demand) and ending with the specific requirements (specific demands). The ability to meet a requirement is given by a design function. The requirements decompositions and their relationships to the corresponding functions are represented in a tree-diagram (Fig. 3.9), where specific requirements are mapped into specific functions.

3.4.2.2 Constraint Parameter Decomposition

The parameters describe the features (quantitative or qualitative data) of the product, while the constraints define the ranges of values assigned to parameters that are defined by product requirements. The problem structure is represented in an incidence matrix [34]. The incidence matrix is decomposed by grouping all non-empty elements in blocks at the diagonal. It is preferable that the blocks be mutually separable (independent). In some cases, overlapping between variables or constraints may occur. The design of a ball bearing is used to illustrate the decomposition [34].

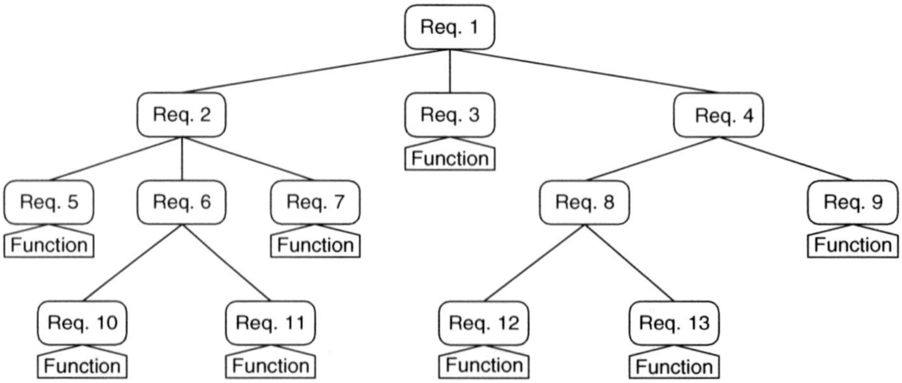

Fig. 3.9 Requirements decomposition

Table 3.1 Ball bearing design parameters

Parameter	Description	Parameter	Description
d_e	Pitch diameter	β_f	Free contact angle
d_o	Outer-race diameter	r_o	Outer-race curvature
d_i	Inner-race diameter	r_i	Inner-race curvature
P_d	Diametrical clearance	P_e	Free endplay
D	Rolling-element diameter	s	Shoulder height
I	Race conformity ratio	θ	Shoulder angle height
R	Race curvature radius	R	Curvature sum
B	Total conformity	R_x	x direction effective radius
I_o	Outer-race conformity	R_y	y direction effective radius
I_i	Inner-race conformity	Γ	Curvature difference
D	Race curvature distance	β	Contact angle

Table 3.2 Ball bearing design

C_1	$d_e = \frac{1}{2}(d_o + d_i)$	C_7	$P_e = 2D\sin\beta_f$
C_2	$P_d = d_o - d_i - 2d$	C_8	$s = r(1 - \cos\theta)$
C_3	$f = r / d$	C_9	$\frac{1}{R} = \frac{1}{R_x} + \frac{1}{R_y}$
C_4	$B = f_o + f_i - 1$	C_{10}	$\Gamma = R\left(\frac{1}{R_x} - \frac{1}{R_y}\right)$
C_5	$D = Bd$	C_{11}	$R_x = d(d_e - d\cos\beta)/2d_e$
C_6	$\beta_f = \arccos\frac{r_o + r_i - \frac{1}{2}(d_o - d_i)}{r_o + r_i - d}$	C_{12}	$R_y = f_i d/(2f_i - 1$

The parameters are listed in Table 3.1 and the constraints are shown in Table 3.2. The constraint–parameter incidence matrix is shown in Fig. 3.10. The decomposed matrix is shown in Fig. 3.11.

3.4.2.3 Decomposition-Based Design Optimization

The decomposition of a large complex design problem into smaller independent sub-problems facilitates the use of mathematical programming techniques to solve and optimize the sub-problems [20–22]. The solutions are integrated to provide an overall solution. The objective is to decompose a complex system into multi-level subsystems in a hierarchical form (Fig. 3.12), in which a higher-level subsystem controls or coordinates the subsystems at the lower level. The subsystems are solved independently at the lower level. The objective at the higher level is to coordinate the action of the first level to ensure that the overall solution is obtained.

	d_e	d_o	d_i	P_d	d	l	r	B	l_o	l_i	D	β_l	r_o	r_i	P_e	s	θ	R	R_x	R_y	Γ	β
C_1	*	*	*																			
C_2		*	*	*	*																	
C_3					*	*	*															
C_4								*	*	*												
C_5					*			*			*											
C_6		*	*		*							*	*	*								
C_7											*	*			*							
C_8							*									*	*					
C_9																		*	*	*		
C_{10}																		*	*	*	*	
C_{11}	*				*														*			*
C_{12}					*					*										*		

Fig. 3.10 Ball bearing design constraint–parameter incidence matrix

	l	r	s	θ	d_e	d_o	d_i	β	P_d	r_o	r_i	l_o	l_i	R_y	R	Γ	D	P_e	d	β_l	R_x	B
C_3	*	*																	*			
C_8		*	*	*																		
C_1					*	*	*															
C_{11}					*			*											*		*	
C_2						*	*		*										*			
C_6						*	*			*	*								*	*		
C_4												*	*									*
C_{12}													*	*					*			
C_9														*	*						*	
C_{10}														*	*	*					*	
C_5																	*		*			*
C_7																	*	*		*		

Fig. 3.11 Decomposed constraint–parameter incidence matrix

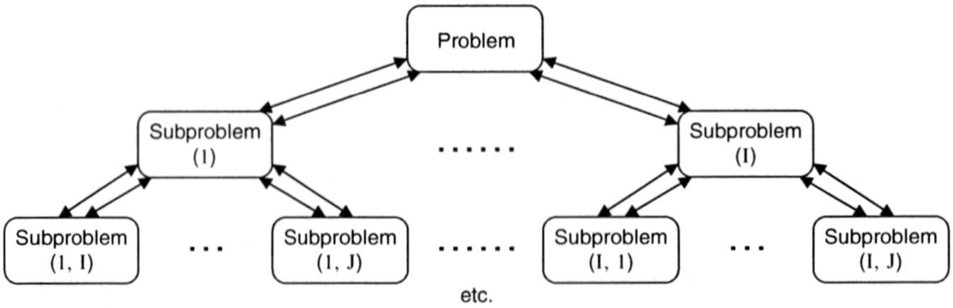

Fig. 3.12 Hierarchical decomposition of a complex system

3.4.3 *Process Decomposition*

Process decomposition is the decomposition of the entire design process, starting with the need recognition and ending with the detail design. The activities in the design process are modeled in a generic manner independent of the specific product being designed. Three perspectives of process decomposition were recognized. These are *product flow perspective*, *information flow perspective*, and *resource perspective*.

3.4.3.1 Product Flow Perspective

Design activities required to translate customer requirements into a detailed design of products are the focus of this perspective. The design activities are modeled as blocks with identified inputs and outputs (the output of one activity becomes the input of another activity). The decomposition tries to eliminate redundant activities and reorganize other activities to be performed concurrently, which will eventually reduce the product development time.

3.4.3.2 Information Flow Perspective

Analysis of the precedence constraints between the design activities is the main concern of this perspective. Precedence constraints are utilized to generate the required information needed to build supporting databases and communication networks and to schedule design activities, all concurrently.

3.4.3.3 Resource Perspective

The resources provide activities with a mechanism for transforming inputs to outputs. In this perspective two types of constraints are considered:

- *External resource constraints*, in which the resource used by the activity is generated by an activity or resource that is external to the design process.
- *Internal resource precedence constraints*, in which the resource is developed in the design process and us ed by other activities.

3.5 Survey of Some Modularity Techniques

There are different forms in which modularization may be applied. Different modularization scenarios have different impacts on the life-cycle characteristics of a product. A product can be modularized to enhance assembly, reusability, etc. The effect of modularization on functions can be represented by functional interactions among components in terms of exchanges of materials, energy and signals, or spatial and geometric relationships. The geometric relationship includes attachment, positioning, motion, and containment.

3.5.1 Product Representation for Modular Design

There are different methods that have been used to represent modular designs, most of them based on matrices. This methods are often included as part of the design methods.

3.5.1.1 Dependence Matrix

Sosale et al. [23] developed a matrix that represents the single dependence relationship on a product, as well as some measurements of this dependence. The matrix relations are established using a scale from 0 to 10, which shows the dependence relationships, as well as the weight of the interactions. The columns and rows are spatial and geometric interactions in the product. Among those physical interactions there are [23]:

- Attachment: physical contacts, joints, fasteners, welds, couplings, etc.;
- Positioning: relative distance or angle between components, alignment, etc.;
- Motion: CAM-controlled objects, trajectory of joints, end effectors, etc.;
- Containments: e.g. components contained in the same housing; as well as
- Life-cycle issues.

3.5.2 Dependence and Similarity

3.5.2.1 Huang and Kusiak

Huang and Kusiak [24] used two matrices to represent a modular design. Their interest was to determine the interactions and the suitability of the matrices by measuring the similarity of the functional interactions. To do so, for the second matrix they used a scale for 0–10 to establish the frequency of the interactions, while the suitability of inclusion of the components in a module was measured using an a, e, o, and u scale to determine if it was strongly desired (a) or strongly undesired (u).

3.5.2.2 Gershenson et al.

Gershenson et al. [11, 14, 15] also used two matrices in some aspects in the same direction as the previous authors (dependencies and similarities) but with the difference that in this case both matrices represent the interactions between components, not functions. The rows in the matrices represent each component, and the columns represent each components and each life-cycle task that the product undergoes, like manufacturing, assembly, service, etc. The scale used to fill these matrices is from 1 to 5, according to the strength of the dependency or similarity. One of the concerns of the author is that since the products might have different levels of detail and at some points it can and cannot be modular, it is important to define the level of detail that is going to be used at the product's structure. To solve this problem, the authors used the component trees, a tool that describes the levels of detail of a product (components and subassemblies). Another problem with this method is that since there is no formal way to fill the matrices, the ratings are the primary source of error in this representation.

3.5.2.3 System Decomposition

Used by Pimmler and Eppinger [25], it is a form of design structure matrix to represent a modular design. To use it, it is necessary to specify the overall product concept in terms of its functional and/or physical elements. According to the authors, the decomposition of the product architecture should be done to one level of detail further than the one required.

3.5.2.4 Requirements and Functions

Proposed by Kusiak and Szczerbicki [10], it is a method that determines the type of material, energy or informational flow required to develop a digraph that can be used to establish the connections between components. It is based on functions and requirements. The digraph represents strongly connected components as vertex, and the edges represent cluster to cluster connections. Its adjacency matrix represents the number of paths of length l leading from vertex i to vertex j in the digraph. A reachability matrix R is also constructed, and the product of it and its transpose has to be calculated. If the result is 0 it means that the object is disconnected; else if it is 1, then the product is only one cluster and it cannot be decomposed.

3.6 Design Methods

There are several methods that have been used to design a product while applying modularity but only a few of them are systematic: *function structure heuristic method (FSHM)* and *modular function deployment (MFD)*. There are also different categories in which design methods can be classified: *checklist methods*, which are usually simplistic and inefficient; *design rules*, which can be easily used but are not too specific to complete a design; *matrix manipulations*, that allow component/module manipulations and are very detailed; and finally, *step-by-step measure and re-design methods*, that are often included as part of the matrix methods.

3.6.1 Systematic Methods

3.6.1.1 Function Structure Heuristic Method

FSHM is a functional decomposition block diagram of the entire product's functions and material, energy, and information flows between them [26, 27]. They use function structure to identify: dominant flows, branching flows, and conversion–transmission flows. Each of them is used to identify dependencies between modules. The idea of this method is to begin with a function structure of a single product, and from that, determine all the many possible modules that can be created by group functions. The objective was to develop a method that would assist the designers in grouping components to form easily detachable modules, as well as create modules that can be reused or remanufactured. In the case were recycling is a concern, it is very important to consider the material compatibility and the ease of disassembly. There are two ways in which the modules are designed [14]:

1. To form modules individually, considering the objectives separately. Once they are established then it is possible to make trade-off between the configurations until one achieve the desired combination.
2. To modularize a product based on a weighted average objective.

The goal of the heuristic is to define possible modules for the product but this method does not determine if those modules are the right ones for design and production. The selection of the module has to be made by the designer, according to the needs of the company. The modularization criteria that is considered in the selection of the modules are functionality and modules interfaces. There are three phases to use this design method [14, 28]:

1. *Problem definition*: identify the type and characteristics of the design problem, decompose the problem into sub-problems, and determine the objectives of modularization.
2. *Interaction analysis*: a matrix is developed to evaluate the interaction between components for the objective of the heuristic. Values are assigned to each objective according to a scale from 0 to 10, and a weighted average is the calculated for the interactions for any two components.
3. *Module formulation*: implement the algorithm to group the components into modules by maximizing the component–component interaction. The algorithm generates random clusters in the search for the best configuration. The case were two components cannot be separated is known as a single component. The search period is controlled by specifying the duration of the search. Based on the probability of change, the objective function values are calculated for each new configuration so that it replaces old configurations.

One of the characteristics of this method is that it has being designed to accommodate multiple life-cycle elements at the same time. Another one is that it can be applied to single products and product family of similar products. Although the goal of this method is to maximize the interactions within a module and with the use of the heuristic it is possible to achieve it, there are some deficiencies of this method:

- It only takes into consideration the maximization of the interactions within modules, but it does not consider the minimization of the interactions between modules
- Does not consider the notion of similarity
- Does not consider the incompatibility between components
- It is not easily quantified
- The final decision is made by the designer, which is based on experience.

3.6.1.2 Modular Function Deployment

Compared to the previous two methods, there is not much information available about this third one. Its main difference compared to the previous two is that it considers several other drivers besides functionality to identify new modules. There is not a consensus on the number of drivers that this method uses, but initially 12 were proposed [13]: *carryover*, i.e. a specific function will carry over to different products and no technology changes are expected; *technology evolution* and *planned product changes*, both drivers take into consideration expected and unexpected changes; *different specification* enables product variation; *styling* considers how the modularity choice would affect the appearance of the product; *common unit* is similar to Zamirowski and Otto's [29] common function heuristic in the function structure heuristic method; *process and/or organization*, *separate testing*, *supplier availability*, and *service and maintenance* are related to the organizational effects of modularization; *upgrading* allows future additions to the product; and *recycling* considers the afterlife of the product. The modules are grouped based on the functions that receive the highest summed scores; and, according to this method, the functions dominated by the same drivers are good candidates for a module. MFD estimates that the ideal number of modules is the square root of the number of parts or assembly operations. There is no step sequence to apply this method, but it is important to have an interface design step that considers from, fixation principles, number of contact surfaces and attachments, number of energy connection points, material flow, and signals. As the other two methods, it relies on designers and engineers to select the modules, which depending on several factors like experience, can be beneficial or detrimental for the company.

3.6.2 Other Methods

Holonic product design (HPD), developed by Marshall et al. [9], is a checklist method that focuses on cellular manufacturing and on the interactions of cells to manufacturing and the implications for the life-cycle stages beyond manufacturing. Its goal is to ensure that the goals of a company and its product requirements are taken into considerations well as the modularity decisions. The objectives are:

- Clarify reasons to change to a modular product architecture;
- Clarify business strategies and corporate objectives;
- Define the organization of the company;
- Provide a platform for the HPD methodology;
- Examine existing and future products and their features for suitability to modularity; and
- Provide guidance on the level of modularity suited to the product and the company.

The deficiency about this method is that although it is very detailed and easy to follow, it is not a method. It is simply a guideline.

Cluster identification method developed by Kusiak and Chow is a simple algorithm used to group a machine-part incidence matrix [30]. Its goal is to use group technology to group manufacturing machines and processes into subsystems, in an efficient way. The way the algorithm works is by crossing out rows and columns of the matrix to cluster all the machines with even indirect connections. The deficiencies of the method are:

- It does not yield optimal results, although it leads to significantly improved modules faster.
- There is no guarantee that it will continue to move a design toward a more modular product.
- It leads to a final configuration that may or may not be feasible since the algorithm does not have intermediate steps to move the algorithm toward a better design.

Fishbone diagram for disassembly is used by Allen and Carlson-Skalak for the development of feature changes [7]. It is an information source that can be very useful if combined with other methods because of its simplicity. The steps are:

- Identify the modules of previous product generations;
- Identify the function structure with respect to the company's structure;
- Develop a product's system function structure; and
- Calculate a metric to indicate the modularity of the product.

Three heuristics methods are developed by Otto et al. to group functions in a module to manage variety [31–33]. Steps are:

- Analyze the need and product uses of the client
- Identify functions within the family structure
- Cluster functions in relation to heuristics
- Select physical modules, in technological and physical dimensions
- Compare and select product architectures.

3.7 Design Structure Matrix

Product development can be viewed as a process that transforms input information about customer needs into output information corresponding to manufacturable designs and tooling for production. Each individual development activity is an information-processing unit that receives information from previous activities and transforms it into information suitable for use by subsequent tasks. Information needs create dependencies between tasks that determine the product development structure and the most appropriate task sequencing. Tasks can be classified [16] – according to information dependencies – into (Fig. 3.13):

- *Sequential*: if there is a uni-directional information flow,
- *Parallel*: if there is no information exchange and thus can be performed simultaneously,
- *Coupled*: if they are mutually dependent and the information flows both ways.

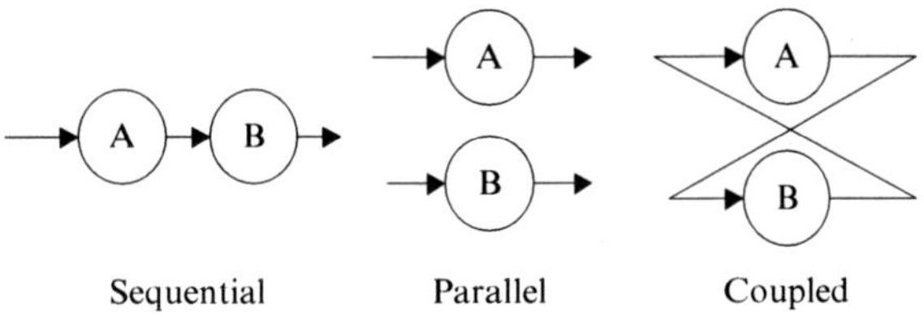

Fig. 3.13 Information dependency types

Fig. 3.14 DSM matrix

	A	B	C	D	E	F
A	•	×				
B		•				
C	×	×	•			
D			×	•		×
E				×	•	
F				×		•

These information dependencies constitute the foundations of building and understating the Design Structure Matrix (DSM).

3.7.1 *The Design Structure Matrix Approach*

Steward [34] introduced the Design Structure Matrix (DSM) as a method that can be used to manage the design of complex systems based on the information flow that occur within the system, identify the dependencies between tasks, and sequence the development process.

The DSM (Fig. 3.14) consists of an $N \times N$ matrix where the design tasks to be performed are listed on the x-axis and then in the same order on the y-axis. The relations between the different tasks are then listed in the matrix, a mark "X" means that a relation exists between the tasks, and a blank cell means that there is no relation between the tasks. A mark in row i, column j means i has the predecessor j. That is, going across the rows shows what precedes, going down the columns shows what follows. Marks below the diagonal represent feed forward, and marks above the diagonal represent feedback.

The primary goal of DSM analysis is to find a sequence of these design tasks that allow this matrix to become lower triangular.

3.7.2 *DSM Types*

There are four types of DSMs, the distinction between the four matrices is based on the type of dependency each matrix represent. These types are as follows.

3.7.2.1 Component-Based DSM: System Architecture

This type of DSM is used to represent system architecture in terms of the relationships between the components that form the system. The way a system is decomposed into its basic subsystems or components affects the success of system integration. An effective decomposition strategy should allow for ease of integration, that is the decomposition step should be performed in a way that prepares the identified components for "future integration." The goal of a component-based DSM is to reduce the coordination complexity by decomposing and then clustering components into subsystems where interactions occur within subsystems rather than between subsystems. The interactions (marks) captured in this type of DSM can be classified into four broad types [35]:

1. *Spatial*: Association of physical space and alignment; needs for adjacency or orientation between two elements.
2. *Energy*: Need for energy transfer/exchange between two elements (ex. power supply).
3. *Information*: Needs for data or signal exchange between two elements.
4. *Material*: Needs for material exchange between two elements.

Four separate DSMs can be built to represent the four types of interaction, and the most important DSM (based on the nature of the system being analyzed) will be used to rearrange the system and complete the DSM analysis.

3.7.2.2 Team-Based DSM: Organization Structure

Team-based DSMs document the information flow between different organizational elements, i.e. individuals and/or teams. The information flow captured could be between multiple organizational levels, although it is recommended that a single DSM information flow is captured between a single organizational level. The goal of an organizational DSM is to identify the key organization interfaces and group teams where information flow is most frequent. An organizational DSM is time-dependent, meaning a single DSM represents a snap shot of the organization information flow at a given point of time, it may be necessary to build several DSMs to capture information flows in different times. Interactions or information flow between teams can be characterized in several ways [36]:

1. *Level of detail*: sparse (documents) to rich (models, face-to-face)
2. *Frequency*: low (one-shot, batch) to high (piece-by-piece, on-line, intensive)
3. *Direction*: one-way, two way, to broadcast
4. *Timing*: early (preliminary, begins the process) to late (completed work, ends the process).

3.7.2.3 Activity-Based DSM

An activity-based DSM describes the input/output relationships between activities that form a certain project. The goal of such a DSM is to find a better sequence of activities to shorten the project time. Activity-based DSM gives a visual representation of project's activities that will allow for identification of feedbacks and loops that exists within the project. Understating the activity-based DSM relies on understanding the concept of iterations and the difference between serial, parallel, and coupled activities. Iterations exist in most product development processes in two ways: (1) intentional iterations which allows a coupled design process to converge to a desirable solution, (2) unintentional iterations which results from new information arriving late in the process.

3.7.2.4 Parameter-Based DSM

Design activities are usually targeted toward determining the values of parameters that constitute the lowest level of design. Those parameters are determined based on other parameters, and the parameter-based DSM objective is to capture the relationships between those parameters and provide a more appropriate arrangement of the sequence by which parameters are determined.

3.7.3 Building the DSM

Building DSMs require selecting an appropriate decomposition methodology capable of identifying all subsystems or components that form the system, and documenting all the relationships that exists between the subsystems. The method used to decompose a certain system into its basic components or modules that comprise that system depends on the nature of the system under study. In the case of "hardware" systems such as mechanical products, a physical decomposition method can be used. Physical decomposition is a hierarchal decomposition technique where the system is broken down into smaller subsystems and those in turn are broken down into components that forms the basic elements that comprises the system. Other systems may be analyzed by asking a group of experts from different functional areas related to the system under study to collectively list all subsystems/components/elements/tasks that comprise the system [37]. Such a method can be used for project management or team-based systems where the system analysis investigates the operations of the project or the dependencies between teams. After identifying the system components or elements, they are listed in the DSMs as row and column in the same order. The elements within the matrix are then identified by asking the appropriate manager/expert in the group for the minimum set of parameters (taken from the list) that influence their own subsystem and contribute to its behavior. In a task-based DSM, this can be the minimum set of activities that need to be performed before the activity under questioning can be started. In a parameter-based DSM, the rows and columns are design parameters that drive the design or define the system and the managers/experts can then be asked to define the precedence relationships between the listed parameters. These tasks/parameters/elements are marked in the DSM by an 'X'. A formal approach to building DSMs was proposed by Dong [38], this approach is presented in the next section.

3.7.3.1 Step 1: Define the System and Its Scope

This step determines the system boundaries so that effort can be focused on the system itself.

3.7.3.2 Step 2: List All the System Elements

Existing project plans or system definition could be used as a starting point in defining the system elements. Those system elements may need to be modified later in the process.

3.7.3.3 Step 3: Study the Information Flow Between System Elements

This step investigates the information flow between system elements by reading the design documents and interviewing experienced engineers who are working on the particular product.

Steps 2 and 3 are highly iterative. Deeper understanding of the system usually results in modification of the initial system elements.

3.7.3.4 Step 4: Build a Matrix to Represent the Information Flow

Initially, a binary DSM can be built to represent the basic dependency structure and information flows between various system elements. A binary DSM serves as a good start for preliminary analysis; however, a better understanding of the system (or project) might require the use of a numerical DSM that will provide better system understanding and allow for more detailed analysis.

3.7.3.5 Step 5: Partition the Matrix

Refer to Sect. 3.7.5 for information about partitioning

3.7.3.6 Step 6: Optimize the Matrix

Enhancing the matrix could be done by tearing, coupling, and de-coupling. More information is presented in the following sections.

3.7.3.7 Step 7: Give the Matrix to Engineers and Managers to Comment On and Use

Since the DSM can provide both managers and engineers with a better understanding of the design process, it is important that they (managers and engineers) get a chance to verify that the DSM clearly represents the system or the process understudy.

3.7.4 *Partitioning the DSM*

An interaction between two elements or components in the DSM is represented by a non-blank cell in the matrix and is called a "mark." The DSM should be reordered – the sequence of elements is kept on both rows and columns – so that the matrix is a lower triangular (all the marks are either on or below the diagonal), this will ensure that whenever a variable or element is being determined all the required predecessors are already known or available. Figure 3.15 shows an example of a DSM where the matrix was reordered. This reordering process is called partitioning.

Most systems have cycles or feedbacks between their elements or components; this makes it difficult to reach an ordering where a lower triangular form of the matrix can be achieved. The partitioning process (reordering) can be used to rearrange the matrix to produce a "block triangular" matrix (Fig. 3.16) rather than "lower triangular." A block triangular matrix is a matrix where the marks appear either below the diagonal or within square blocks on the diagonal. These square blocks represent the smallest set of elements/variables that must be developed/or determined together. There are many techniques that can be used to partition a DSM, most of these techniques are similar, but they do differ in the manner by which they identify loops (circuits/feedbacks). In general, partitioning techniques or algorithms proceed as follows [34]:

a

	A	B	C	D	E	F
A	•			X		
B	X	•	X		X	
C			•		X	
D				•		
E	X				•	
F		X	X			•

Original DSM

b

	D	A	E	C	B	F
D	•					
A	X	•				
E		X	•			
C			X	•		
B		X	X	X	•	
F				X	X	•

Re-Ordered DSM

Fig. 3.15 Lower triangular DSM

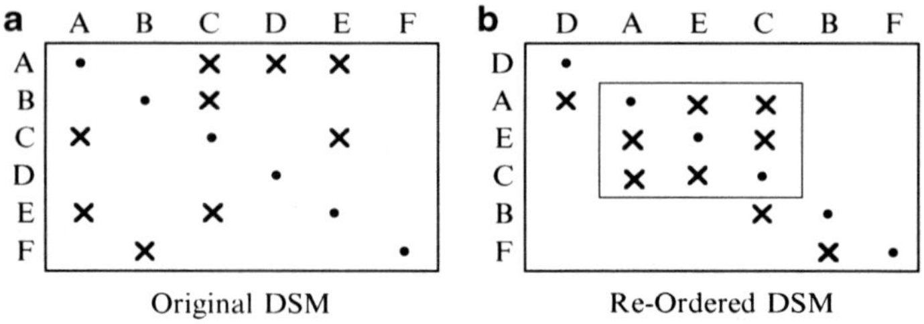

Fig. 3.16 Block triangular DSM

Step 1: Identify tasks or elements that do not require any input or interaction from any other element in the DSM, i.e. identify elements with an empty row. Then place that element or tasks at the top of the matrix (re-arrange). Once an element is rearranged it should be removed from the matrix with all its corresponding marks.
Step 2: Repeat step 1 on all remaining elements.
Step 3: Identify tasks or elements that do not provide any output or interaction with the rest of the elements in the DSM, i.e. identify elements with an empty column. Then place that element or task at the bottom of the matrix (re-arrange). Finally, remove the element with all its corresponding marks from the matrix.
Step 4: Repeat step 3 on all remaining elements.
Step 5: If there are no remaining elements, then the DSM is completely partitioned; Stop. Otherwise loops/circuits/feedbacks exist and must be analyzed; proceed to step 6.
Step 6: Select any remaining element/task (arbitrary). Call this task (task 1), then trace its dependence until "task 1" is reached again. All tasks encountered during the tracing are said to be in a loop and should be "collapsed" together to form a single composite task.
Step 7: Redraw the DSM with the new composite task, and go to step 1.

3.7.5 Tearing the DSM

After a DSM is partitioned, we may be left with some "marks" above the diagonal. These marks mean that some needed information is still unknown, and exists within a loop or a circuit. A guess is needed to estimate the value of those marks [34]. The "marks" or unknowns to be estimated should be selected in a way that will render the matrix lower triangular after removing the estimated marks. The process of choosing the marks to be removed is called "tearing," and the removed marks are called "tears." The goal of tearing is to work within the loops of information flow identified by the partitioning step. Each loop defines a set of tasks which are in a block and which must be considered simultaneously. However, a loop can be temporarily broken to initialize iteration if one or more tasks are performed without all of its necessary information.

The aim of the tearing procedure is then to determine which task or tasks provide the best opportunity to initialize the loops of information flow. There is no optimal method for tearing, but the following general guidelines can be used when tearing: (1) minimize the number of tears, (2) start tearing in the smallest blocks along the diagonal, and (3) tear where good estimates can be made.

3.7.6 Comments on DSM and Product Development Improvement

Product design and development is a complicated and challenging task due to many reasons such as:

1. Product can be technologically complex
2. A complete design process may involve millions of tasks
3. All tasks are coupled in some manner
4. Iterations are an inherent characteristic of design.

In order to improve the product design and development process, one must first model the process to understand it. There are several methods for modeling machining and to a lesser degree assembly, but there is not any single convincing and comprehensive modeling tool that can be used to fully analyze the design process and allow for systematic improvements. The Design Structure Matrix is a step in the development of such a modeling tool. The DSM assumes that the basic elements of a design process are tasks that require input information, take time to execute, and produce decisions or output information for input to other tasks. It further assumes that a major route to process improvement, other than making each task more efficient, is to re-sequence individual tasks or groups of tasks so that required information is available sooner and available information is used sooner.

The DSM can be used as a display of the existing design process or of the designers' view of the process. The matrix graphically displays all the existing information flows and makes it easy to see the difficulties in the form of coupling and unnecessary delays. The DSM can also be used as a management tool to redirect engineering efforts to tasks involved in key iterations. Also, design reviews and progress assessment can be based on the matrix, with managers assuring that available information was actually used, required information was actually transmitted and received, and tasks started as soon as they could.

3.8 Modularity Advantages

Many benefits have been identified with the implementation of modularity. Having the ability to create sub-components that might be assembled in different ways to build a product, or the chance to interchange parts in order to satisfy customer requirements, or been able to reduce costs because of standardization, are few of the many benefits that have been written about modularity. The benefits of modularity are [5, 12, 15]:

1. Component economies of scale due to the use of components across product families,
2. Ease of product updating due to functional modules,
3. Increased product variety from a smaller set of components,
4. Decreased order lead-time due to fewer components,
5. Ease of design and testing due to decoupling of product functions,
6. Ease of service due to differential consumption.

In terms of the designer, modularity allows to control the degree of changes, simplifies the product architecture, and eases the design of product components because it uses the same ones. It also provides flexibility for the manufacturing of the product, allowing postponement which enables the company to respond in a more effective way to changes in demand by delaying the differentiation of the product to the last stages of its production. Flexibility also allows independent development of each module and enables product variations. By having several modules and standardization it reduces the number of processes and repetitions, thus reducing the life-cycle costs. Modularity promotes interchangeability, hence increases flexibility and reduces time cycle. Its products can be more reusable than the non-modular ones. For customers, modularity gives the opportunity to mix and match, to create a product according to their needs; and for the manufacturer, it gives them the tools to satisfy the customer requirements and offer customization. In summary, impacts of the modularity are as follows.

3.8.1 Reduction in Product Development Time

Modularity relies on dividing a product into components with clear definition of the interfaces. These interfaces permit the design tasks to be decoupled. This decoupling results in a reduction in the design complexity and enables design tasks to be performed concurrently, which will eventually reduce the product development time.

3.8.2 Customization and Upgrades

Modular products accomplish customer requirements by integrating several functional components interacting in a specific manner. This integration allows products to be improved and upgraded by using more efficient components that can perform the required functions effectively. In addition, components can be replaced by custom-made ones to fulfill different functions.

3.8.3 *Cost Efficiencies Due to Amortization*

Modular components are used in several product lines, which means that their production volumes are higher. This will allow the amortization of the development expenses over a large number of products.

3.8.4 *Quality*

Modularity allows production tasks to be performed simultaneously. Thus, independent components can be produced and tested separately before they are integrated into a modular product. This will help build quality into the product.

3.8.5 *Design Standardization*

Modular design facilitates design standardization by identifying the component functions clearly and minimizing the incidental interactions between a component and the rest of the product.

3.8.6 *Reduction in Order Lead-Time*

Modular products can be made by combining standardized and customized components. This allows standard components to be inventoried, and then customization can be focused on the differentiating components. Also, modular products can be a combination of standard components, that is, the same standard components (usually kept in inventory) are integrated in different ways to form a variety of products that can respond to customer requirements.

3.9 Summary

Modular design is a design technique that can be used to develop complex products using similar components. Modular design can be viewed as the process of producing units that perform discrete functions, then connecting the units together to provide a variety of functions. The product development life cycle can be defined as a sequence of all the required activities that a company must perform to develop, manufacture and sell a product. These activities include marketing, research, engineering design, quality assurance, manufacturing, and a whole chain of suppliers and vendors. This chapter provides the concept of modular design, modularity, product modularity representation, the modular systems development process, the product development process using the DSM, and finally a complete description for DSM building.

3.10 Review Questions

3.1. Define the modular design.
3.2. What is the modularity? Discuss it types.
3.3. Discuss in details product modularity representation.
3.4. Describe the modular systems development process.
3.5. Discuss briefly some of modularity techniques.
3.6. How can the product development process using the design structure matrix be used?
3.7. Describe how DSM can be built.
3.8. What are the advantages of modularity?

References

1. Kamrani AK (1997) Modular design methodology for complex parts. In: Industrial engineering research conference, Miami Beach, FL
2. Shirly GV (1992) Modular design and the economics of design for manufacturing. In: Susman G (ed) Integrating design and manufacturing for competitive advantage. Oxford University Press, UK
3. Jose A, Tollenaere M (2005) Modular and platform methods for product family design: literature analysis. J Intell Manuf 16:371–390
4. Sosa ME, Eppinger SD, Rowles CM (2000) Designing modular and integrative systems. In: Asme 2000 international design engineering technical conferences and computers and information in engineering conference. ASME, Baltimore, MA, pp 1–9
5. Galsworth GD (1994) Smart, simple design, Essex junction. Oliver, Vermont
6. Wilhelm B (1997) Platform and modular concepts at Volkswagen – their effects on the assembly process. In: Shimokawa K, Juergens U, Fujimoto T (eds) Transforming automobile assembly – experience in automation and work organization. Springer, Berlin, p 146
7. Allen KR, Carlson-Skalak S (1998) Defining product architecture during conceptual design. In: Proceedings of the 1998 ASME design engineering technical conference, Atlanta, GA. The American Society of Mechanical Engineers, New York
8. Newcomb P 3rd, Bras B, Rosen DW (1998) Implications of modularity on product design for the life cycle. J Mech Des Trans As/Vie 120(3):483–491
9. Marshall R, Leaney PG, Botterell P (1998) Enhanced product realisation through modular design: an example of product/process integration. J Integr Des Process Technol 3(4):143–150
10. Kusiak A, Szczerbicki E (1993) Transformation from conceptual to embodiment design. IIE Trans 25(4):6–12
11. Gershenson JK, Prasad GJ, Allamneni S (1999) Modular product design: a life-cycle view. J Integr Des Process Sci 3(4):13–26
12. Ulrich K, Tung K (1991) Fundamentals of product modularity. In: Proceeding of the 1991 Asme design engineering technical conferences – conference on design/manufacture integration, Miami, FL

13. Holtta KM, Salonen MP (2003) Comparing three different modularity methods. ASME 9(3):189–206
14. Gershenson JK, Prasad GJ, Zhang Y (2004) Product modularity: measures and design methods. J Eng Des 15(1):33–51
15. Gershenson JK, Pasad GJ, Zhang Y (2003) Product modularity: definitions and benefits. J Eng Des 14(3):295–313
16. Ulrich TK, Eppinger SD (2000) Product design and development, 2nd edn. McGraw-Hill, Inc., New York
17. Roozenburg NFM (1995) Product design: fundamentals and methods. Wiley, Chichester
18. Pahl G, Beitz W(1996), Engineering design: a systematic approach. Springer, London
19. Ulrich K, Tung K (1991) Fundamentals of product modularity. Issues Des Manuf Integr DE-Vol. 39, ASME
20. Finger S, Dixon JR (1989) A review of research in mechanical engineering design, part I: descriptive, prescriptive, and computer-based models of design processes. Res Eng Des 1:51–67
21. Finger S, Dixon JR (1989) A review of research in mechanical engineering design, Part II: Representation, analysis, and design for the life cycle. Res Eng Des 1:121–137
22. Johnson RC, Benson RC (1984) A basic two-stage decomposition strategy for design optimization. Trans ASME 106 Please provide page ranges for Ref. [31].
23. Sosale S, Hashemian M, Gu P (1997) Product modularization for reuse and recycling. Concurrent Product Design and Environmentally Conscious Manufacturing. ASME, De-94/Med 5, pp 195–206
24. Huang CC, Kusiak A (1998) Modularity in design of products and systems. IEEE Trans Syst Man Cybern A 28(1):66–77
25. Pimmler TU, Eppinger SD (1994) Integration analysis of product decompositions. In: Proceedings of the 1994 ASME design engineering technical conferences – 6th international conference on design theory and methodology, Minneapolis, MN
26. Pahi G, Beitz W (1984) Engineering design: a systematic approach. Springer, Berlin
27. Stone RB, Wood KL, Crawford RH (1998) A heuristic method to identify modules from a functional description of a product. In: Proceedings of the 1999 ASME design technical conferences – 11th international conference on design theory and automation, Las Vegas, NV
28. Jose A, Tollenaere M (N.D.) Optimization of platforms for product family development. Gilco Inpg 1–8 Please update Ref. [10].
29. Zamirowski EJ, Otto KN (1999) Identifying product portfolio architecture modularity using function and variety heuristics. In: Proceedings of the 1999 ASME design engineering technical conferences – 11th international conference on design theory and automation, Las Vegas, NV
30. Kusiak A, Chow WS (1987) Efficient solving of the group technology problem. J Manuf Syst 6(2):117–124
31. Otto K (2000) Modular platform-based product family design. In: Proceedings of Detc'oo Asme design engineering technical conferences and computers and information in engineering conference, Baltimore, MD, 10–13 Sept
32. Otto K (2001) Identifying product family architecture modularity using function and variety heuristics. Paper Work Center for Innovation on Product Development, Massachusetts Institute of Technology, Cambridge, MA
33. Otto K, Sudjianto A (2001) Modularization to support multiple brand platforms. In: Proceedings of Detc: Asme design engineering technical conferences. DETC 2001, DTM-21695, Pittsburgh, PA, 9–12 Sept
34. Steward D (1981) The design structure matrix: a method for managing the design of complex systems. IEEE Trans Eng Manag EM-28(3):71–74
35. Pimmler TU, Eppinger SD (1994) Integration analysis of product decompositions. Design theory and methodology – DTM'94, DE-Vol. 68, ASME
36. Clark KB, Wheelwright SC (1993) Managing new product and process development. Harvard Business School Press, Boston

37. Kusiak A, Larson N (1995) Decomposition and representation methods in mechanical design. Trans ASME 117
38. Sabbaghian N, Eppinger S, Murman E (1998) Product development process capture & display using web-based technologies. In: Proceedings of the IEEE international conference on systems, Man, and Cybernetics, San Diego, CA, 11–14 Oct, pp 2664–2669
39. Dong Q (1999) Representing information flow and knowledge management in product design using the design structure matrix. ME Dept SM Thesis, MIT

Chapter 4
Design for Modularity

Abstract This chapter presents a three-phase methodology, which is proposed for the development of complex products using the modularity concept. The proposed methodology matches the criteria set by the design for functionality, assembly, and manufacture. A detailed discussion of needs analysis, quality function deployment, product requirement analysis, product/concept analysis, product physical analysis, product functional analysis, product/concept integration, identification of the Impact of system-level specifications on general functional requirements, similarity index, and optimization-based solution methodology for grouping components into modules is presented in this chapter. Finally, genetic algorithm-based solution methodology and algorithm-based solution methodology for grouping components into modules are described for understanding the presented concepts.

Keywords Modularity concept • Quality function deployment • Product requirement analysis • Genetic algorithm

4.1 Introduction

In this chapter, a three-phase methodology is proposed for the development of complex products using the modularity concept. The proposed methodology matches the criteria set by the design for functionality, assembly, and manufacture. Figure 4.1 illustrates an overview of the product design sequence in the proposed methodology.

Some of the major benefits associated with this methodology include:

- Increased design accuracy and efficiency and the reuse of existing design for new programs.
- Potential for integration of the developed methodology and technology into the engineering design activities.
- Modular product design and the process of planning the production are integrated in one overall engineering process in which product features are mapped into their feasible process(es) in a one-to-one correspondence.

A.K. Kamrani and E.A. Nasr, *Engineering Design and Rapid Prototyping*,
DOI 10.1007/978-0-387-95863-7_4,

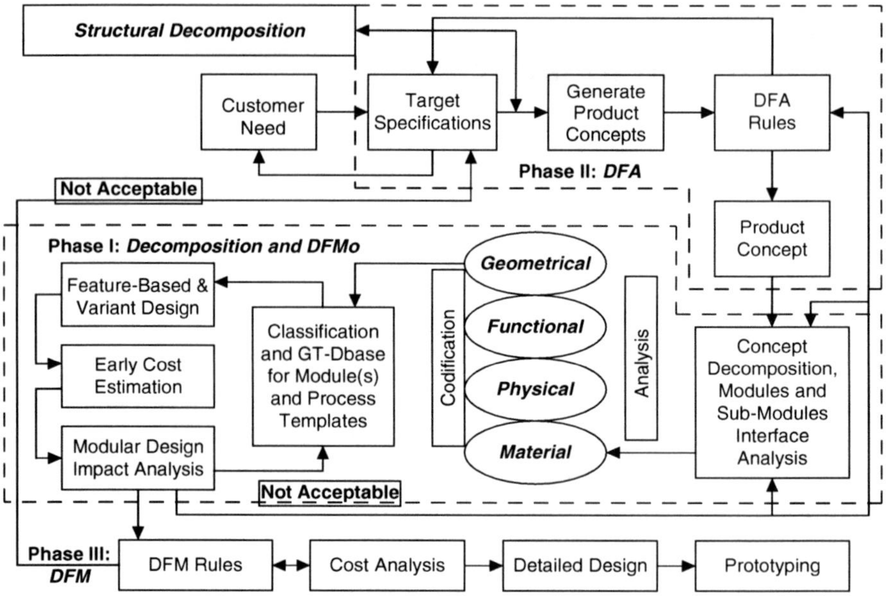

Fig. 4.1 Overview of the proposed design environment

The steps associated with this methodology include:

Phase I: Decomposition Analysis: Design for Modularity and Classification (DFMo)

- Product and problem decomposition
- Structural and modular decomposition
- Associativity analysis between the components and specifications
- Application of group technology classification system
- Construction of the associativity measure matrix
- Optimum module selection

Phase II: Product Analysis: Design for Assembly and Functionality Analysis

- Identify the components that could be produced and assembled separately
- Determine the order of disassembly and assembly for each subcomponent module
- Establish the interfaces based on the analysis of the design features
- Determine the order in which the subassemblies are assembled to produce the final product

Phase III: Process Analysis: Design for Manufacture

- Family identification and template retrieval
- Determination of the logical order of GT codes for the process of modules
- Machine and process parameter calculation
- Variant process planning
- Phase I of this methodology (Fig. 4.2) is the focus of this chapter

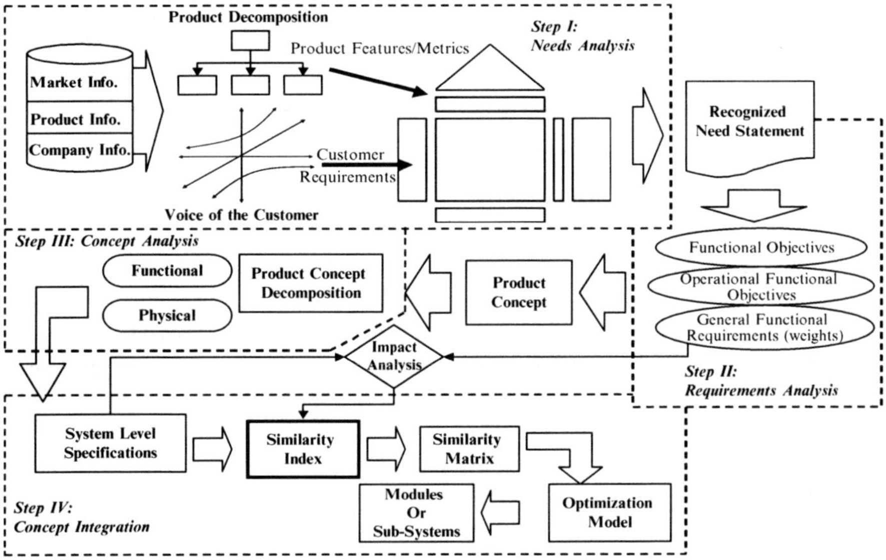

Fig. 4.2 Design for modularity

Further specifications associated with this phase include:

- Needs analysis
- Requirements analysis
- Concept analysis
- Concept integration

4.2 Needs Analysis

The design engineer is usually given an ill-defined problem. In many situations, the designer has to respond to the mere suggestion that there is a need for a product to perform a certain function. One of the main tasks is to find out precisely what the needs are and what customers really want. An important step in the design is to describe the product fully in terms of functional needs and physical limitations. These functional needs and physical limitations will form the product specifications.

Several sources of information can be used to identify the needs [8]. Such sources include potential customers, the company for which the design is being made, the competition, and any authorities that can impose restrictions on the product (standards, safety, etc.).

Information required to identify customer needs can be collected by surveying prospective purchasers or customers. This can be done by conducting a marketing study that begins by establishing target markets and customers. Then customers' wants and needs can be obtained by using several methods such as interviews and

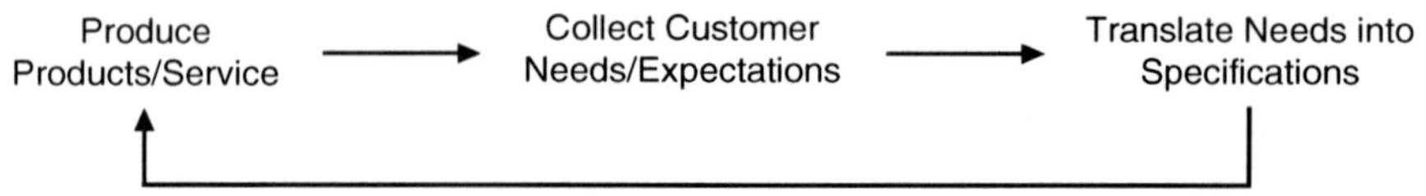

Fig. 4.3 Customer satisfaction process

questionnaires. Also, similar products (competitive products) are investigated to find possible improvement opportunities by focusing on weakness points and features desired by customers.

Next, customer wants and needs are arranged into groups and prioritized according to their importance. Needs analysis usually results in a statement of recognized needs and the expected manner in which those needs should be met. The resulting information is used to identify the requirements.

Satisfying the customer needs is a critical task, since the customer is becoming the center of successful product development efforts, that is, organizations are trying to explore what the customer needs and to deploy these needs into their products. Furthermore, organizations are trying to exceed customer needs and expectations to the point that the customer is delighted. The focus on customer needs is stimulated by the high competition resulting from the globalization of organizations operating in a globalized market place. This makes understanding the customer needs and expectations a valuable tool to maintain current business and to compete effectively to gain new business.

Satisfying the customer is not an easy task. It involves the process of communicating with customers to understand what they really need and then to translate these needs into specifications that can be met by products or services provided by the organization. Next, customer reaction to these new products/services is collected and incorporated into the development effort to improve the products/services and thus increase customer satisfaction.

The process of satisfying the customer can be viewed as an ongoing process (Fig. 4.3). The process should be performed continually as long as the organization is in business.

Customer satisfaction is an important step toward gaining customer loyalty and ensuring stable and growing market share. It is important for organizations to realize that only highly satisfied customers are most likely to become loyal customers. Customer loyalty can be measured by the retention rate that an organization has. Also, customer loyalty differs depending on how much the customer is satisfied. A "very satisfied" customer is more likely to repurchase a product than a "satisfied" customer [5]. The importance of satisfying customers can be substantiated by the difference in cost between maintaining existing customers and attracting new customers. It costs about five times more to attract a new customer than to maintain a current customer [7].

It can be argued that market share is a direct result of both customer satisfaction and loyalty. An organization with highly satisfied customers, and thus loyal customers, will have higher retention rates and can gain more market share.

4.2.1 Kano's Model of Customer Satisfaction

Kano [6] presented a conceptualization model of customer satisfaction, shown in Fig. 4.4 [2]. In this model, three major types of product requirement that affect customer satisfaction are recognized. These are as follows:

- Must-be requirements
- Performance requirements
- Attractive requirements

Must-Be Requirements

The "must-be" requirements are the basic need for a product. They constitute the main reason that the customer needs the product. That is, they fulfill the abstract need of the customer. The customer expects these requirements, and the manufacturer gets no credit if they are there. On the other hand, if these requirements are not fulfilled, the customer will be extremely dissatisfied (the customer will have no reason to buy the product). Fulfilling the "must-be" requirements alone will not satisfy the customer; they will lead only to a case of a "not dissatisfied" customer.

Must-be requirements are unspoken and nonmeasurable; they are either satisfied or not satisfied. They are shown in Kano's model by the lower right curve of the figure.

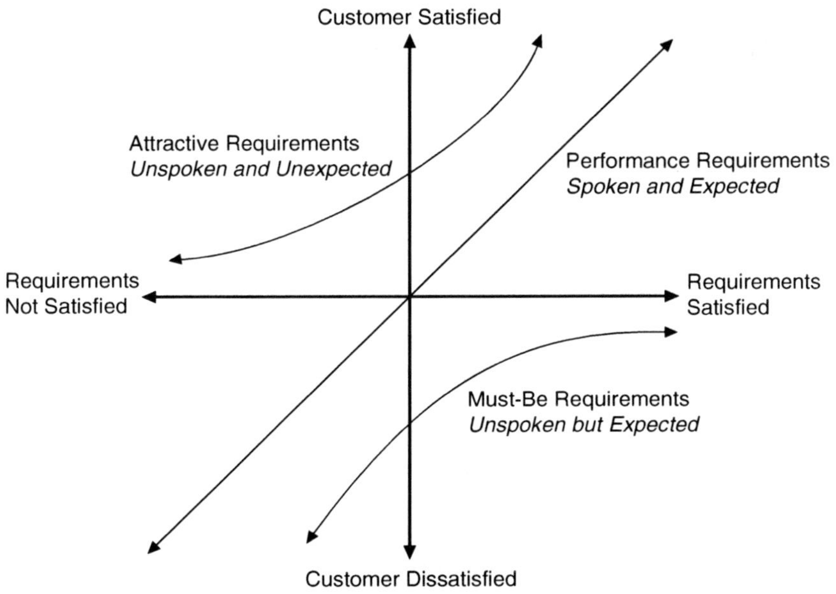

Fig. 4.4 Kano's model

Performance Requirements

Performance requirements are spoken needs. They can be measured for importance as well as for a range of fulfillment levels. The customers explicitly tell us or answer our question about what they want. Performance requirements include written or spoken requirements and are easily identified and expected to be met. Usually, customer satisfaction increases with respect to the degree to which these requirements are fulfilled. Also, performance requirements serve as differentiating needs and differentiating features that can be used to benchmark product performance. The diagonal line in Kano's model depicts these requirements.

Attractive Requirements

Attractive requirements are future oriented and usually high-tech innovations. These requirements are unspoken and unexpected because the customer does not know that they exist. Usually, these requirements come as a result of the creativity of the research and development effort in the organization. Because they are unexpected, these creative ideas often excite and delight the customer and lead to high customer satisfaction. These requirements are shown by the curved line in the upper left corner of the figure. It should be mentioned here that these requirements quickly become expected.

4.2.1.1 Customer Requirements

Classification of Customer Requirements

Four major groups of customer requirements exist in Kano's model. These groups are as follows:

- Expected requirements
- Spoken requirements
- Unspoken requirements
- Unexpected requirements

Expected Requirements

These are the basic requirements that customers expect to exist self-evidently. An organization must satisfy these requirements to be competitive and to remain in business. These requirements are satisfied through basic product/service characteristics that customers assume are part of the product or service; that is, they expect them as standard features. Fulfilling these expected requirements does not create much satisfaction and mostly remains unnoticed; however, lack of it causes considerable dissatisfaction. It is important for an organization to recognize the minimum acceptable level of the expected requirements, because a lower level increases dissatisfaction rapidly, and a higher level raises costs without much increase in customer satisfaction.

Unspoken Requirements

Unspoken requirements are product features that customers do not talk about and, though silent, are important and cannot be ignored. Unspoken requirements are like expected requirements in the sense that customers do not feel obligated to tell the developer about them, either because they feel they are clear and obvious or because they do not know that they exist. These requirements are usually defined by means of a thorough investigation of customers' needs. An organization must try to explore why the customer needs a certain product and how customers historically behaved toward new and innovative features. Meeting these requirements can lead to high customer satisfaction, especially if customers did not realize that they actually needed these product features.

Spoken Requirements

Spoken requirements are specific product/service features customers say they want in a product. An organization must be willing to provide these features in their products to satisfy their customers. Most of the time, these requirements are related to technical issues that can be measured, and they can be used to compare and benchmark product performance with respect to the competition. Customer satisfaction increases proportionally as these requirements are being met. The cost of meeting these requirements can be high, and thus an organization must determine what the acceptable levels of meeting these requirements are. That is, an organization should balance meeting these requirements with the cost of the product and offer a product at competitive cost and performance. Usually, trade-offs in the design process are made on the product features that fulfill these spoken requirements.

Unexpected Requirements

Unexpected requirements are unspoken features of a product that make the product unique and distinguish it from the competition. Unexpected requirements define product attributes or features that are a pleasant surprise to customers when they first encounter them. Surprising the customer with these unexpected features will lead to customer delight. However, if these features are not present, then the customers will not be dissatisfied, since they will be unaware of what they are missing. Organizations usually make large investments in research and development departments in order to be able to make big breakthroughs that will surprise the customer and offer unexpected features, guaranteeing high customer satisfaction. However, these unexpected requirements will become expected after a short period of time.

Recording Customer Requirements

There are many techniques that can be used to investigate customer preferences about products. The selection of the appropriate techniques depends on (1) the amount and type of information required, (2) the availability of information, and (3) the time and cost allocated to collect the information. Information sources can be grouped into two large groups: *existing information* and *new information*.

Existing Information

- Sales data
- Customer complaints
- Warranty data
- Publications from the government and trade journals
- The company's designers, engineers, and managers
- Benchmarked products

New Information

- Surveys, including mail, telephone, and comment cards
- Focus groups
- Interviews, group or individual
- Service calls and trade shows
- Direct visits with the users

Analyzing Customer Requirements

Information collected using the methods mentioned in the previous section should be analyzed to identify customer preferences of product features and performance. It can be noted that there are two types of requirement data, qualitative and quantitative, or linguistic and numerical. Qualitative data are subjective and exploratory in nature and are represented in the form of labels. On the other hand, quantitative data are objective, specific, and number-oriented information. Quantitative data are represented by numbers, which provide metrics that measure the degree of achievement of a certain requirement. Requirement analysis begins by identifying the qualitative information of interest to the customer. This will establish a fundamental understanding of customer requirements. Then quantitative information is mapped into qualitative information. The assumption here is that the customer's requirements, represented in qualitative labels, can be satisfied through achieving some quantitative targets. For example, a customer may say that he/she desires an economical car. This can be mapped to measurable quantitative information, e.g., miles per gallon.

4.3 Quality Function Deployment

4.3.1 Introduction

Quality function deployment (QFD) is a team-based technique that provides a means of identifying and translating customer requirements into technical specifications for product planning, design, process, and production. It is a planning tool used to fulfill customer expectations. QFD helps in defining the product in terms of the customers' requirements. In addition, it takes customer requirements and provides means of converting them into product specifications.

In general, QFD addresses the following:

- Customer wants and needs (WHATs)
- Importance of the requirements to the customer (WHYs)
- Product features needed to satisfy the customer's needs (HOWs)

4.3.2 *The House of Quality*

The house of quality (HOQ) is the nerve center and the engine that drives the entire QFD process. It is a conceptual map that provides the means for cross-functional planning and communication. The HOQ is a large matrix that contains seven different elements [11], as shown in Fig. 4.5. Using these seven elements, the HOQ becomes a repository of information that can be used as a mechanism for applying commonsense engineering.

4.3.2.1 Customer Requirements/Needs

Customer requirements represent the wants and needs of the customer. Collecting these requirements is not part of QFD; they must be collected before QFD begins. QFD is used to deploy customer requirements.

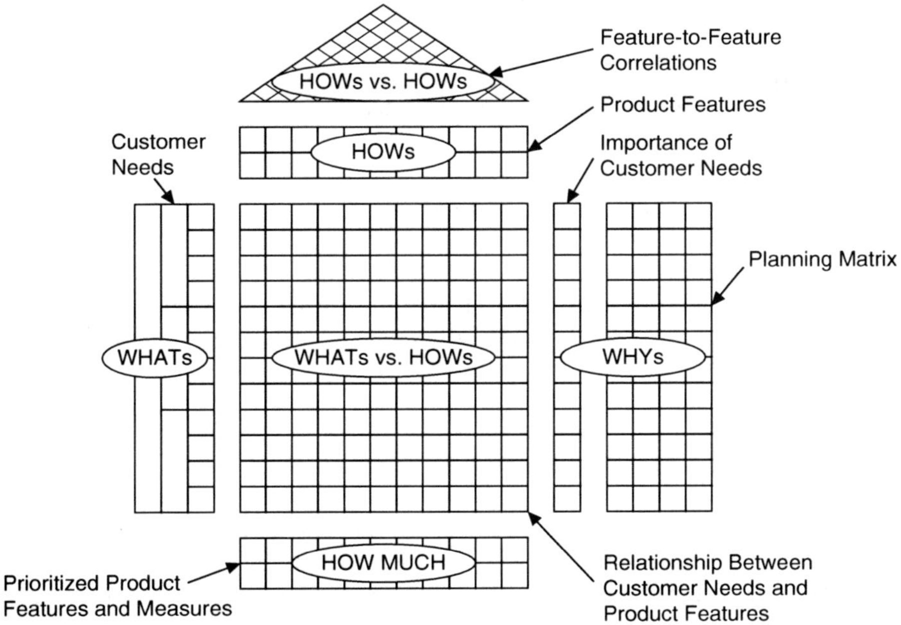

Fig. 4.5 The house of quality

4.3.2.2 Product Features

Product features are also called design requirements or engineering characteristics. The product features will become the criteria used to determine the degree to which customer needs are met. That is, product features are proposed by engineering to respond to the customer needs identified by marketing. Product features must be stated in measurable and benchmarkable terms.

4.3.2.3 Importance of Customer Needs

Not only do we need to know what the customer wants, but also how important those needs are.

4.3.2.4 Planning Matrix

The planning matrix comprises a competitive analysis of the company's product with respect to each customer need, in comparison with major competitors' products. It contains measures of how much improvement is needed in the current product, how much sales leverage may result from such improvement, and a final overall score for each customer need.

4.3.2.5 Relationship Between Customer Needs and Product Features

The inter-functional teams use the body matrix to indicate how much each product feature (engineering characteristic) affects each customer need.

4.3.2.6 Feature-to-Feature Correlation

The effect of changes in one feature on other features is recorded in the correlation matrix. Sometimes a desirable change in one feature results in a negative effect on another feature. This correlation aids in identifying such impact and trade-offs.

4.3.2.7 Prioritized Product Features or Targets

This section is a summation of the effects of all prior variables on each product feature. It may also contain target measures for engineering characteristics, as well as a competitive analysis of other manufacturers' measures for the same variables.

4.3.3 *Building the House of Quality*

4.3.3.1 Identify Customer Requirements (WHATs)

QFD starts by asking "What does the customer want?" The answer to this question will provide a primary list of customer needs or expectations in a particular product. This list is usually very general and vague and needs to be refined into secondary and perhaps tertiary levels of customer requirements. This will result in a hierarchy of primary, secondary, and tertiary customer requirements that can be addressed by the engineering staff.

4.3.3.2 Identify Product Features (HOWs)

The QFD team must identify product features (engineering characteristics) that will be needed to achieve the customer requirements. Each product feature must directly affect a customer expectation and be expressed in measurable terms.

4.3.3.3 Build a Relationship Matrix Between WHATs and HOWs

The QFD team now fills the inside of the HOQ (the relationship matrix). In this step, customer requirements and product features are compared with respect to each other, and their respective relationships are determined. Usually, symbols are used to represent the degree of relationship between the customer requirements and the product features. For example:

- ⊙ represents a strong relationship
- O represents a medium relationship
- Δ represents a weak relationship

The box is left blank if no relationship exists.

Next, symbols used to define the relationships are replaced with numbers; for example:

- ⊙ = 9
- O = 3
- Δ = 1

After the relationship matrix is completed, it should be evaluated for empty rows or columns. An empty row indicates that a customer need is not being met since that need is not being addressed by any of the product features. In this case, additional product features must be considered to satisfy that customer need. An empty column indicates that a particular product feature does not affect any of the customer requirements and may be removed from the HOQ.

4.3.3.4 Construct Correlation Matrix Between HOWs

Interrelationships between product features are identified in this step and recorded in the roof of the HOQ. Symbols can be used to describe the strength of the interrelationships. For example:

- ⊙ represents a strong positive relationship
- O represents a positive relationship
- Δ represents a negative relationship
- ⚠ represents a strong negative relationship

The correlation matrix identifies which product features support one another and which are in conflict. Conflicting product features could be the result of conflicting customer requirements and represent points at which trade-offs must be made.

4.3.3.5 Customer Competitive Assessment

Customer competitive assessment is a summary of the top competitive products' features in comparison with the product being developed. Also, customer competitive assessment contains an appraisal of where an organization stands relative to its major competitors in term of each customer requirement.

4.3.3.6 Technical Competitive Assessment

Technical competitive assessment is a benchmarking study that compares the competitors' specifications for each of the product features and the proposed specification to either meet or exceed the customer requirements.

4.3.3.7 Prioritize Customer Requirements

Customer requirements are prioritized in this step. The prioritization can be divided into the following:

- ***Importance of Customer Needs Rating***: Customer needs are rated according to their importance with respect to each other. The question being answered here is "How important is this feature to the customer?"
- ***Target Value***: The QFD team sets values for each product feature depending on whether they (the team) want to keep the feature unchanged, improve it, or make the feature better than the competition.
- ***Scale-Up Factor***: The scale-up factor is the ratio of the target value to the product rating given in the customer competitive assessment. The higher the value, the more effort is needed.
- ***Sales Point***: Marketing uses the sales point to determine if they will get any leverage out of any improvement. That is, the sales point answers the question "Taking into consideration the importance of this feature to the customer and the

effort needed to make the change, if we change this feature, can marketing get any leverage from it?"

- ***Absolute Weight***: The absolute weight is calculated using the following formula:

  Absolute weight = (Importance Rating) × (Scale-Up Factor) × (Sales Point)

4.3.3.8 Prioritized Product Features

The prioritization of product features includes:

The Degree of Technical Difficulty: A measure of the organization's capability of making a certain product feature

Target Value: An objective measure that defines values that must be achieved in order to fulfill customer requirements

Absolute Weight: The absolute weight for the *j*th product feature is:

$$a_j = \sum_{i=1}^{n} R_{ij} C_i \tag{4.1}$$

Where:
a_j = absolute weight for product features ($j = 1,\ldots, m$)
R_{ij} = weight assigned to relationship matrix ($i = 1,\ldots, n, j = 1,\ldots,$ m)
C_i = degree of importance of customer requirements ($i = 1,\ldots, n$)
m = number of product features
n = number of customer requirements

4.3.4 Implementing Quality Function Deployment

The QFD process starts with thoroughly understanding customers and their needs and the relative importance of those needs. The matrices used in QFD are used to document and organize the needs so that they are manageable and actionable. A simplified four-phase methodology for implementing QFD was described by Sullivan [13], in which each phase, or matrix, represents a more specific aspect of the product's requirements. Binary relationships between elements are evaluated for each phase. Only the most important aspects from each phase are deployed into the next matrix. The four phases are as follows:

4.3.4.1 Phase 1: Product Planning

Led by the marketing department, phase 1, or product planning, is also called the HOQ. Many organizations get through only this phase of a QFD process. Phase 1 documents customer requirements, warranty data, competitive opportunities, product measurements, competing product measures, and the technical ability of the organization to meet each customer requirement. Getting good data from the customer in phase 1 is critical to the success of the entire QFD process.

4.3.4.2 Phase 2: Product Design

The engineering department leads phase 2. Product design requires creativity and innovative team ideas. Product concepts are created and part specifications are documented during this phase. Parts that are determined to be most important to meeting customer needs are then deployed into process planning, or phase 3.

4.3.4.3 Phase 3: Process Planning

Process planning comes next and is led by manufacturing engineering. During process planning, manufacturing processes are flowcharted and process parameters (or target values) are documented.

4.3.4.4 Phase 4: Production Planning

Finally, in production planning, performance indicators are created to monitor the production process, maintenance of schedules, and skills training for operators. Also, in this final phase, decisions are made as to the risk involved in not controlling a process. Phase 4 is usually led by the quality assurance department in concert with manufacturing.

4.3.5 Benefits of Quality Function Deployment

- QFD emphasizes customer requirements and uses them as the driving force for the development effort.
- QFD helps in discovering conflicting design requirements in the early planning stages prior to production, reducing the implementation time by reducing engineering changes, retooling, and operator training.
- QFD is based on teamwork, which will increase both efficiency and productivity.
- QFD provides documentation for future use.

4.4 Product Requirements Analysis

Results of the needs-analysis step are used to identify the product requirements. The development group begins by preparing a list of *functional objectives* needed to meet the customer's primary needs. Further analysis of customer needs reveals *operational functional requirements* that impose both functional and physical constraints on the design. Secondary customer requirements will be categorized as *general functional requirements*; they are ranked secondary because they will not affect the main function of the product. That is, a product may lack one or more GFR and still be considered as a functional product that meets the intended function. GFR should be weighted with respect to their importance.

Following is a description of functional objectives, operational functional requirements, general functional requirements, and weights associated with general functional requirements.

4.4.1 Functional Objectives

Functional objectives are an abstraction of the product function required to satisfy customer needs. They provide information about what the device/product under investigation is supposed to do. Functional objectives can be thought of as the basic operations or transformations that must be performed by the system to satisfy customers' primary needs. Customers' primary needs are those needs for which the customer will buy the product. Usually, customers do not explicitly state these primary needs. Customers assume that these needs are obvious and there is no need to indicate them. For example, the primary need of a car brake system is to stop the car as needed. This need is so clear that most customers may not mention it in an interview. It is the responsibility of the development team to find the primary needs critical to the product usage and translate it into functional objectives. Functional objectives are often somewhat general, but they should always employ action phrases such as "to transform," "to support," or "to lift." Table 4.1 is a list of several products along with their functional objectives:

4.4.2 Operational Functional Requirements

Needs analysis will identify the operational conditions and the physical limitations of the product under investigation, which should be translated into operational functional

Table 4.1 Functional objectives

Product	Functional objectives
Eraser	To erase marks made by a pencil
Binder	To hold paper together
Clipboard	To hold and support paper while writing
Knife	To cut meat
Cup	To hold liquids
Power cable	To transmit electric power

Table 4.2 Operational functional requirements

Product	Operational functional requirements
Cup	⇒ The cup inside diameter should be between 2.75 and 3.25 in.
	⇒ The cup handle radius should be about 2 in.
	⇒ The cup height should be about 4 in.
Power cable	⇒ The cable should transmit 110–120 V
	⇒ The cable should transmit 10 A

requirements giving quantitative data wherever possible. Operational functional requirements are detailed and contain specific information representing a set of constraints that the design must possess in order to satisfy the product's intended function. Operational functional requirements can be the result of an integrated effort incorporating marketing staff, design engineers, manufacturing engineers, service personnel, suppliers, and customers and are usually presented in the form of ranges. For example, the operational functional requirements of the CPU speed of a computer can be a Pentium 90 MHz to Pentium 166 MHz. Possible operational functional requirements of several products are listed in Table 4.2.

4.4.3 General Functional Requirements

General functional requirements are the criteria set by the designer, based on the needs analysis, to evaluate the resulting design. They are those requirements that satisfy the customers' secondary needs, which could be a critical factor for the customer when comparing different competitive products that accomplish the same function. The general functional requirements will differ based on the prospective markets or users of the product and can be related to qualitative features of the product such as appearance and color. Also, maintenance needs and assembly can be termed as general requirements. They are usually stated in general terms. The following are some examples of common general functional requirements:

- Ease of manufacture
- Ease of service or replacement of components
- Ease of operations
- Safety of use

4.5 General Functional Requirements' Weights

Several general functional requirements may exist for a product, and some are more important than others; therefore, different weights should be assigned to different requirements. Customer needs are considered as an essential factor in weight assignment. Weight assignment can be made by using a benchmarking study of competitive products. Alternatively, it can be derived from the input of the design team based on previous knowledge of the importance of such requirements.

4.6 Product/Concept Analysis

Product/concept analysis is the decomposition of the product into its basic functional and physical elements. These elements must be capable of achieving the product's functions. Functional elements are defined as the individual operations and transformations that contribute to the overall performance of the product. Physical elements are the parts, components, and subassemblies that ultimately implement the product's function.

Product concept analysis consists of product physical decomposition and product functional decomposition. In product physical decomposition, the product is decomposed into its basic physical components that, when assembled together, will accomplish the product function. Physical decomposition should result in the identification of basic components that must be designed or selected to perform the product function. Product functional decomposition describes the product's overall functions and identifies component functions. Also, the interfaces between functional components are identified. Both functional and physical elements can be presented in a function–structure diagram that will illustrate the relationship between them (Fig. 4.6) [10].

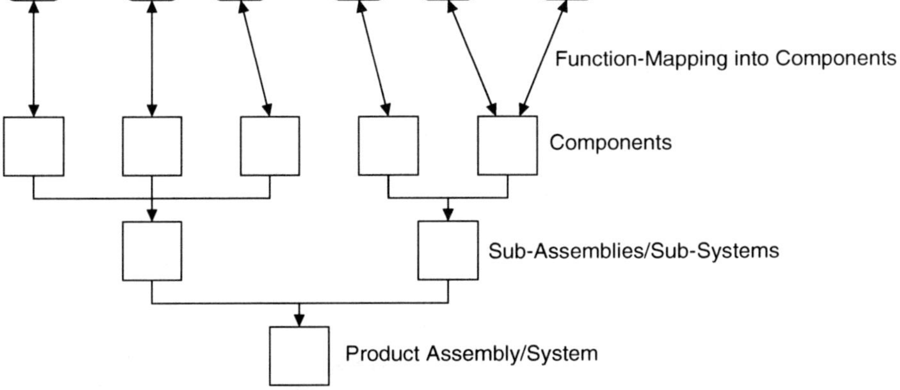

Fig. 4.6 Function–structure diagram

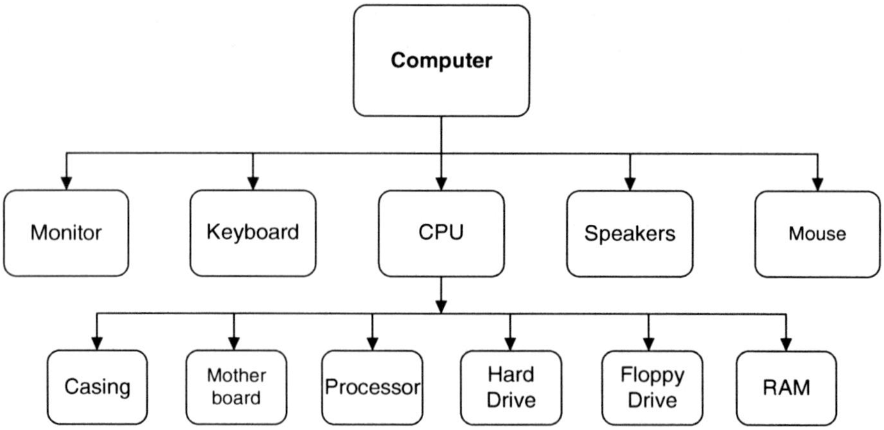

Fig. 4.7 Computer physical decomposition

4.7 Product Physical Decomposition

The product is decomposed into subsystems and/or subassemblies capable of achieving the product function. The decomposition process should continue until basic physical components are reached. Figure 4.7 shows the physical decomposition of a computer.

4.8 Product Functional Decomposition

Functional decomposition should aim at representing the intended behavior (the functions) of a product and its parts. A function can be implemented by a single physical element (component) or by a combination of elements arranged in a specific manner. Functional components are arranged according to several logical considerations that will ensure the accomplishment of their intended combined function. The logical arrangement is called a working principle [10], which defines the mode of action that the product/system will perform on the inputs to reach the output state.

To analyze the product function, the overall function of the product should be conceptualized into an action statement (verb–noun form) [1, 3, 4, 9]. Then, the overall function is broken into subfunctions, and those are further decomposed into lower level functions. This function breakdown is continued until a set of functions that can be achieved by available components is reached. At this point, functions are mapped into components, and components are arranged to form subassemblies leading to an overall assembly that will ultimately accomplish the overall function.

A function flow diagram can be used to represent how the function is being implemented in the system. The function flow diagram will show how each component supports the product's overall function. A component function should tell what the reason is for using the component. At this point, two categories of functions are identified: primary functions and auxiliary functions.

- ***Primary functions*** are those functions that support the overall function (the functional objectives) in a direct manner.
- ***Auxiliary functions*** are those functions that support the overall function indirectly, or they are those functions that support the overall function by supporting the primary functions.

To draw a function flow diagram, first the input and output parameters are determined, then the overall function needed to transform the input into output is identified and represented in a simple flow diagram that will also show the relationship between the inputs and outputs as general constraints (Fig. 4.8).

The next step is to focus on the processes that inputs go through until they reach the output. This can be done by following the inputs' flows through the components and identifying the contribution of each component in transforming the inputs into outputs. A function flow diagram for all the components' functions is prepared. The output from one component function should be the input to another component function. Some functions may have more than one input or output. Changes or alterations of the inputs should be noted and documented as a relationship between inputs and outputs to further explain the function (Fig. 4.9 and Table 4.3).

This analysis will determine the functions implemented by each physical component and show how different functions are integrated to accomplish the overall function. This will also allow the use of different physical components that can perform the same function.

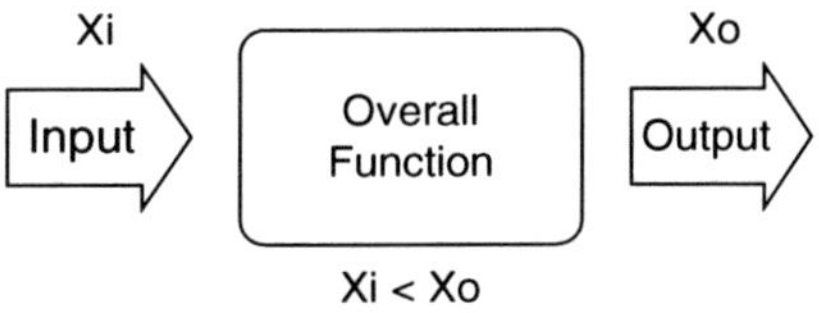

Fig. 4.8 Overall function flow diagram

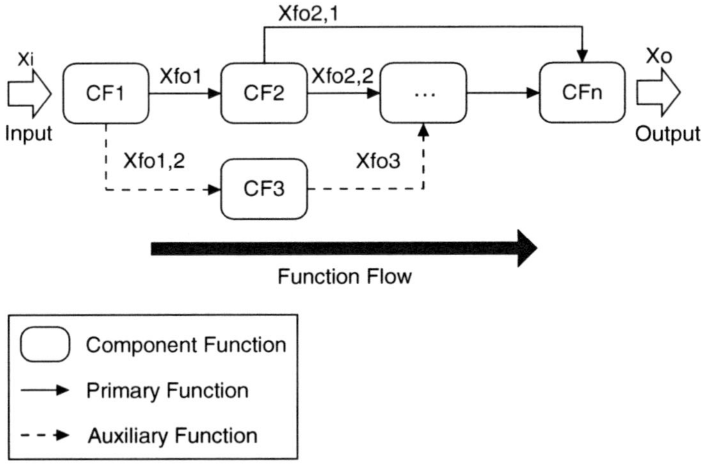

Fig. 4.9 Function flow diagram

Table 4.3 Relationships between component functions

Component function	Input	Output	I/O constraints	Function type
CF1	Xi	Xfo1,1	Xi < Xfo1,1	Primary
		Xfo1,2	Xi = Xfo1,2	
CF2	Xfo1	Xfo2,1	Xfo1 < Xfo2,1	Primary
		Xfo2,2	Xfo1 = Xfo2,2	
CF3	Xfo1,2	Xfo3	Xfo1,2 = Xfo3	Auxiliary
–	–	–	–	–
CFn	Xfo2,1	Xo	Xfo2 < Xi	Primary

4.9 Product/Concept Integration

Basic components resulting from the decomposition process should be arranged in modules and integrated into a functional system. The manner by which components are arranged in modules will affect the product design. The resulting modules can be used to structure the development teams needed. Following are the steps associated with product integration.

4.9.1 Identify System-Level Specifications

System-level specifications are the one-to-one relationships between components with respect to their functional and physical characteristics. Functional characteristics are a result of the operations and transformations that components perform in order to contribute to the overall performance of the product. Physical characteristics are a result of the components' arrangements, assemblies, and geometries that implement the product function. A general guideline for identifying the relationships can be presented as follows.

4.9.2 Functional Characteristics

- Identify the main function(s) based on the functional decomposition
- Identify the required operations and transformations that must be performed in order to achieve the function based on the function flow diagram
- Document the operations and transformations
- Categorize operations and transformations into a hierarchy structure

4.9.3 Physical Characteristics

- Identify any physical constraints imposed on the product based on the requirement analysis

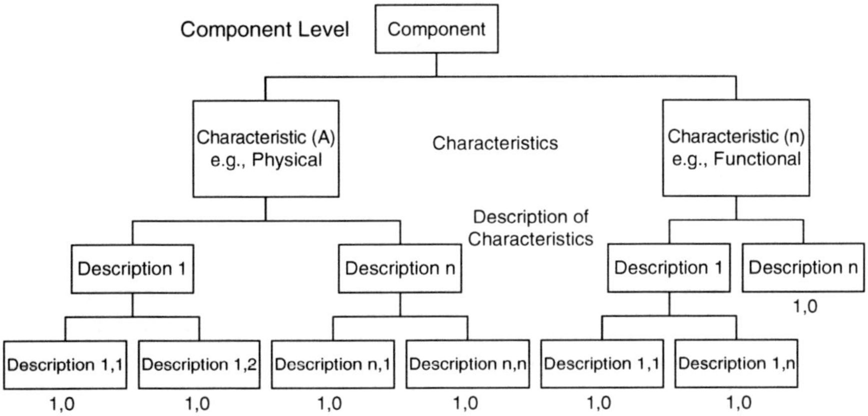

Fig. 4.10 System-level specification decomposition hierarchy

- Identify possible arrangements and/or assemblies of the components based on previous experiences, previous designs, engineering knowledge, or innovative designs/concepts
- Document possible arrangements and/or assemblies
- Categorize arrangements and assemblies into a hierarchy structure

Physical and functional characteristics, forming the system-level specifications, are arranged into a hierarchy of descriptions that begins with the component at the top level and ends with the detailed descriptions at the bottom level. Bottom-level descriptions (detailed descriptions) are used to determine the relationships between components, 1 if the relationship exists and 0 otherwise. This binary relationship between components is arranged in a vector form, the system-level specifications vector (SLSV). Figure 4.10 illustrates the hierarchy structure of the physical and functional characteristics.

4.10 Identify the Impact of System-Level Specifications on General Functional Requirements

System-level specifications identified in the previous step affect the general functional requirements in the sense that some specifications may help satisfy some GFR, while other specifications might prevent the implementation of some desired GFR. The impact of the SLS on GFR should be clearly identified, which will help in developing products that will meet, up to a satisfactory degree, the GFR stated earlier. The impact will be determined based on

- −1 = Negative impact
- 0 = No impact
- +1 = Positive impact

The user can set this number with any range. A negative impact represents an undesired effect on the general functional requirements, such as limiting the degree to which

Table 4.4 GFR vs. SLS

	General functional requirements		
System-level specifications	FR(1)	FR(2)	FR(m)
SLS(1)	−1	1	0
SLS(2)	–	–	–
SLS(*n*−1)	–	–	–
SLS(*n*)	1	0	1

the product will meet the general requirement or preventing the product from implementing the general requirement. Inasmuch as a positive impact represents a desired effect that the SLS will have on the general requirements, such SLS will ensure that the product will satisfy the requirements and result in customer satisfaction. An SLS is said to have no impact if it neither prevents the implementation of the GFR nor helps satisfying the GFR. An example of the SLS impact on the GFR is given in Table 4.4.

For example, the SLS(1) have a negative impact on the FR(1), positive impact on FR(2), and no impact on the FR(m).

4.11 Similarity Index

The degree of association between components should be measured and used in grouping components into modules. This can be done by incorporating the GFR weights, in addition to the system-level specifications vectors and their impacts on the general functional requirements, to provide a similarity index between components. The general form of the similarity index is as follows:

SLSV (C_1 & C_2) SLS & FRs Weights for FRs

$$(S)_{1\times 1} = (1 \quad 0 \ldots a_n)_{1\times n} \begin{pmatrix} 1 & . & . & . & b_{1,m} \\ 0 & . & . & . & . \\ . & . & . & . & . \\ . & . & . & . & . \\ b_{n,1} & . & . & . & b_{n,m} \end{pmatrix}_{n\times m} \begin{pmatrix} 1 \\ 0.9 \\ . \\ . \\ c_{m\times 1} \end{pmatrix}_{m\times 1} \tag{4.2}$$

The similarity indices associated with components are arranged in a component vs. component matrix as shown below:

	C_1	C_2	C_3	–	–	C_n
C_1	X	$S_{1\times 2}$	$S_{1\times 3}$	–	–	$S_{1\times n}$
C_2		X	$S_{2\times 3}$	–	–	$S_{2\times n}$
C_3			X	–	–	$S_{3\times n}$
–				X	–	–
–					X	–
C_n						X

4.12 Optimization-Based Solution Methodology for Grouping Components into Modules

Components with a high degree of association should be grouped together in design modules. This can be accomplished by using an optimization model that maximizes the sum of the similarities. The optimization model will identify independent modules that can be designed simultaneously. Several models are available for use (Chap. 3). The p-median model will be used to cluster components into modules [12]. The model is as follows:

Notation:

C = Number of components
M = Number of modules
S = Similarity Index

Maximize the sum of all the similarities:

$$\sum_{p=1}^{C}\sum_{q=1}^{C} S_{pq} X_{pq} \tag{4.3}$$

Where

$$X_{nm} = \begin{Bmatrix} 1, \text{if component } n \text{ belongs to module } m \\ 0, \text{otherwise} \end{Bmatrix}$$

Subject to:
Each component is assigned to exactly one module:

$$\sum_{q=1}^{C} X_{pq} = 1, \quad \forall p$$

Components are assigned to a predefined number of modules:

$$\sum_{q=1}^{C} X_{pq} = M$$

Components are assigned to modules that have a median component:

$$X_{pq} \le X_{qq}, \quad \forall p,q$$

The model is an integer model:

$$X_{pq} = 0/1, \quad \forall p,q$$

4.13 Genetic Algorithm-Based Solution Methodology

Many combinatorial optimization problems from the manufacturing systems world are very complex and hard to solve using conventional optimization techniques. As it has been suggested, enumeration techniques are usually obsolete if a problem is extremely big. Since real-life problems are in most cases enormous, new techniques have to be developed and used to solve these large and difficult problems. Many algorithms have been proposed since the development of the approximation techniques. These techniques include simulated annealing, tubule search, and genetic algorithms. The rapid development of computer science along with the rapid development of artificial intelligence techniques has created a great interest among engineers to solve difficult problems. Evolutionary algorithms are stochastic optimization and are similar to the simulation of the natural evolutionary process of human beings. The idea is that by simulating the thought process that the human uses to solve difficult problems, the algorithm will be able to solve the same kind of problems. Moreover, and unlike the normal local search algorithm, evolutionary algorithms (and more specifically genetic algorithms) will try to take into account a wider range of possible solutions, which will then in turn reduce the probability of not finding an optimal solution to a given combinatorial problem.

4.13.1 Genetic Algorithms

Genetic algorithms are stochastic search techniques. The process start with a set of random solutions called population. Each individual or solution in the population is called a chromosome. A chromosome is a string of symbols (usually, a binary bit string), which represents a solution to the problem being solved. The initial population has to be determined by the user. Each combinatorial optimization problem is different, so special attention has to be given to the definition of a solution. Solutions can be represented as binary bit strings, number strings, word strings, among many other options. However, if a representation does not accurately represent what the solution actually means, the algorithm will perform poorly, and no useful information will be obtained. Chromosomes evolve via successive iterations called generations. During each iteration, the chromosomes are evaluated using a fitness function that will eventually decide if a chromosome passes to the next generation or not. After an iteration of the algorithm, a new generation is created with new chromosomes called offspring. They are formed by either merging two chromosomes from the current generation using the crossover operation or by modifying a single chromosome using the mutation operation. The new generation is then formed by selecting some of the chromosomes according to their fitness value and by rejecting others that do not qualify as valuable individuals. Figure 4.11 illustrates the basic process of a genetic algorithm.

The algorithm shown below is highly adaptive and many combinatorial optimization problems can be solved using this basic outline:

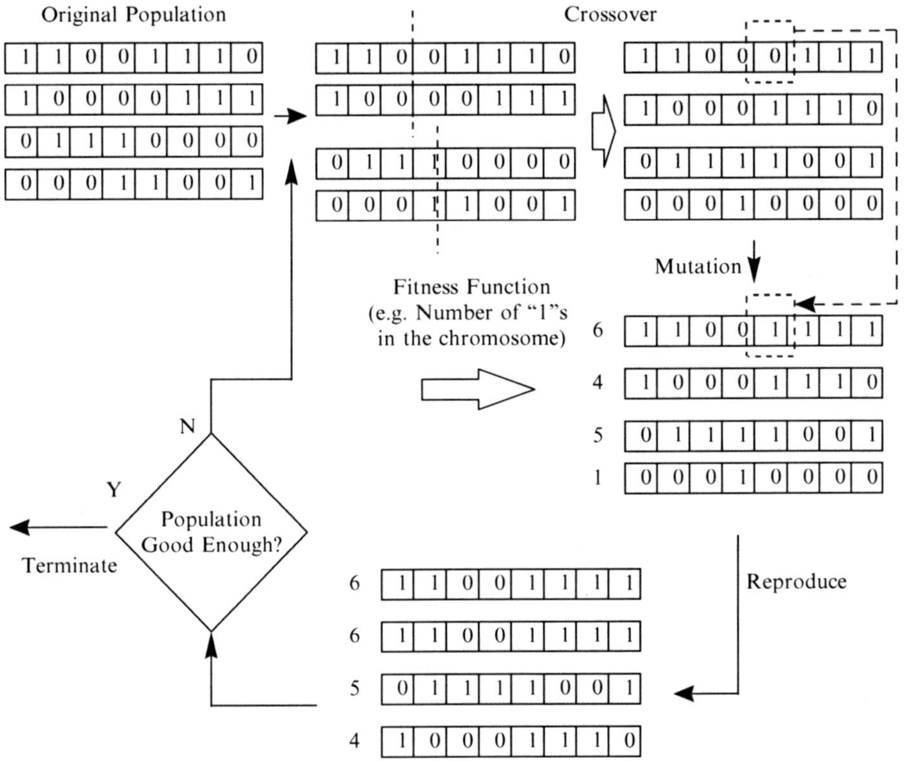

Fig. 4.11 Basic structure of a genetic algorithm

Procedure: *Genetic Algorithm*

[*Initialization*] Generate random population of *n* chromosomes.

[*Evaluation*] Evaluate the fitness *f*(x) of each chromosome *x* in the population.

[*New population*] Create a new population by repeating the following steps until the new population is complete:

- [*Selection*] Select two parent chromosomes from a population according to their fitness (the better fitness, the bigger chance to be selected).
- [*Crossover*] Given a crossover probability, crossover the parents to form new offspring. If no crossover is performed, offspring is an exact copy of parents.
- [*Mutation*] With a mutation probability, mutate new offspring at given positions in a chromosome.
- [*Accepting*] Place new offspring in a new population.
- [*Replace*] Use new generated population for a further run of algorithm (new generation).
- [*Test*] If the end condition is satisfied (usually, required fitness value is obtained), STOP, and return the best solution in current population.

[*Loop*] Otherwise go to step 2 (a new generation will be created).

For combinatorial optimization problems such as the modularity problem, the evaluation process must be carefully defined. The performance of the fitness function is crucial for the algorithm's performance. If the evaluation technique is too general, the algorithm may never reach an optimal solution because the fitness function will not measure the chromosomes properly. On the other hand, if the process is too strict, the algorithm will be unable to find any kind of optimal solution. This may be very costly if one takes into account the cost associated with the use of genetic algorithms in combinatorial optimization problems. The three most important steps in a genetic algorithm are presented below.

4.13.2 Proposed Model

The need of developing an efficient algorithm that will produce acceptable answers for a wide variety of combinatorial optimization problems has been stressed. An efficient genetic algorithm is developed to solve the modular design problem. The algorithm developed can be used to solve a vast variety of combinatorial optimization problems, as the differences among these problems lie only on the chromosome representation of the answers. The objective of the problem can be different as well, but a few modifications to the algorithm would have to be made. These changes, however, would not affect the actual algorithm. They would only affect the complexity of the developed user interface for the problem. The objective of the modularity problem is to maximize the sum of all the similarities between components, so that they can be grouped together to form a modular product. Developing the genetic algorithm involves various steps. The proposed algorithm consists of ten steps that are broken into four major points:

Initialization:

- Define the chromosomes and their representation.
- Determine the population size, crossover probability, and mutation probability.
- Generate initial population of solutions that satisfy the problem constraints.

Reproduction:

- Compute the fitness value of each chromosome.
- Calculate the total fitness of the population.
- Find the reproduction probability for each chromosome.
- Calculate the cumulative reproduction probability for each chromosome.
- Choose the best chromosomes for the next generation.

Crossover:

- Do crossover on the selected chromosomes based on a crossover probability defined in the initialization procedure.

Mutation:

- Perform mutation on the chromosomes based on a mutation probability defined in the initialization procedure.

As the objectives and the nature of each problem are different, one has to pay close attention to each of the steps to develop the method that will enhance the performance and the usability of the algorithm.

4.13.2.1 The Chromosomal Representation

The first step in the implementation of the algorithm involves the representation of the problem to be solved with a finite-length string called chromosome. Each element of the chromosome (each gene) represents a decision variable, feature, or parameter of the problem. The chromosome representation involves two related decisions. The first decision has to do with the problem of finding an effective way of encoding the problem in terms of a string chromosome. This decision can lead to the determination of the size of each chromosome, which is completely dependent on the nature of the problem. The second decision involves the selection of a string format for each gene in the chromosome. The value of a gene can be either a binary number or an integer. The binary format, which is the genetic one, has been widely used. The integer format (used in this work) is domain specific, and has also been widely used in the areas such as facility layout, traveling salesman problems, scheduling, among many other problems.

For the problem of modularity, the integer format is used. Each gene will have an integer value, which will represent the classification of components into groups depending on their similarities. Encoding for this problem is done as follows: the chromosome strings consist of n integer genes, each of which represent a component in the family of components ($1, \ldots, n$). Each value in a chromosome stands for the module number to which a part is assigned. For example, if there are ten parts to be divided into three modules, then the chromosome can be represented as (1, 1, 3, 2, 1, 2, 2, 3, 3, 3). Here, this chromosome indicates that module #1 has parts 1, 2, and 5; module #2 has parts 4, 6, and 7; and module #3 has parts 3, 8, 9, and 10. The size of the chromosome depends only on the size of the matrix being studied, so if a problem has 20 parts/components, the size of the chromosome will be 20 (20 genes). Also, since the similarities between components can and will be different, the integer value of each gene is allowed to be any number between 1 and the number of components being grouped (n), which will basically mean that any group from 1 to n could be selected in the process.

4.13.2.2 Determination of Systems Parameters

Some parameters, including population size, maximum number of generations, and the probability of crossover and mutation, should be decided before the algorithm starts to find a solution. These parameters are very sensitive to the computational performance of the algorithm, and although researchers proposed some standard values for these parameters, it is usually with extensive experimentation that these parameters are best set.

The population size directly affects the computation time of the algorithm, as having a large population size will mean that the algorithm will have to perform more calculations. It is important to note that a higher population size will give the algorithm a higher chance of success, as the solution space searched will be larger. However, if good solutions are required in a reasonable amount of time, the population size should be smaller. This will allow the algorithm to quickly find a good solution to the problem being studied. If the crossover probability is set to be 0.65, this basically means that about 65% of the chromosomes in each generation will experience crossover. If the crossover probability is set to 100%, then all offspring will be created by crossover. If it is 0%, the whole new generation will be made from exact copies of chromosomes from old populations. Crossover is made in hope that new chromosomes will have good parts of old chromosomes and maybe the new chromosomes will be better. However, it is always good to allow some chromosomes to survive without change in the next generation, an occurrence called elitism. If the mutation probability is set to be 0.20, this means that about 20% of the genes in a population will experience mutation. If the mutation probability is set to a 100%, all of the chromosomes will be changed. On the other hand, if it is set to 0%, nothing will change.

Mutation occurs to prevent falling into local optima, but it should not occur very often, as the genetic algorithm will become just a random search method. The number of generations is directly affected by the performance of the algorithm. A maximum number of generations can be set, but the algorithm will usually find a solution to the problem before this number is ever reached. This means that many extra computations will be performed after the algorithm has already found a solution. The algorithm developed examines the fitness of each population, and stops whenever the global fitness cannot be greater than what it already is. This means that the algorithm has found an optimal solution, at which point, the solution is given to the user while the algorithm starts another run to investigate the grouping of components with a different number of groups.

4.13.2.3 Generation of Initial Populations

After the problem is represented, a set of initial solutions (chromosomes) called population has to be determined. This first population will represent the first generation, which will then evolve until a good solution to the problem being studied is found. In order to obtain feasible solutions, each chromosome has to satisfy the problem constraints. The *replacement policy* is a way of making sure that the chromosomes created for the first generation are valid. Validity can be asserted in many ways, but one easy method checks the problem constraints given by the problem with each created chromosome. Two different methods have been widely used. The first method is called the variable restriction method, which only chooses the chromosomes that meet the feasible region of constraints. The second method is called the penalty function method, which allows chromosomes to violate the constraints by giving them a penalty, which will then in turn lower their probability of survival to zero. In this model, a modification of the variable restriction method is utilized.

A random number generator creates the initial population by randomly generating numbers between 1 and n (where n is the size of the chromosome). After each chromosome is created, the algorithm checks which groups were created, and with this information, the algorithm checks the validity of the chromosome by comparing it with the specified problem constraints. If the chromosome is valid, no changes are made, but if it violates one of the constraints specified in the problem, the necessary changes are made to ensure the validity of the chromosome. Due to the chromosomal representation chosen for the problem, the only problem that can occur is the assignment of components to modules that have not been created. The problem is easily resolved by creating the modules that are represented within the chromosome. For example, if the modules selected are 2, 5, and 8, and the chromosome is (5, **2**, 8, 8, **2**, 5, 5, **8**, 2, 2), one has to make sure that position 2 in the chromosome has a 2, position 5 in the chromosome has a 5, and position 8 in the chromosome has an 8. In this case, the chromosome would not be valid, and the algorithm would change the fifth position value from a 2 to a 5 to make the chromosome a valid one. This modification of the variable restriction method is used after each population is created, so that only valid chromosomes are studied and manipulated.

4.13.2.4 Selection and Use of the Fitness Function

In order to find better solutions, a fitness function is needed for evaluating and selecting good generations of chromosomes. The fitness function should give domain-specific information about the value of each chromosome. For this reason, it is usually a good idea to define it in the form of a mathematical formulation, either by maximization or minimization of some parameters and constraints. In the case of this model, the objective function of the p-median model is selected as the fitness function. Each chromosome, created by either the initialization of the genetic algorithm, or by the creation of a new generation, will have a fitness value. Since the objective of this model is to maximize the sum of the similarities, the fitness values are continuously increasing, until an optimal solution is found. Once the fitness function is defined and used for the first time (after the first generation of chromosomes are created), the algorithm starts the selection and reproduction process. Using the fitness values of each chromosome, the roulette wheel selection process begins by finding the total fitness of the population.

Using the roulette principles, a reproduction probability is assigned to each chromosome (based on their fitness values), and the roulette is filled using the respective cumulative probabilities of each chromosome. Since the space on the wheel is totally dependent on the fitness value of each chromosome, fitter chromosomes will have a larger space in this bias roulette, increasing their chances of survival. The algorithm then randomly generates a number between 0 and 1, and a chromosome is chosen as a parent based on that random number and the cumulative reproduction probability. At the end of the reproduction cycle, the new group of chromosomes is ready for crossover and mutation. The new chromosomes are the basis of a new generation; a generation that will be altered by crossover and mutation, and that will be examined after the two genetic operators have done their part in introducing new areas of unvisited solution space to the problem.

4.13.2.5 Crossover

After the better fitted chromosomes are selected, each pair of chromosomes is selected sequentially to exchange information according to the crossover probability previously determined. A random number is generated, and if the number lies below the crossover probability, crossover is performed. A mode of single-point crossover is used in this method; a mode for which the cut point for doing crossover is randomly selected at each instance. All but one chromosome are subject to crossover. After reproduction gives a new generation of chromosomes, the best chromosome is assured to be a part of the next generation unchanged, as it is always a good idea to keep the best solution within a population. Elitism ensures that the best chromosome will not suffer crossover and/or mutation. The algorithm will also benefit from elitism, as it will have a better chance of finding good solutions to the problem without losing the best solution found so far.

After all chromosomes have been exposed to crossover, the new generation is subjected to mutation. A crossover cycle is over whenever each pair of sequential chromosomes have either experienced crossover (because their probability of crossover fell within the $P_{cr}\%$), or have been passed without change because their probabilities fell outside the crossover probability. If a pair of chromosomes experience crossover, the next chromosomes to be sequentially selected will not include any of these two chromosomes, as it is not desired to double crossover a single chromosome.

4.13.2.6 Mutation

After crossover, mutation in each chromosome occurs according to previously set mutation probability. The mutation operator occasionally alters genes within a chromosome by changing the value of a single gene within a chromosome. The genetic algorithm in this case alters the genes based on the mutation probability as well as on the groups or modules that have already been selected for each chromosome. This ensures that new chromosomes will be explored, and thus, the solution space will be broader.

It is important to note that the changes due to mutation lie within the groups selected for each chromosome. In essence, this means that once a gene is randomly selected for mutation, the change will only occur within the number of groups being used. This ensures that the number of groups within a population will not increase, as this is a highly undesirable occurrence when the algorithm is studying a number of groups. Once all chromosomes are subjected to mutation (except the best one), the new generation of chromosomes is set. The new run will not require an initialization procedure, as the generation that just experienced reproduction, crossover, and mutation acts as the new generation. However, another mode of mutation is introduced at this point, as the chromosomes of this new generation are subjected to a validity check. An illegal chromosome should never be passed to the next generation (replacement policy), and so the required changes to each chromosome are made in order to ensure their validity.

The new generation of chromosomes starts the process of reproduction, crossover, and mutation all over again. New fitness values are found, and thus a new population fitness is calculated. The population fitness measures the improvement of the genetic algorithm. After the first run, these values are compared with the previous population fitness. This serves as a test that tells whether the new population of individuals is better than the previous population. If the number does not increase after a few runs, or if after a few runs the best population fitness is reached, the algorithm will stop and give the best solution found. An optimal solution is reached whenever the chromosomes in a population are all the same. This basically means that the fitness of each chromosome will be the same, and thus, the population fitness will have reached a maximum point, at which point the algorithm stops and returns the best chromosome within the final population. The algorithm will give various arrangements for the different number of groups studied, and the one with highest fitness will be regarded as the optimal solution. This will mean that if the components are grouped in that manner, the modular design will be optimized. The algorithm developed in this section can be used in a wide variety of combinatorial optimization problems, as it is flexible enough to achieve good results using simple, yet powerful tools such as crossover and mutation. Genetic algorithms can handle bigger and more complex problems that may be impossible to solve using conventional methods. As it has been stressed throughout this section, GAs also give the user good solutions to problems that may take too long to solve. The best solution is not always possible, and genetic algorithms can give plausible solutions to combinatorial optimization problems that are NP-Hard and that may require an infinite number of calculations.

4.14 Algorithm-Based Solution Methodology for Grouping Components into Modules

Components with a high degree of association should be grouped together in design modules. This can be accomplished by using an optimization model that maximizes the sum of the similarities. The optimization model will identify independent modules that can be designed simultaneously. The model is Np-Hard. Heuristic algorithms have been extensively used as an alternative technique for solving Np-Hard problems. Modularity decomposition problem could benefit from these algorithms for finding solution in less time. This method for decomposing similarity matrices in modularity is based on network flows and optimization. The main reason for the application of network optimization as base structure is its ability in coupling deep intellectual contents with a remarkable range of applicability. Many combinatorial problems that are innately hard can be solved by transforming to network concepts. This specification makes us to define our problem as a graph acceptable in network flows rules to have the opportunity to solve this decomposition by Network algorithms. After calculation of similarity measure, nodes and edges are defined as follows:

- ***Node***: each component represents a node in our proposed graph.
- ***Edge***: relationship between any two components represents an edge such that if component *i* has similarity index more than 0 with component *j*, there exists an edge between them.
- ***Flow of each edge***: the similarity index between each two components is the flow of the edge between them (see Fig. 4.12).

The objective function associated with this network is to find modules, such that each module has the maximum amount of collective similarity indexes. It means that it is necessary to find modules created by group of components that are connected with each other by the largest similarity indexes. Each component can be assigned to just one module and each module must contain the most similar components to each other. Based on this objective, Max-Flow algorithm is used for solving this problem. In a capacitated network, it is required to send as much flow between start and end node. This concept has been applied in other engineering applications such as matrix rounding problem and feasible flow problems. In this problem, the maximum flow for the main graph is determined. This establishes the solution for the first module that contains components (nodes) in maximum flow. Components that are assigned to first module will be removed and a new graph will be created and this graph for maximum flow is solved to find the second module. The algorithm will continue until all possible modules are assigned. For the Max-Flow algorithm, augmenting path algorithm or Preflow-push algorithm will be used based on network specifications. The Preflow-push algorithm is one of the most powerful and flexible algorithms for max-flow problems. The Preflow-push algorithms maintain a preflow at each intermediate stage. Active nodes in this algorithm are nodes that have positive

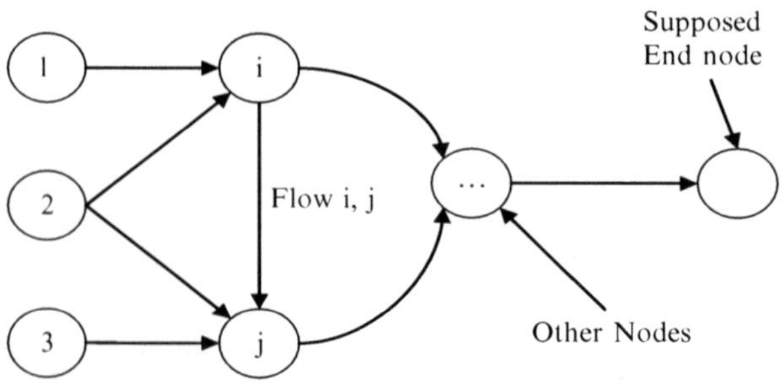

Fig. 4.12 Graph for a decomposed product

excess. Because this algorithm attempts to achieve feasibility, the presence of active nodes indicates that the solution is infeasible and the basic operation of this algorithm is to select an active node and try to remove its excess by pushing flow to its excess. A maximum preflow is defined as a preflow with the maximum possible flow into the sink. This algorithm is a polynomial algorithm. Sample of the pseudo-codes for the solution methodology are as follows:

```
Network Set Begin
x = 1
n = number of components
While each component has assigned to a module
Design graph of components.
Use pre flow push algorithm to find the maximum flow from each node to node t.
Assign components in the maximum flow with similarity more than 1 to a module
named module x.
            Similarity of module = Sum (similarity of components)
                        x = x + 1.
            End while.
            Use selecting Procedure

End Procedure.
Selecting Procedure Begin
y = 1
While n <> 0
            Choose the module with biggest similarity
            Eliminate components in that module from total components.
                        n = n – Eliminated components
                        y = y + 1
                        End while
            Y = number of modules
            Write the selected modules as final modules.

End Procedure.
```

Matlab® software is used for the implementation of the Preflow push max flow algorithm.

4.15 Summary

A three-phase methodology is proposed for the development of complex products using the modularity concept. The proposed methodology matches the criteria set by the design for functionality, assembly, and manufacture. A detailed discussion of needs analysis, QFD, product requirement analysis, product/concept analysis, product physical analysis, product functional analysis, product/concept integration, identification of the impact of SLS on GFR, similarity index, optimization-based solution methodology for grouping components into modules is presented in this chapter. Finally, genetic algorithm-based solution methodology and algorithm-based solution methodology for grouping components into modules are described for understanding the presented concepts.

4.16 Review Questions

4.1. Discuss the three-phase methodology proposed for the development of complex products using the modularity concept.
4.2. Describe customer satisfaction process.
4.3. Define Kano's diagram of customer satisfaction.
4.4. What are the four major groups of customer requirements that exist in Kano's model?
4.5. How can the customer requirements prioritize?
4.6. Define Quality Function Deployment (QFD).
4.7. How can the House of Quality (HOQ) be built?
4.8. What are the benefits of Quality Function Deployment?
4.9. Discuss briefly how can the impact of system-level specifications on GFR be identified.
4.10. Select any product in the market and do all the required analysis that you learned in this chapter.

References

1. Akiyama K (1991) Function analysis: systematic improvement of quality and performance (trans: Dillon A P), Productivity Press, Cambridge, MA
2. Besterfield DH, Besterfield-Michna C, Besterfield GHet al (1995) Total quality management. Prentice-Hall, Englewood Cliffs, NJ
3. Chandrasekaran B, Goel AK, Iwasaki Y (1993) Functional representation as design rationale. Computer 26:48–56
4. Fasal JH (1972) Practical value analysis methods. Hayden Book Co., New York
5. Finkelman DP, Goland AR (1990) How not to satisfy your customers. McKinsey Q 18(1) (1998):25–38
6. Kano N, Seraku N, Takahashi F, Tsuji S (1984) Attractive quality and must-be quality. Hinshitsu (Quality, The Journal of the Japanese Society for Quality Control),14(2):39–48
7. Matzler K, Hinterhuber HH (1998) How to make product development projects more successful by integrating Kano's model of customer satisfaction into quality function deployment. Technovation 18(1):25–38

8. Middendorf WH (1990) Design of devices and systems, 2nd edn. Marcel Dekker, New York
9. Miles LD (1972) Techniques of value analysis and engineering, 2nd edn. McGraw-Hill, New York
10. Pahl G, Beitz W (1996) Engineering design: a systematic approach, Springer, New York
11. Shillito LM (1994) Advanced QFD: linking technology to market and company needs. Wiley, New York
12. Singh N, Rajamani D (1996) Cellular manufacturing systems: design, planning, and control. Chapman & Hall, London
13. Sullivan LP (1986) Quality function deployment. Qual. Prog. pp. 39–50

Chapter 5
DFMo Case Study: Four-Gear Speed Reducer Design

Abstract This chapter presents a design for modularity approach which is tested and validated using a test product. The selected test product is of moderate complexity to ensure that effort is focused on applying and validating the proposed approach rather than on attempting to understand a complex product. Maintaining moderate complexity will also show the potential for using the proposed approach in designing complex products or systems.

Keywords Modularity • Complexity

5.1 Introduction

In this chapter, the design for modularity approach is tested and validated using a test product. The selected test product is of moderate complexity to ensure that effort is focused on applying and validating the proposed approach rather than on attempting to understand a complex product. Maintaining moderate complexity will also show the potential for using the proposed approach in designing complex products or systems.

5.2 Problem Description

A need to design a speed reducer is identified. The speed reducer is part of the power-transmission system of a pump. The power is generated by an electric motor that operates at a fairly high speed while the pump must rotate more slowly. Figure 5.1 shows an overview diagram of the system.

A.K. Kamrani and E.A. Nasr, *Engineering Design and Rapid Prototyping*,
DOI 10.1007/978-0-387-95863-7_5, © Springer Science+Business Media, LLC 2010

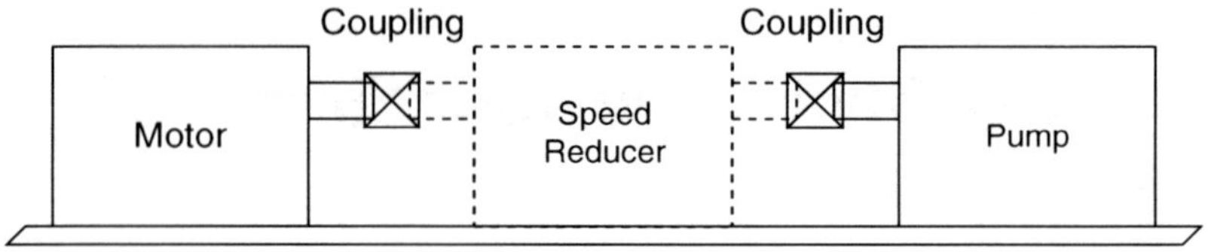

Fig. 5.1 System diagram

5.3 Needs Analysis

5.3.1 Recording the Voice of the Customer

Customer needs and requirements are collected and categorized according to Kano's model. The following table shows a sample of those requirements.

No.	Customer needs	Expected	Spoken
1.	15 hp must be transmitted from the motor to the pump	Yes	Yes
2.	The speed delivered to the pump must be reduced from 2,000 rpm (motor speed) to 300 rpm	Yes	Yes
3.	The motor output shaft and the pump input shaft must remain in-line	Yes	Yes
4.	The space available to install the device is 20″ × 20″	Yes	Yes
5.	The height of the product should not exceed 24″	Yes	Yes
6.	The component of the product should have enough space between them to allow easy maintenance	Yes	No
7.	The product should have small size and weight	Yes	No
8.	The product should be capable of operating at temperatures ranging from 0° to 130°F	Yes	Yes
9.	The product should have a moderate cost	No	No
10.	The product should be safe to use	Yes	No
11.	The product should be easy to operate	No	No
12.	The product should use standard materials and components	No	No

5.3.2 Product Features/Metrics Identification

The product to be designed is decomposed into basic components and the main features/metrics are identified. Following is a partial list of the identified features/metrics.

Component	Feature description	Feature symbol
Gear	Outside Diameter	G_{OD}
	Inside Diameter	G_{ID}
	Number of teeth	t
Shaft	Center line Height	H1
	Outside Diameter	Sh_{OD}
	Length	L

5.3.3 *Building the House of Quality*

Customer requirements and product features are used in building the house of quality as follows.

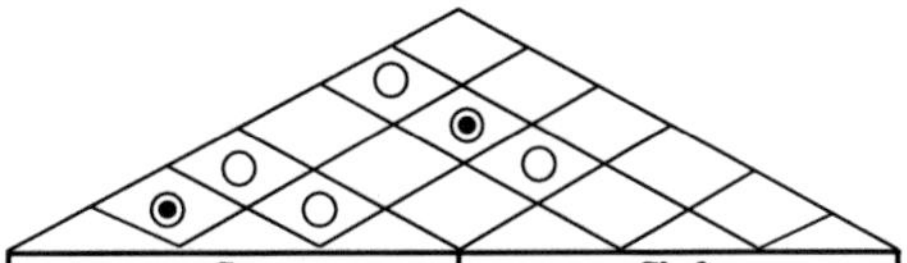

No	Customer Needs	Gear: Outside Diameter G_{OD}	Gear: Inside Diameter G_{ID}	Gear: Number of teeth t	Shaft: Center line Height H1	Shaft: Outside Diameter Sh_{OD}	Shaft: Length L	
1	15 hp must be transmitted from the motor to the pump.	3	3	3	1	1	1	9
2	The speed delivered to the pump must be reduced from 2,000 rpm (motor speed) to 300 rpm.	3	3	9	0	0	0	9
3	The motor output shaft and the pump input shaft must remain in-line.	0	0	0	9	1	1	9
4	The space available to install the device is 20" X 20".	9	9		9	3	9	3
5	The height of the product should not exceed 24".	9	3	3	9	9	0	3
		3	3	9	3	3	9	

5.4 Product Requirements Analysis

The needs listed in the previous step are translated into requirements and presented as follows.

5.4.1 *Functional Objectives*

- To receive power from an electric motor through a rotating shaft.
- To transmit power through machine elements that reduces the rotational speed to a desired value.
- To deliver the power at the lower speed to an output shaft that ultimately drives the pump.

5.4.2 *Operational Functional Requirements*

- The reducer must transmit 15.0 hp.
- The input is from an electric motor at a rotational speed of 2,000 rpm.
- The output delivers the power at a rotational speed range of 290–300 rpm.
- The input and output shafts must be in-line.
- The reducer must be installed on a square surface 20″ × 20″, with a maximum height of 24″.

5.4.3 General Functional Requirements

- *Performance*: The degree to which the design meets or exceeds the design objectives.
- *Compactness*: Small size and weight.
- *Ease of Service*: Components should be arranged in such a way that they are accessible for maintenance and replacement.

5.4.4 General Functional Requirements' Weights

It is assumed that the three general functional requirements (GFRs) listed have equal importance; therefore, equal weight of 1 is assigned to each of them.

5.5 Product Concept Analysis

It is assumed that concept generation and concept selection were performed and resulted in the selection of a four-gear speed reducer to accomplish the required function and meet the requirements stated in the previous step. Figure 5.2 shows the selected conceptualized speed reducer.

5.5.1 Product Physical Decomposition

The overall system is decomposed into four physical subsystems that include the speed reducer. Then the speed reducer is decomposed into its basic physical components as shown in Fig. 5.3.

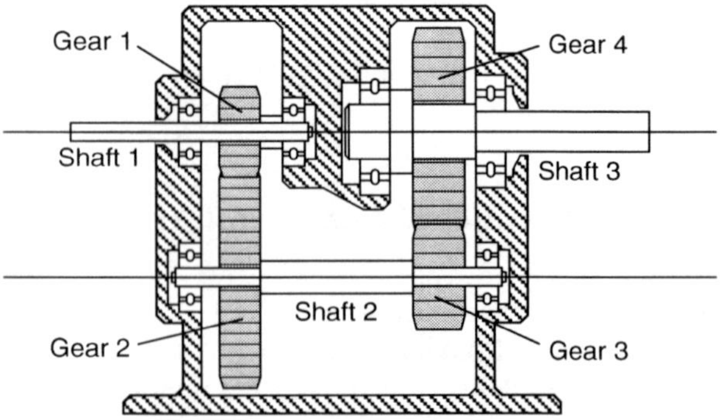

Fig. 5.2 Four-gear speed reducer

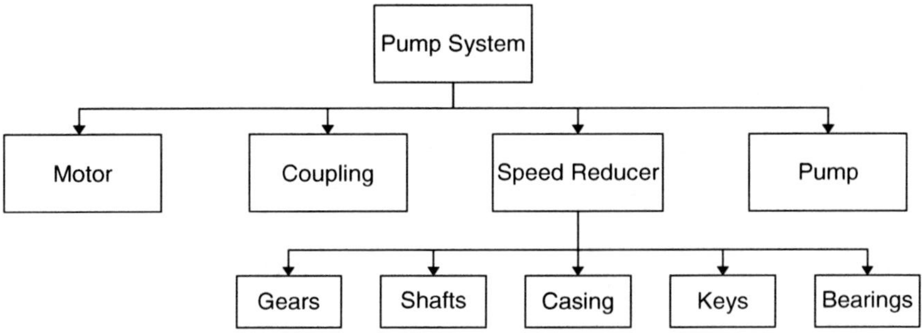

Fig. 5.3 Physical decomposition of pump system

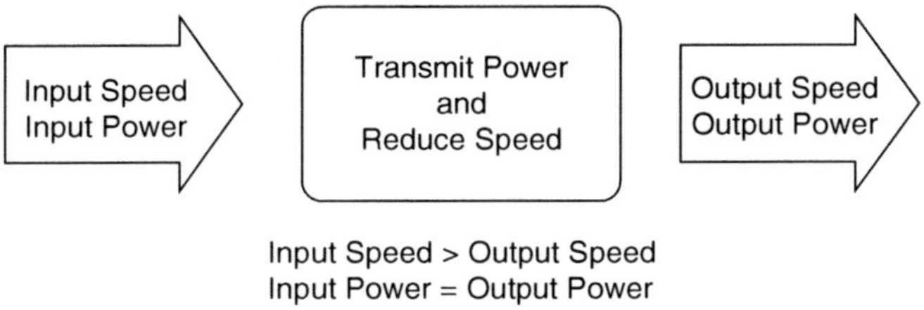

Fig. 5.4 Overall function of the speed reducer

5.5.2 Product Functional Decomposition

The product overall function is conceptualized into an action statement, "to transmit power and reduce speed," and represented in a function block diagram (Fig. 5.4).

Next, the components' functions are analyzed as shown in Fig. 5.5 and Table 5.1.

5.6 Product/Concept Architecture

5.6.1 System-Level Specifications

System-level specifications (SLS) are determined based on the functional and physical decomposition. The resulting decomposition description is arranged in a hierarchy structure as in Fig. 5.6.

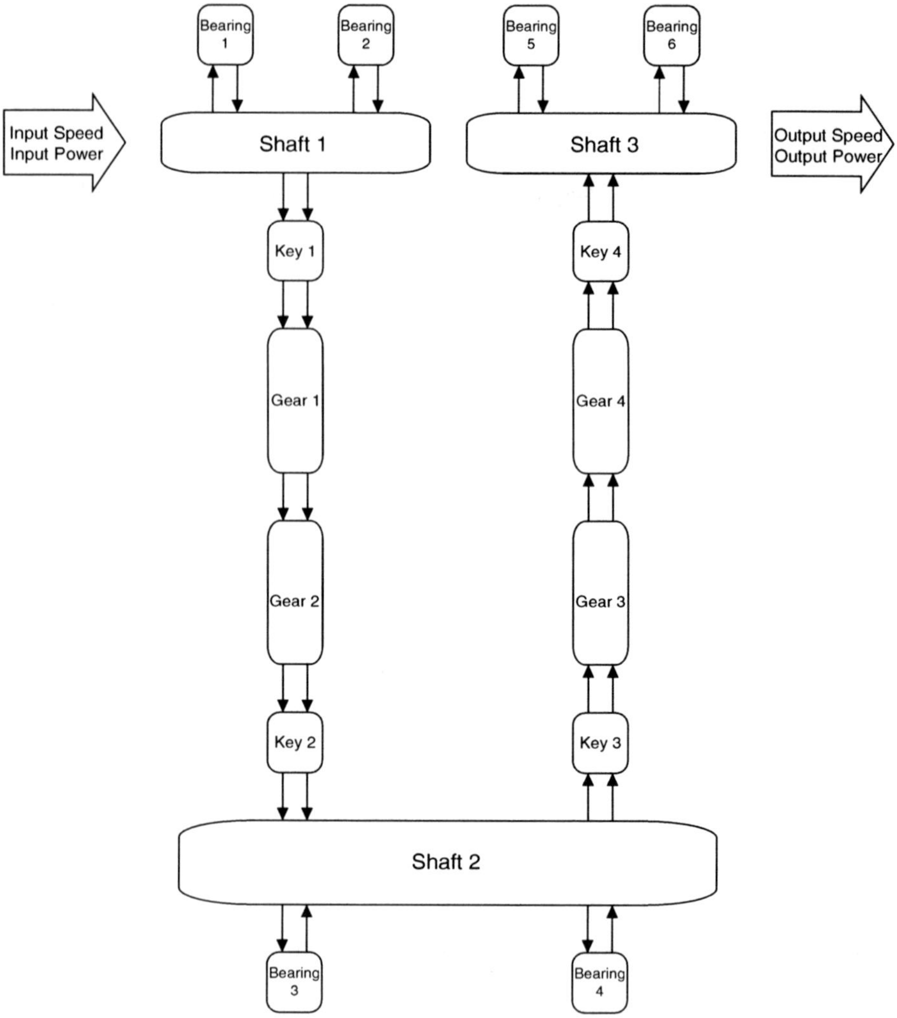

Fig. 5.5 Components' functions

The one-to-one relationships between components (system-level specifications) are determined based on the detailed descriptions of the system-level specifications. These are located in the bottom level of the hierarchy. The relationship will be assigned 1 if it exists and 0 otherwise. A partial listing of the resulting system-level specifications is given in Table 5.2 (the complete SLS are listed in the Appendix). From Table 5.2, it can be shown that gear 1 and gear 2 have a parallel arrangement and direct contact. Also, the functional characteristics show that the power is transmitted directly between them and their speed is different.

Table 5.1 Relationships between components' functions

Component	Input	Output	I/O constraints	Function type
B1	S_{Sh1}	S_{B1}	$S_{Sh1} = S_{B1}$	Auxiliary
B2	S_{Sh1}	S_{B2}	$S_{Sh1} = S_{B2}$	Auxiliary
Sh1	S_{in}, P_{in}	S_{Sh1}, P_{Sh1}	$S_{in} = S_{Sh1}$ $P_{in} = P_{Sh1}$	Primary
K1	S_{Sh1}, P_{Sh1}	S_{K1}, P_{K1}	$S_{Sh1} = S_{K1}$ $P_{Sh1} = P_{K1}$	Primary
G1	S_{K1}, P_{K1}	S_{G1}, P_{G1}	$S_{K1} = S_{G1}$ $P_{K1} = P_{G1}$	Primary
G2	S_{G1}, P_{G1}	S_{G2}, P_{G2}	$S_{G1} > S_{G2}$ $P_{G1} = P_{G2}$	Primary
K2	S_{G2}, P_{G2}	S_{K2}, P_{K2}	$S_{G2} = S_{K2}$ $P_{G2} = P_{K2}$	Primary
SH2	S_{K2}, P_{K2}	S_{Sh2}, P_{Sh2}	$S_{K2} = S_{Sh2}$ $P_{K2} = P_{Sh2}$	Primary
B3	S_{Sh2}	S_{B3}	$S_{Sh2} = S_{B3}$	Auxiliary
B4	S_{Sh2}	S_{B4}	$S_{Sh2} = S_{B4}$	Auxiliary
K3	S_{Sh2}, P_{Sh2}	S_{K3}, P_{K3}	$S_{Sh2} = S_{K3}$ $P_{Sh2} = P_{K3}$	Primary
G3	S_{K3}, P_{K3}	S_{G3}, P_{G3}	$S_{K3} = S_{G3}$ $P_{K3} = P_{G3}$	Primary
G4	S_{G3}, P_{G3}	S_{G4}, P_{G4}	$S_{G3} > S_{G4}$ $P_{G3} = P_{G4}$	Primary
K4	S_{G4}, P_{G4}	S_{K4}, P_{K4}	$S_{G4} = S_{K4}$ $P_{G4} = P_{K4}$	Primary
SH3	S_{K4}, P_{K4}	S_{Sh3}, P_{Sh3}	$S_{K4} = S_{Sh3}$ $P_{K4} = P_{Sh3}$	Primary
B5	S_{Sh3}	S_{B5}	$S_{Sh3} = S_{B5}$	Auxiliary
B6	S_{Sh3}	S_{B6}	$S_{Sh3} = S_{B6}$	Auxiliary

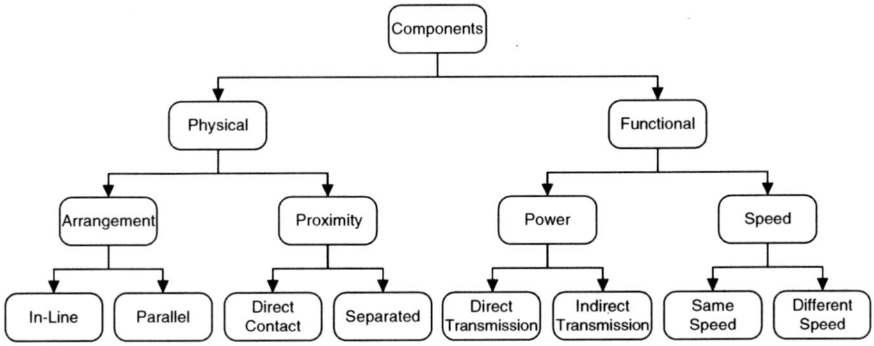

Fig. 5.6 System-level specification hierarchy structure

Table 5.2 System-level specifications

		Components							
		Physical				Functional			
		Arrangement		Proximity		Power		Speed	
Comp. 1	Comp. 2	In-line	Parallel	Direct contact	Separated	Direct transmission	Indirect transmission	Same speed	Different speed
Gear 1	Gear 2	0	1	1	0	1	0	0	1
	Gear 3	0	1	0	1	0	1	0	1
	Gear 4	1	0	0	1	0	1	0	1
	Shaft 1	1	0	1	0	1	0	1	0
	Shaft 2	0	1	0	1	0	1	0	1
	Shaft 3	1	0	0	1	0	1	0	1
	Bearing 1	1	0	0	1	0	0	1	0
	Bearing 2	1	0	0	1	0	0	1	0
	Bearing 3	0	1	0	1	0	0	0	1
	Bearing 4	0	1	0	1	0	0	0	1
	Bearing 5	1	0	0	1	0	0	0	1
	Bearing 6	1	0	0	1	0	0	0	1
	Key 1	0	1	1	0	1	0	1	0
	Key 2	0	1	0	1	0	1	0	1
	Key 3	0	1	0	1	0	1	0	1
	Key 4	0	1	0	1	0	1	0	1

5.6.2 *Impact of the System-Level Specifications on the General Functional Requirements*

The impact of SLS on GFR is determined by using the needs analysis and previous knowledge of the system under design. In determining the impact, the designer might ask questions such as follows:

- What will happen to the GFR if we did/did not have this specific SLS?
- What kind of effect will this SLS have on the GFR?

Answers to these questions should identify the impacts. If it is a desired impact, it will be assigned (1). If the impact is undesired or if it will prevent the product from achieving its functions, it will be assigned (−1). If the impact is not significant or it does not affect the GFR, it will be given a (0).

Applying the above general guidelines to the product under study will reveal the following:

- In-line arrangements of the components will facilitate the serviceability of the product because they are easier to reach than intersecting or inclined arrangements. Also, in-line arrangements usually maintain the compactness of the system while having no effect on the system's performance. The same thing can be said about the parallel arrangement.

- Direct contact between components will have no effect on the serviceability because it is a common practice to disassemble the components during maintenance. Direct contact has a desired effect on the compactness of the product. The closer the components are, the more compact the product will be. Also, direct contact will support the performance because power and speed are transmitted through contacting parts only.
- Separation of components from each other will make the maintenance process easier, since component disassembly will become easier. Separated components, however, will lead to a larger size and thus will have a negative impact on the compactness of the product. Also, separated components will not help in transmitting power and speed, which will lead to a negative impact on the performance.
- Power-transmission (direct/indirect) and speed-ratio (same/different) specifications will not affect the ease of service and compactness requirements, because these requirements are affected by physical specifications, and power transmission and speed ratio are function specifications.
- Direct power transmission will support the overall function of the product and thus have a desired impact. Because the impact of indirect power transmission is not significant on the performance of the power, it will neither support the overall function in a direct manner nor prevent the overall function from being accomplished.
- As for the speed ratio, a different speed ratio between the components will support the overall function (reduce speed), while a steady ratio (same speed) will not have any positive or negative impacts on the overall function.

The resulting impacts are given in Table 5.3.

5.6.2.1 Similarity Matrix

The similarity index is used to determine the degree of association between the different components. The indexes are arranged in a matrix. Three types of similarity indexes are analyzed: (1) functional similarity indexes, (2) physical similarity indexes, and (3) combined similarity indexes. Functional similarity indexes will be based on the functional SLS alone, while physical similarity indexes will be based

Table 5.3 Impact of SLS on GFR

System-level specifications	General functional requirements		
	Ease of service	Compactness	Performance
In-line	1	1	0
Parallel	1	1	0
Direct contact	0	1	1
Separated	1	−1	−1
Direct transmission	0	0	1
Indirect transmission	0	0	0
Same speed	0	0	0
Different speed	0	0	1

on the physical SLS. Combined similarity indexes will use both the functional and physical SLS. The combined similarity results are shown in Fig. 5.7.

5.6.2.2 Grouping Components into Modules Using p-Median

The p-median model is used to rearrange the similarity matrices into independent modules. The independent modules are found as follows:

- Construct an optimization model for each of the similarity matrices (physical, functional, and combined).
- Solve each model by increasing the number of independent modules.
- Compare the results and select the solution that has the maximum objective value.

The best solution has three independent modules, as shown in Fig. 5.8.

The optimization model identifies independent modules. These modules can be designed simultaneously. The solution also illustrates interactions do exit between the independent modules. These interactions are represented by the similarity between different components that belong to independent modules. For example, gear 1 and gear 2 have a similarity factor of 1 in the functional modules and a similarity factor of 4 in the physical modules. This indicates that association exists between gear

	Gear 1	Gear 2	Gear 3	Gear 4	Shaft 1	Shaft 2	Shaft 3	Bearing 1	Bearing 2	Bearing 3	Bearing 4	Bearing 5	Bearing 6	Key 1	Key 2	Key 3	Key 4
Gear 1		5	1	1	6	1	1	2	2	1	1	1	1	6	1	1	1
Gear 2	5		2	1	1	6	1	1	1	2	2	1	1	1	6	3	2
Gear 3	1	2		2	1	6	1	1	1	2	2	1	1	1	1	6	1
Gear 4	1	1	2		1	1	6	1	1	1	1	5	5	1	1	1	6
Shaft 1	6	1	1	1		1	1	5	5	1	1	1	1	6	1	1	1
Shaft 2	1	6	6	1	1		1	1	1	5	5	1	1	1	6	6	1
Shaft 3	1	1	1	6	1	1		1	1	1	1	5	5	1	1	1	6
Bearing 1	2	1	1	1	5	1	1		2	1	1	1	1	2	1	1	1
Bearing 2	2	1	1	1	5	1	1	2		1	1	1	1	2	1	1	1
Bearing 3	1	2	2	1	1	5	1	1	1		2	1	1	1	2	2	1
Bearing 4	1	2	2	1	1	5	1	1	1	2		1	1	1	2	2	1
Bearing 5	1	1	1	5	1	1	5	1	1	1	1		2	1	1	1	2
Bearing 6	1	1	1	5	1	1	5	1	1	1	1	2		1	1	1	2
Key 1	6	1	1	1	6	1	1	2	2	1	1	1	1		1	1	1
Key 2	1	6	1	1	1	6	1	1	1	2	2	1	1	1		2	1
Key 3	1	3	6	1	1	6	1	1	1	2	2	1	1	1	2		1
Key 4	1	2	1	6	1	1	6	1	1	1	1	2	2	1	1	1	

Fig. 5.7 Combined similarity matrix

	Gear 1	Shaft 1	Bearing 1	Bearing 2	Key 1	Gear 2	Gear 3	Shaft 2	Bearing 3	Bearing 4	Key 2	Key 3	Gear 4	Shaft 3	Bearing 5	Bearing 6	Key 4
Gear 1		6	2	2	6	5	1	1	1	1	1	1	1	1	1	1	1
Shaft 1	6		5	5	6	1	1	1	1	1	1	1	1	1	1	1	1
Bearing 1	2	5		2	2	1	1	1	1	1	1	1	1	1	1	1	1
Bearing 2	2	5	2		2	1	1	1	1	1	1	1	1	1	1	1	1
Key 1	6	6	2	2		1	1	1	1	1	1	1	1	1	1	1	1
Gear 2	5	1	1	1	1		2	6	2	2	6	3	1	1	1	1	2
Gear 3	1	1	1	1	1	2		6	2	2	1	6	2	1	1	1	1
Shaft 2	1	1	1	1	1	6	6		5	5	6	6	1	1	1	1	1
Bearing 3	1	1	1	1	1	2	2	5		2	2	2	1	1	1	1	1
Bearing 4	1	1	1	1	1	2	2	5	2		2	2	1	1	1	1	1
Key 2	1	1	1	1	1	6	1	6	2	2		2	1	1	1	1	1
Key 3	1	1	1	1	1	3	6	6	2	2	2		1	1	1	1	1
Gear 4	1	1	1	1	1	1	2	1	1	1	1	1		6	5	5	6
Shaft 3	1	1	1	1	1	1	1	1	1	1	1	1	6		5	5	6
Bearing 5	1	1	1	1	1	1	1	1	1	1	1	1	5	5		2	2
Bearing 6	1	1	1	1	1	1	1	1	1	1	1	1	5	5	2		2
Key 4	1	1	1	1	1	2	1	1	1	1	1	1	6	6	2	2	

Fig. 5.8 Combined modules

1 and gear 2, i.e., the design of gear 1 will affect the design of gear 2 and vice versa. Although gear 1 and gear 2 are associated according to their similarities, they were not grouped in the same module, which shows that the model used lacks the ability to handle the interaction effectively. These interactions are critical to the development process and must be studied for it.

The association between components belonging to different modules can be used to identify the types of interactions and thus identify the information required to explain this interaction. For example, gear 1 and gear 2 belong to different modules, but there exists a functional association between them. This association can be explained by examining the function flow diagram. It can be shown that a major contribution to the function occurs (i.e., the speed is reduced). The same can be concluded for gear 3 and gear 4.

For physical association, all components are physically interacting with each other, but components within the same module have a strong association with a major part that will determine the physical arrangement. For example, in module 1, all components have a strong relationship with shaft 1, which means that the physical location of shaft 1 will guide the arrangement of the other components. The same can be argued for shaft 2 in module 2 and shaft 3 in module 3.

From the above discussion, it can be concluded that the design of the speed reducer using the developed modules will be guided by two important interactions: (1) the speed reduction ratio between gears 1 and 2 and gears 3 and 4 and (2) the physical location of the three shafts. This includes modules that will be identified based on the speed of the gears and the physical specifications of the shaft.

5.7 Grouping Components into Modules Using Genetic Algorithm Model

The genetic algorithm begins by randomly generating the initial population of solutions for the similarity matrix. Since the first generation of solutions group all the components into one group, the algorithm just finds the column on the matrix that yields the highest fitness values. The fitness value for the best chromosome is 48. The grouping for this case does not give the user any valuable information as all the components are grouped into one family. After eight generations, the algorithm reaches the best solution for this case, a solution with a total fitness (for the generation) of 960, which is the sum of all the fitness values of each chromosome in a generation composed of 20 individuals. Figure 5.9 shows the improvement made by the GA after every generation. A maximum of 80 generations is allowed, but in this case, the algorithm only needs eight generations mainly because of the simplicity of the problem being studied (classification of components into one module).

The second run groups the components into two separate modules. The result is shown in Fig. 5.10.

At the end of this run, the fitness values of each chromosome should be greater than the ones given in the previous run (48 was the best chromosome for the one

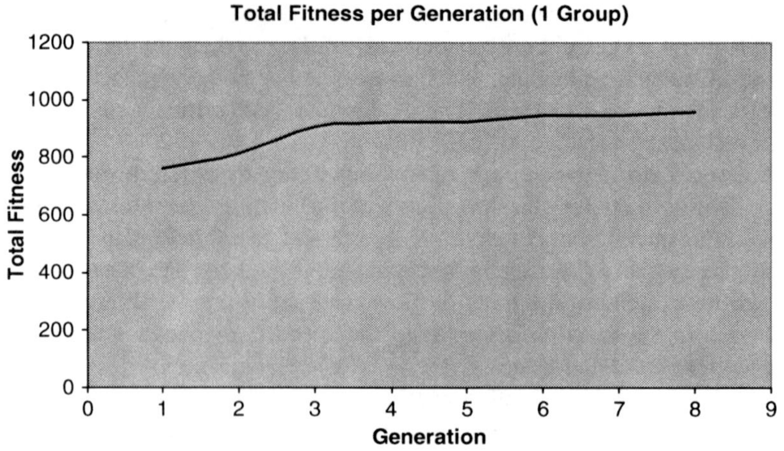

Fig. 5.9 GA response for one-module solution

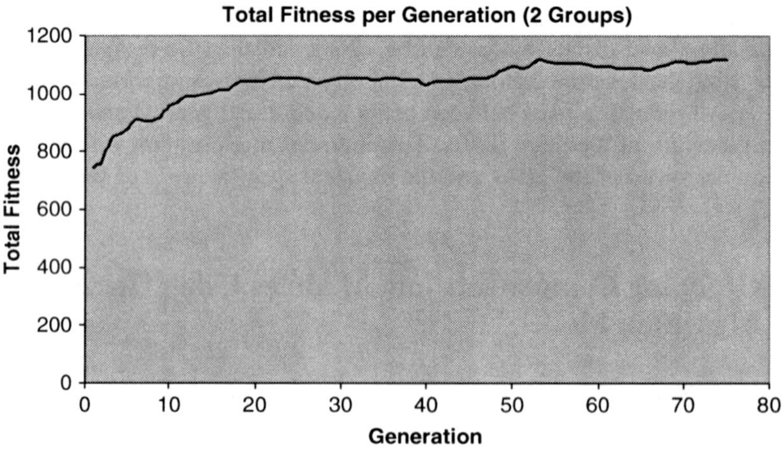

Fig. 5.10 GA response for two-module solution

group run). However, if the groupings do not yield a better fitness value, this will mean that grouping the components into two groups does not give the best overall optimal solution. It is important to note that some groupings, even if they are not the optimal ones, will give companies what they are looking for, as they may be satisfied with good solutions that will improve their operations.

As shown in Fig. 5.10, the genetic algorithm gives a better solution in this case. The fitness function value is 56, and the best grouping is (4, 6, 6, 4, 4, 6, 4, 6, 6, 6, 6, 4, 4, 6, 6, 6, 4). This basically means that, for example, components 1 and 4 are together in one module (positions one and four are both fours), and components 2 and 3 are together in another module. The rest of the groupings for this solution

	SS1C1	SS1C4	SS2C1	SS2C3	SS3C5	SS3C6	SS4C4	SS1C2	SS1C3	SS2C2	SS3C1	SS3C2	SS3C3	SS3C4	SS4C1	SS4C2	SS4C3
SS1C1		2	5	2	2	2	2	6	2	2	1	1	2	2	5	2	2
SS1C4	2		2	5	4	4	5	2	1	2	2	2	2	2	2	2	2
SS2C1	5	2		2	2	2	2	2	2	2	4	4	2	2	5	2	2
SS2C3	2	5	2		4	4	5	2	2	2	2	2	2	2	2	2	2
SS3C5	2	4	2	4		1	1	2	2	2	2	2	2	2	2	2	2
SS3C6	2	4	2	4	1		1	2	2	2	2	2	2	2	2	2	2
SS4C4	2	5	2	5	1	1		3	2	2	2	2	2	2	2	2	2
SS1C2	6	2	2	2	2	2	3		1	5	2	2	1	1	2	5	2
SS1C3	2	1	2	2	2	2	2	1		5	2	2	1	1	2	2	5
SS2C2	2	2	2	2	2	2	2	5	5		2	2	4	4	2	5	5
SS3C1	1	2	4	2	2	2	2	2	2	2		1	2	2	1	2	2
SS3C2	1	2	4	2	2	2	2	2	2	2	1		2	2	1	2	2
SS3C3	2	2	2	2	2	2	2	1	1	4	2	2		1	2	1	1
SS3C4	2	2	2	2	2	2	2	1	1	4	2	2	1		2	1	1
SS4C1	5	2	5	2	2	2	2	2	2	2	1	1	2	2		2	2
SS4C2	2	2	2	2	2	2	2	5	2	5	2	2	1	1	2		1
SS4C3	2	2	2	2	2	2	2	2	5	5	2	2	1	1	2	1	

Fig. 5.11 Grouping for two modules

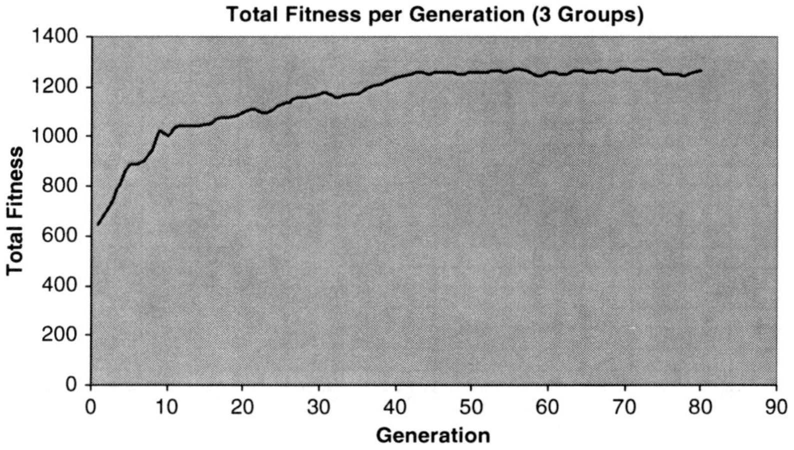

Fig. 5.12 GA response for a three-module solution

follow the physical proximity of the components being studied. A total of 75 generations were needed this time, as the problem got more interesting and complicated. The solution given by the GA for two modules is a good solution, but it is not the optimal one, as there are several high similarities left out for the two groups (an occurrence that may imply that the solution is not optimal). It is apparent from the grouping that the addition of at least another group would improve the arrangement of components, as the assignment of the components into two groups can be very restrictive. Figure 5.11 shows the final groupings.

The third run of the genetic algorithm groups the components into a total of three modules. This new run of the genetic algorithm brings a new improvement with respect to the previous runs. In this case, the genetic algorithm yields a fitness value of 64, the highest one so far, and a total fitness of 1,263. As Fig. 5.12 shows,

	SS1C1	SS2C1	SS3C1	SS3C2	SS4C1	SS1C2	SS1C3	SS2C2	SS3C3	SS3C4	SS4C2	SS4C3	SS1C4	SS2C3	SS3C5	SS3C6	SS4C4
SS1C1		5	1	1	5	6	2	2	2	2	2	2	2	2	2	2	2
SS2C1	5		4	4	5	2	2	2	2	2	2	2	2	2	2	2	2
SS3C1	1	4		1	1	2	2	2	2	2	2	2	2	2	2	2	2
SS3C2	1	4	1		1	2	2	2	2	2	2	2	2	2	2	2	2
SS4C1	5	5	1	1		2	2	2	2	2	2	2	2	2	2	2	2
SS1C2	6	2	2	2	2		1	5	1	1	5	2	2	2	2	2	3
SS1C3	2	2	2	2	2	1		5	1	1	2	5	1	2	2	2	2
SS2C2	2	2	2	2	2	5	5		4	4	5	5	2	2	2	2	2
SS3C3	2	2	2	2	2	1	1	4		1	1	1	2	2	2	2	2
SS3C4	2	2	2	2	2	1	1	4	1		1	1	2	2	2	2	2
SS4C2	2	2	2	2	2	5	2	5	1	1		1	2	2	2	2	2
SS4C3	2	2	2	2	2	2	5	5	1	1	1		2	2	2	2	2
SS1C4	2	2	2	2	2	2	1	2	2	2	2	2		5	4	4	5
SS2C3	2	2	2	2	2	2	2	2	2	2	2	2	5		4	4	5
SS3C5	2	2	2	2	2	2	2	2	2	2	2	2	4	4		1	1
SS3C6	2	2	2	2	2	2	2	2	2	2	2	2	4	4	1		1
SS4C4	2	2	2	2	2	3	2	2	2	2	2	2	5	5	1	1	

Fig. 5.13 Grouping for a three-group solution

	SS1C1	SS1C2	SS1C4	SS1C3	SS2C2	SS3C3	SS3C4	SS4C3	SS2C1	SS2C3	SS3C1	SS3C2	SS3C6	SS4C1	SS3C5	SS4C2	SS4C4
SS1C1		6	2	2	2	2	2	2	5	2	1	1	2	5	2	2	2
SS1C2	6		2	1	5	1	1	2	2	2	2	2	2	2	2	5	3
SS1C4	2	2		1	2	2	2	2	1	5	2	2	4	2	4	2	5
SS1C3	2	1	1		5	1	1	5	2	2	2	2	2	2	2	2	2
SS2C2	2	5	2	5		4	4	5	2	2	2	2	2	2	2	5	2
SS3C3	2	1	2	1	4		1	1	2	2	2	2	2	2	2	1	2
SS3C4	2	1	2	1	4	1		1	2	2	2	2	2	2	2	1	2
SS4C3	2	2	2	5	5	1	1		2	2	2	2	2	2	2	1	2
SS2C1	5	2	1	2	2	2	2	2		2	4	4	2	5	2	2	2
SS2C3	2	2	5	2	2	2	2	2	2		2	2	4	2	4	2	5
SS3C1	1	2	2	2	2	2	2	2	4	2		1	2	1	2	2	2
SS3C2	1	2	2	2	2	2	2	2	4	2	1		2	1	2	2	2
SS3C6	2	2	4	2	2	2	2	2	2	4	2	2		2	1	2	1
SS4C1	5	2	2	2	2	2	2	2	5	2	1	1	2		2	2	2
SS3C5	2	2	4	2	2	2	2	2	2	4	2	2	1	2		2	1
SS4C2	2	5	2	2	5	1	1	1	2	2	2	2	2	2	2		2
SS4C4	2	3	5	2	2	2	2	2	2	5	2	2	1	2	1	2	

Fig. 5.14 Grouping for a four-module solution

this run uses 80 generations to get the optimal solution for the problem, a solution that will end up being the overall optimal solution for the speed reducer.

The result of this run is shown in Fig. 5.13.

The grouping given by the genetic algorithm is (5, 6, 6, 7, 5, 6, 7, 5, 5, 6, 6, 7, 7, 5, 6, 6, 7). The solution for the three-module classification is in fact, the optimal solution. The total fitness (as well as the fitness of the chromosomes) of the new run for four-module grouping goes down to almost 900, while for the previous run it was over 1,200. Moreover, the fitness value of the best chromosome goes down from 64 to just 47, which is actually lower than the fitness value of the first run, in which the algorithm grouped all components into one group. This sudden drop in fitness, along with the poor grouping of the components (1, 1, 6, 1, 5, 6, 5, 5, 5, 6, 6, 15, 5, 5, 15, 6, 15), gives the user a "preview" of what is about to happen. Figure 5.14 lists the solution for the four-module classification.

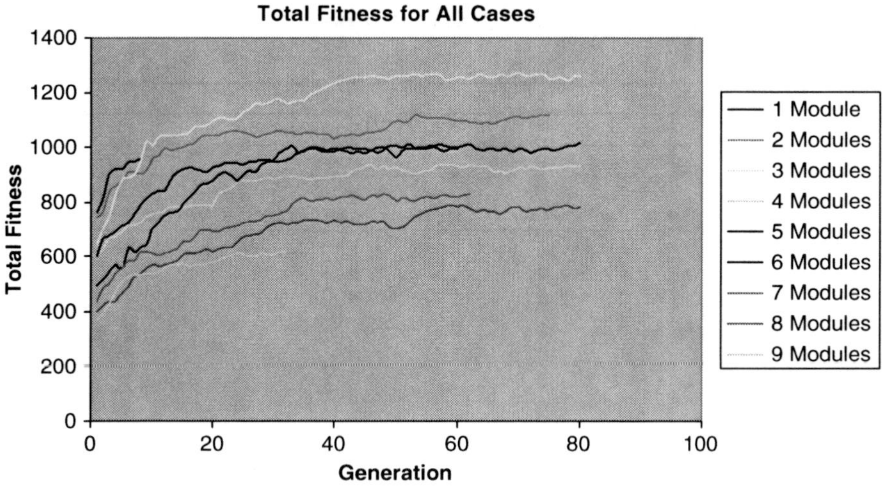

Fig. 5.15 Summary results of the GA process

The grouping shown is a more specific grouping of the components. However, as it can be seen on the matrix, there are high similarity values outside the groupings. This implies that the solution is not optimal, although it may give the user a much more specific arrangement of components. Figure 5.15 illustrates the summary of the runs for this study. As it is shown, the total sum of the fitness value decreases after the three modules solution.

5.8 Summary

In this chapter, a complete case study for understanding the presented concepts described in the previous chapter is presented. The design for modularity approach is tested and validated using a test product. The selected test product is of moderate complexity to ensure that effort is focused on applying and validating the proposed approach rather than on attempting to understand a complex product. Moreover, maintaining moderate complexity is also shown in designing complex products or systems.

5.9 Review Question

5.1. Select any product/system in the market and do all the required analysis that you learned in this chapter.

5.10 Engineering Design Specifications

		Components							
		Physical				Functional			
		Arrangement		Proximity		Power		Speed	
Comp. 1	Comp. 2	In-line	Parallel	Direct Contact	Separated	Direct transmission	Indirect transmission	Same Speed	Different Speed
Gear 1	Gear 2	0	1	1	0	1	0	0	1
	Gear 3	0	1	0	1	0	1	0	1
	Gear 4	1	0	0	1	0	1	0	1
	Shaft 1	1	0	1	0	1	0	1	0
	Shaft 2	0	1	0	1	0	1	0	1
	Shaft 3	1	0	0	1	0	1	0	1
	Bearing 1	1	0	0	1	0	0	1	0
	Bearing 2	1	0	0	1	0	0	1	0
	Bearing 3	0	1	0	1	0	0	0	1
	Bearing 4	0	1	0	1	0	0	0	1
	Bearing 5	1	0	0	1	0	0	0	1
	Bearing 6	1	0	0	1	0	0	0	1
	Key 1	0	1	1	0	1	0	1	0
	Key 2	0	1	0	1	0	1	0	1
	Key 3	0	1	0	1	0	1	0	1
	Key 4	0	1	0	1	0	1	0	1
Gear 2	Gear 3	1	0	0	1	0	1	1	0
	Gear 4	0	1	0	1	0	1	0	1
	Shaft 1	0	1	0	1	0	1	0	1
	Shaft 2	1	0	1	0	1	0	1	0
	Shaft 3	0	1	0	1	0	1	0	1
	Bearing 1	0	1	0	1	0	0	0	1
	Bearing 2	0	1	0	1	0	0	0	1
	Bearing 3	1	0	0	1	0	0	1	0
	Bearing 4	1	0	0	1	0	0	1	0
	Bearing 5	0	1	0	1	0	0	0	1
	Bearing 6	0	1	0	1	0	0	0	1
	Key 1	0	1	0	1	0	1	0	1
	Key 2	0	1	1	0	1	0	1	0
	Key 3	0	1	0	0	0	1	1	0
	Key 4	0	1	0	0	0	1	0	1
Gear 3	Gear 4	1	0	0	1	0	1	1	0
	Shaft 1	0	1	0	1	0	1	0	1
	Shaft 2	1	0	1	0	1	0	1	0
	Shaft 3	0	1	0	1	0	1	0	1
	Bearing 1	0	1	0	1	0	0	0	1
	Bearing 2	0	1	0	1	0	0	0	1
	Bearing 3	1	0	0	1	0	0	1	0
	Bearing 4	1	0	0	1	0	0	1	0
	Bearing 5	0	1	0	1	0	0	0	1
	Bearing 6	0	1	0	1	0	0	0	1
	Key 1	0	1	0	1	0	1	0	1
	Key 2	0	1	0	1	0	1	0	1
	Key 3	0	1	1	0	1	0	1	0
	Key 4	0	1	0	1	0	1	0	1

(continued)

(continued)

		Components							
		Physical				Functional			
		Arrangement		Proximity		Power		Speed	
Comp. 1	Comp. 2	In-line	Parallel	Direct Contact	Separated	Direct transmission	Indirect transmission	Same Speed	Different Speed
Gear 4	Shaft 1	1	0	0	1	0	1	0	1
	Shaft 2	0	1	0	1	0	1	0	1
	Shaft 3	0	1	1	0	1	0	1	0
	Bearing 1	1	0	0	1	0	0	0	1
	Bearing 2	1	0	0	1	0	0	0	1
	Bearing 3	0	1	0	1	0	0	0	1
	Bearing 4	0	1	0	1	0	0	0	1
	Bearing 5	1	0	1	0	0	0	1	0
	Bearing 6	1	0	1	0	0	0	1	0
	Key 1	0	1	0	1	0	1	0	1
	Key 2	0	1	0	1	0	1	0	1
	Key 3	0	1	0	1	0	1	0	1
	Key 4	0	1	1	0	1	0	1	0
Shaft 1	Shaft 2	0	1	0	1	0	1	0	1
	Shaft 3	1	0	0	1	0	1	0	1
	Bearing 1	1	0	1	0	0	0	1	0
	Bearing 2	1	0	1	0	0	0	1	0
	Bearing 3	0	1	0	1	0	0	0	1
	Bearing 4	0	1	0	1	0	0	0	1
	Bearing 5	1	0	0	1	0	0	0	1
	Bearing 6	1	0	0	1	0	0	0	1
	Key 1	0	1	1	0	1	0	1	0
	Key 2	0	1	0	1	0	1	0	1
	Key 3	0	1	0	1	0	1	0	1
	Key 4	0	1	0	1	0	1	0	1
Shaft 2	Shaft 3	0	1	0	1	0	1	0	1
	Bearing 1	0	1	0	1	0	0	0	1
	Bearing 2	0	1	0	1	0	0	0	1
	Bearing 3	1	0	1	0	0	0	1	0
	Bearing 4	1	0	1	0	0	0	1	0
	Bearing 5	0	1	0	1	0	0	0	1
	Bearing 6	0	1	0	1	0	0	0	1
	Key 1	0	1	0	1	0	1	0	1
	Key 2	0	1	1	0	1	0	1	0
	Key 3	0	1	1	0	1	0	1	0
	Key 4	0	1	0	1	0	1	0	1
Shaft 3	Bearing 1	1	0	0	1	0	0	0	1
	Bearing 2	1	0	0	1	0	0	0	1
	Bearing 3	0	1	0	1	0	0	0	1
	Bearing 4	0	1	0	1	0	0	0	1
	Bearing 5	1	0	1	0	0	0	1	0
	Bearing 6	1	0	1	0	0	0	1	0
	Key 1	0	1	0	1	0	1	0	1
	Key 2	0	1	0	1	0	1	0	1
	Key 3	0	1	0	1	0	1	0	1
	Key 4	0	1	1	0	1	0	1	0

(continued)

(continued)

		Components							
		Physical				Functional			
		Arrangement		Proximity		Power		Speed	
Comp. 1	Comp. 2	In-line	Parallel	Direct Contact	Separated	Direct transmission	Indirect transmission	Same Speed	Different Speed
Bearing 1	Bearing 2	1	0	0	1	0	0	1	0
	Bearing 3	0	1	0	1	0	0	0	1
	Bearing 4	0	1	0	1	0	0	0	1
	Bearing 5	1	0	0	1	0	0	0	1
	Bearing 6	1	0	0	1	0	0	0	1
	Key 1	0	1	0	1	0	0	1	0
	Key 2	0	1	0	1	0	0	0	1
	Key 3	0	1	0	1	0	0	0	1
	Key 4	0	1	0	1	0	0	0	1
Bearing 2	Bearing 3	0	1	0	1	0	0	0	1
	Bearing 4	0	1	0	1	0	0	0	1
	Bearing 5	1	0	0	1	0	0	0	1
	Bearing 6	1	0	0	1	0	0	0	1
	Key 1	0	1	0	1	0	0	1	0
	Key 2	0	1	0	1	0	0	0	1
	Key 3	0	1	0	1	0	0	0	1
	Key 4	0	1	0	1	0	0	0	1
Bearing 3	Bearing 4	1	0	0	1	0	0	1	0
	Bearing 5	0	1	0	1	0	0	0	1
	Bearing 6	0	1	0	1	0	0	0	1
	Key 1	0	1	0	1	0	0	0	1
	Key 2	0	1	0	1	0	0	1	0
	Key 3	0	1	0	1	0	0	1	0
	Key 4	0	1	0	1	0	0	0	1
Bearing 4	Bearing 5	0	1	0	1	0	0	0	1
	Bearing 6	0	1	0	1	0	0	0	1
	Key 1	0	1	0	1	0	0	0	1
	Key 2	0	1	0	1	0	0	1	0
	Key 3	0	1	0	1	0	0	1	0
	Key 4	0	1	0	1	0	0	0	1
Bearing 5	Bearing 6	1	0	0	1	0	0	1	0
	Key 1	0	1	0	1	0	0	0	1
	Key 2	0	1	0	1	0	0	0	1
	Key 3	0	1	0	1	0	0	0	1
	Key 4	0	1	0	1	0	0	1	0
Bearing 6	Key 1	0	1	0	1	0	0	0	1
	Key 2	0	1	0	1	0	0	0	1
	Key 3	0	1	0	1	0	0	0	1
	Key 4	0	1	0	1	0	0	1	0
Key 1	Key 2	0	1	0	1	0	1	0	1
	Key 3	0	1	0	1	0	1	0	1
	Key 4	0	1	0	1	0	1	0	1
Key 2	Key 3	0	1	0	1	0	1	1	0
	Key 4	0	1	0	1	0	1	0	1
Key 3	Key 4	0	1	0	1	0	1	0	1

Chapter 6
Design for Manufacture and Assembly

Abstract This chapter presents the concepts of design for manufacturing and assembly. Product design is the first step in manufacturing and is where the critical decisions are made that will affect the final form and cost of the product. Design for manufacturing and assembly concentrates on simplifying designs while also evaluating assembly improvements to further enhance the overall design for manufacturability and quality. DFMA is a product development process and improvement methodology that provides a systematic process to achieve improved product design, robustness, and cost reductions through simplifications of the overall design. Finally, many case studies are shown for the purpose of understanding the concepts presented in this chapter.

Keywords DFMA • product design • manufacturability • quality

6.1 Introduction

The engineer today is unaware of the impact that design changes will have on manufacturing functions when a part is designed and tested only for functionality. When the product engineer has met the program objective, the design is released to the manufacturing engineer for a feasibility review and manufacturability study. This is a sequential process and does not allow for trade-off analysis during the initial stages of the design process. This results in a number of iterations between product and manufacturing engineering, which is time consuming and error prone.

This enforces the need for evaluating what impacts the design specifications have on both manufacturing and assembly and how these impacts come about. Design for manufacturing and assembly (DFMA) is an analytical process that considers all aspects of the design, development, total parts, manufacturability, cost, assembly time, and modularity. DFMA has been in use by industry for several years. It focuses on product enhancements to allow improvements in manufacturing, cost, quality, reliability, time to market, and many other areas.

A.K. Kamrani and E.A. Nasr, *Engineering Design and Rapid Prototyping*,
DOI 10.1007/978-0-387-95863-7_6,

Product design is the first step in manufacturing and is where the critical decisions are made that will affect the final form and cost of the product. Product design has an impact on more than 70% of a product's total cost. Manufacturing and assembly concepts can greatly improve production and development costs while also improving reliability and quality.

DFMA concentrates on simplifying designs while also evaluating assembly improvements to further enhance the overall design for manufacturability and quality.

DFMA is a product development process and improvement methodology that provides a systematic process to achieve improved product design, robustness, and cost reductions through simplifications of the overall design.

6.2 DFMA Methodology

DFMA is a systematic design evaluation process that is used to improve part design and part manufacture early in the design process. Figure 6.1 shows the scope of the DFMA process.

Design for manufacture (DFM) methodology analyzes individual part geometry and process choices for impact on material, manufacturing process, and tooling costs, whereas design for assembly (DFA) is a structured methodology for analyzing product concepts or existing products for simplification of design and assembly processes. Even though DFA can be thought of as a separate philosophy, it is commonly thought of as a central element of DFM.

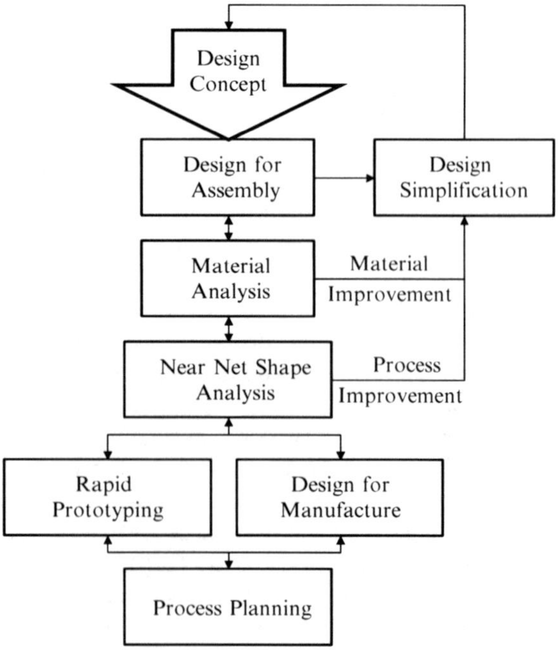

Fig. 6.1 Scope of DFMA

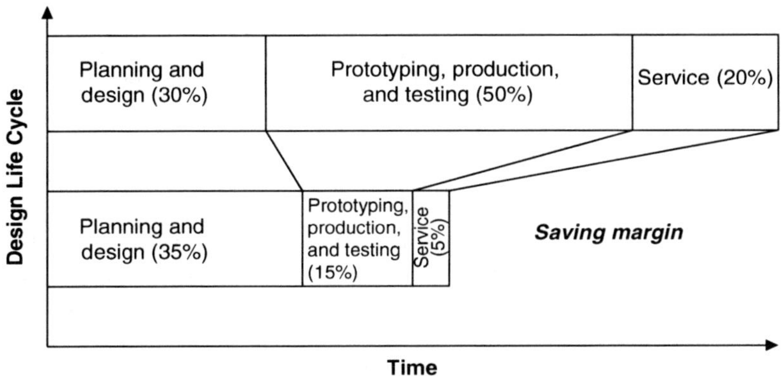

Fig. 6.2 Traditional process vs. concurrent engineering process

The DFMA process is done by a team rather than by a single engineer. The DFMA process team is a cross-functional team from all aspects of design and manufacturing. The major design activities of the concurrent engineering team occur during the conceptual phase of the design life cycle. There is a need for planning extra time early in the design phase for the analysis and the implementation of DFMA. This time is usually recovered during the later phases of development, since the early attention to the detailed design and its impact on manufacturing will eliminate much iteration necessary for design evaluation and improvement. Figure 6.2 shows an estimated saving in percentage of time for product development, comparing traditional design philosophy to concurrent engineering philosophy/DFMA methodology.

As shown in Fig. 6.2, an estimated saving of 45 percent of time to market can be achieved by applying the CE/DFMA methodologies. DFMA is usually experience-based and mostly achieved through hands-on learning. The major goals associated with using the DFMA methodology are based on:

- Profit being the major goal.
- Elimination of unnecessary adjustment.
- Minimizing reorientations during manufacturing and assembly.
- Final-design quality improvements.
- Minimizing the number of parts.
- Maximizing teamwork and consensus decision-making.
- Design for easy service.
- Enhancing ergonomic conditions for the operators.
- A good manual design could lead to a partially or fully automated assembly process.

These principles should be considered during the initial analysis of an existing or new design. A performance evaluation factor known as design efficiency is used to determine how good the current design is, and this value is used as the baseline to improve the design by applying the identified goals. This design efficiency is calculated for all the new design alternatives, and the method of calculation varies depending upon which DFMA method is being used. Even though there are many different methods for evaluating parts for DFMA, there are a number of common steps associated with these techniques. All methods will begin by tearing down the part, and as each part is removed from the assembly, three basic questions are asked to help in the identification of the component that is a candidate (or the components that are candidates) for elimination or combination into other components. These questions are (1) Does the component move relative to the others? (2) Does the component have to be of a different material? and (3) Could combination of these parts prevent assembly/disassembly of other parts? Different methods are used to answer these questions. One such method used in component reduction and design simplification is the "subtract and operate procedure " (SOP) [1]. This simple procedure identifies the components that can be eliminated. The SOP is a four-step procedure, as illustrated in Fig. 6.3.

The SOP is an in-depth analysis of the assembly that consists of basic concepts, and it provides the answer to the fundamental question of *"is this part required?"* The SOP provides a method for exposing components that can be eliminated from the overall assembly. As components are eliminated, a method should be used for

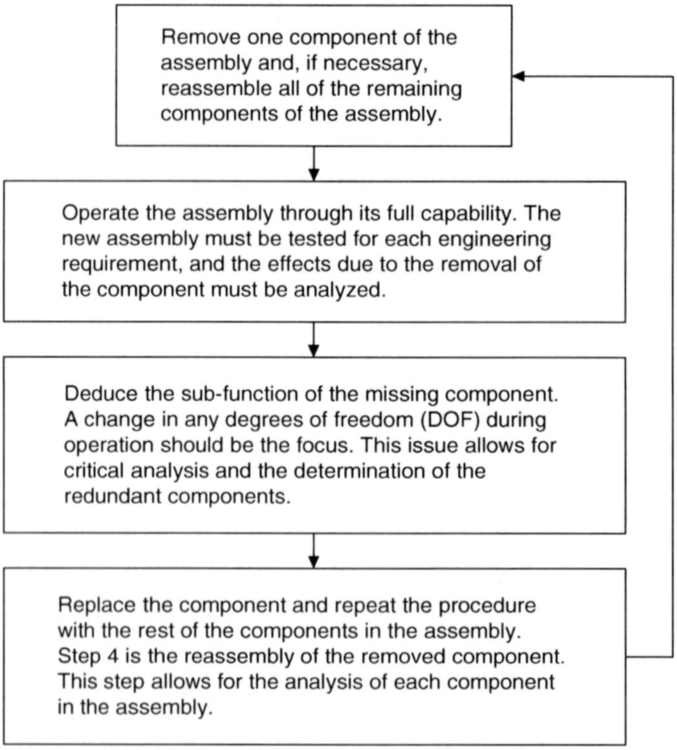

Fig. 6.3 The subtract and operate procedure

the combination of the eliminated components with the other components of the assembly. For evaluating the impact on the combination of the parts, "force flow diagrams " are proposed. Force flow diagrams , like other linear diagrams, represent the transfer of force through a product's components. In this method, components are represented by circles, and forces are represented by arrows that are used to connect the components. The force flow diagram maps the force flow through a product. Using this diagram, the assembly is analyzed to determine which components can be combined. An "R" in this diagram will illustrate a relative between two components.

The diagram is then decomposed into subassemblies based on the "R"s. The components in each subassembly can now be combined if not violating the material, assembly, or disassembly issues. Combination of parts that are not in the same subassemblies becomes more complicated.

One method that is widely used to answer the three major questions and to perform DFMA is using the "functionality flowchart." The functionality flowchart provides the sequence and the logic for the designer to determine if the component is a candidate for elimination or for possible combination with other components. Using this chart, the essential and non-essential components are identified, and, depending on which DFMA methodology is used, the design efficiency is then calculated. The three questions that dominate the analysis of this chart include (1) Material, (2) Component Movement, and (3) Assembly/Disassembly. The design efficiency is a snapshot of how good the design is compared to the three major questions. A good example illustrating the power of the DFMA analysis method is the armrest assembly. The armrest was broken down into 11 components. An estimated 33 assembly operations are required for this part. For the initial design, the designer had used all the background knowledge to develop this part, probably a very robust part. The DFMA team should begin by deleting all the fasteners that are considered to be non-essential components. Since the pivot is considered to be of the same material as the side arm, which is also the same material as the rest of the assembly, the pivot and side arm are determined to be non-essential components and can be combined with the top component. This results in a one-piece assembly design, which is considered to be the optimum theoretical number of an assembly. Injection molding of the part, with a reduction in manufacturing cost of approximately 50%, is the best possible redesign of this part.

6.3 The Boothroyd–Dewhurst Method for Manual Assembly Analysis

One very effective DFA method proposed and used for product DFA analysis was developed by Geoffrey Boothroyd and Peter Dewhurst [2, 3]. The objectives of this method are to address the following issues:

- Determining the appropriate assembly method
- Reducing the number of individual parts that must be assembled
- Ensuring that the remaining parts are easy to assemble

The first step in the Boothroyd–Dewhurst method is to determine the best method of assembly for the product. The team must constantly remind itself that the lowest assembly cost can be achieved by designing the product so that it can be economically assembled by the most appropriate assembly method. The manual DFA method and rules are summarized below.

6.3.1 Manual Assembly

This method includes bench or transfer assembly using only simple tools. If the cross-functional team decides on this assembly method as the most cost-effective method for the product, then the following design rules should be considered:

- Reduce part count and part types
- Strive to eliminate adjustments
- Design parts to be self-aligning and self-locating
- Ensure adequate access and unrestricted vision
- Ensure the ease of handling of parts from bulk
- Minimize the need for reorientation during assembly
- Design parts that cannot be installed incorrectly
- Maximize part symmetry if possible or make parts obviously asymmetrical

6.3.1.1 Design for Manual Assembly Procedure

The procedure will outline the steps necessary for analysis of manual assemblies proposed by the Boothroyd–Dewhurst DFA methodology.

Step 1: Collect and obtain the best information about the assembly. With the current advances in CAD modeling, an exploded design or a prototype would be most desirable. Use a DFMA worksheet [2, 3] to record the information collected in the following steps.

Step 2: Take the assembly apart (if possible) and assign identification numbers as the components are being removed. Record the numbers on the worksheet from the highest to the lowest number in the first column of the worksheet. In the second column, record the number of times that each operation is repeated for removed components.

Step 3: Using the two-digit manual handling code estimate the necessary handling time for the components. Material handling involves the physical transportation and orientation of the part from shipping point to the point where the assembly is performed. The manual handling code is based on the rotational symmetry about the perpendicular axis and rotational symmetry about the insertion axis. Sample analysis is shown in Fig. 6.4a.

This code also takes into consideration the size (length of the longest side of the smallest rectangular/cylinder that can enclose the part), thickness (length of the

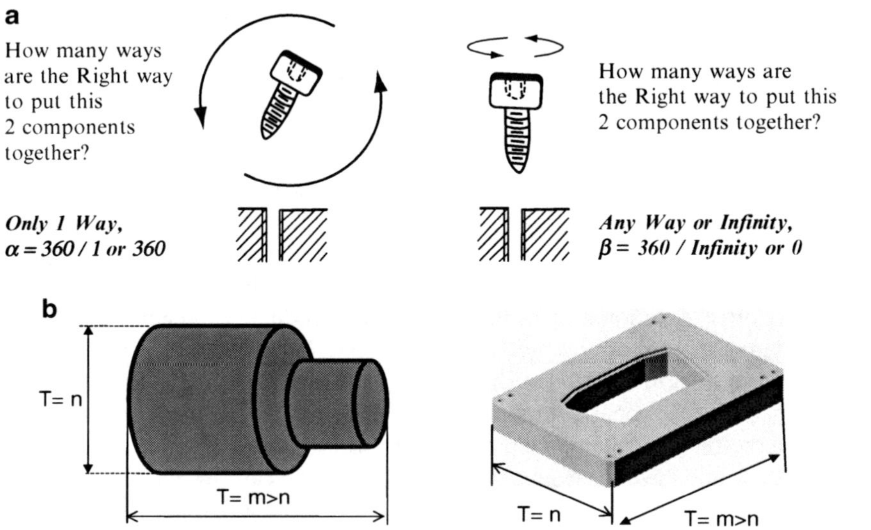

Fig. 6.4 Sample analysis of size and thickness of parts

shortest side/radius of the smallest rectangular prism/cylinder that enclose the part), as shown in Fig. 6.4b, and other factors related to the part extraction from the shipping container.

The code and the time are recorded in the third and fourth columns of the worksheet using the material handling worksheet [2, 3]. The manual insertion worksheet [2, 3] is used for the determination of the insertion code and time, which will be recorded in fifth and sixth columns. This table is classified based on how the part is inserted and how it is secured once the part is added. Each section is then further decomposed based on the complexity of the insertion process. Examples of such complexities include obstructed vision or inadequate access. The total operation time in seconds is calculated by adding the handling and insertion times from columns 4 and 6. This sum is the multiplied by the number of repeated operations. The result is recorded in column 7. The total operation cost is an estimated value. This value is the result of the multiplication of the result from column 7 by the industry standard for assembly rate in cents per second. This is recorded in column 8. The last column, column 9, is detailed on the basis of the team decision-making process. In this column the three major questions are reviewed and answered. This column has a direct impact on the result of the design efficiency that is based on the identification of the candidate for removal from the assembly.

Step 4: Add Columns 7, 8, and 9, respectively, to determine the estimated design efficiency. The result from the column 9 is used as a guide for the redesign of the component. After the redesign, the same procedure is repeated and performed in order to further modify and simplify the overall assembly. For the purpose of redesign, several strategies are outlined:

- Test the need for each part's existence as a separate part
- Eliminate mechanical fasteners
- Eliminate parts that act as conduits on connections (cables, tubing, etc.)
- Design multifunctional parts by getting the most out of individual near-net-shape manufacturing processes, such as injection molding, forging, or casting

One major point that is emphasized is to eliminate the product features or functions that do not add any value to the customer. Once the team has chosen the proper assembly method, the next step in DFA is to reduce the number of individual parts that must be assembled. This is the part-integration phase of the DFA analysis. Once the cross-functional team has studied the product, the team may determine that certain parts are not theoretically necessary. These parts then become candidates for physical integration with either one or more other parts. Even though the resulting multifunctional component may be complex, advanced technologies in injection molding and stamping can often incorporate several different features that otherwise would have been separate parts. This is the case for the armrest bracket. Part integration can provide a company with several benefits, some of which are:

- Part fabrication process accomplishes the assembly of the different features of the part.
- Cost is reduced because integrated parts are often more expensive to fabricate than to replace.
- Product quality is increased due to the control of critical design by the part fabrication rather than by an assembly operation.
- For ease of assembly, the ideal characteristics for a part include:
- Inserted from the top of the assembly (Z-axis assembly)
- Self-aligning and not needing unnecessary orientation
- Assembled in a single, linear motion
- Secured immediately upon insertion

In addition to these characteristics, it is also critical to maximize part symmetry or to emphasize obvious asymmetry. This will have great impact on the handling phase of the assembly process.

In summary, the DFA process makes the difference between a competitive product line and, in the extreme case, products that are not manufacturable or assemblable at all. The following section provides an example of the application of the DFA process to the analysis and redesign of a vehicle fog lamp.

6.4 Case Study: DFA Analysis of a Fog Lamp Design

A fog lamp is a device that improves the visibility of the road during poor driving conditions. It is used to improve road surface illumination in fog, snow, heavy rain, or dusty conditions. These devices are mounted on the exterior with an optical unit mounted either upright on top or suspended beneath the bumper. Their effect depends upon the size of the illuminated area and the focal length of the reflector.

The analysis is initiated with the use of the existing design of the fog lamp. An exploded view of the part is illustrated in Fig. 6.5. The sequence of the assembly operation is then identified using an assembly diagram that helps to define the assembly steps one by one. This is illustrated in Fig. 6.6. Once the sequence is defined, the bill of material necessary to build a fog lamp can be established.

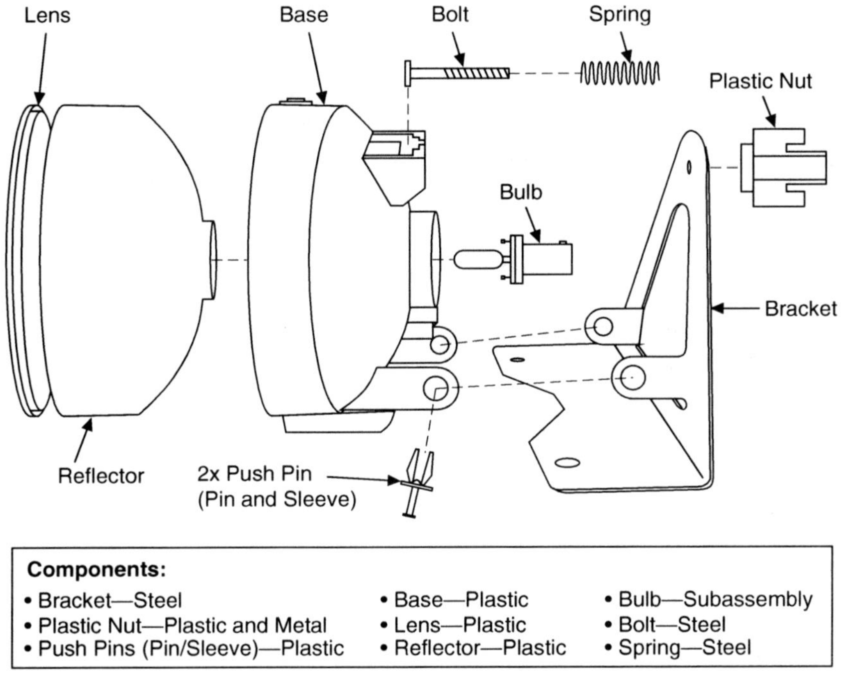

Fig. 6.5 Exploded view of fog lamp (current design)

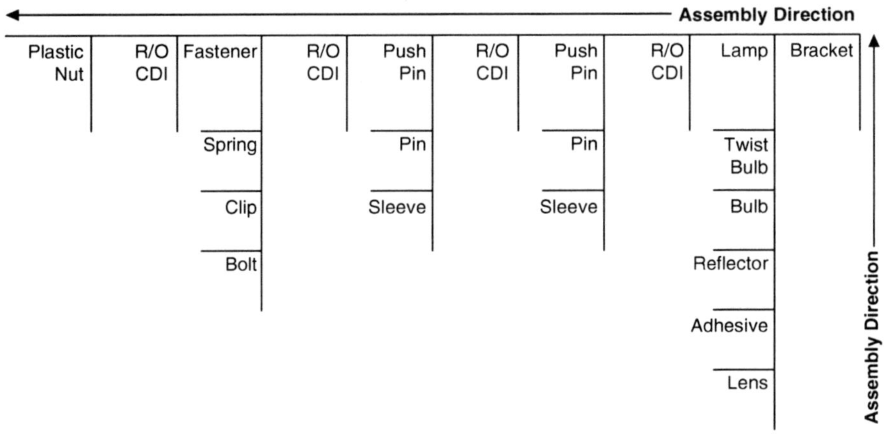

Fig. 6.6 Assembly sequence of current fog lamp design

The next step includes using the functionality flowchart and the DFA worksheet to detail the analysis and determine the design efficiency . This analysis will also provide an estimate of time and cost of the assembly of the fog lamp. Finally, it allows for the identification of opportunities for product structure simplification.

6.4.1 Alternative Fog Lamp Designs

Three alternative designs were generated during the DFA analysis of the fog lamp. A summary of each design is presented in Tables 6.1–6.3 and Figs. 6.7–6.9, respectively. In general, the benefits that can be identified using the DFA process methodology are:

- Understanding the impact of the design on the manufacturing
- Promotes the concept of "right-first-time"
- Structured methodology for product assessment and development
- Promotes creativity and innovation
- Encourages teamwork

It will also improve the product quality through part-count reduction.

***Part-count reduction will promote*:**

- Fewer opportunities for misalignment
- Fewer tolerance stack-up problems
- Fewer adjustments
- Fewer mating points

***Cost is greatly improved due to*:**

- Less material to inventory
- Fewer assembly stations

Table 6.1 Analysis results for the existing design and alternative design 1

Parameter	Current design	Alternative design 1
Total assembly time	73.04	39.35
Number of parts in design	10	4
Total assembly cost	1.22	1.57
Design efficiency (%)	17.91	30

Table 6.2 Analysis results for the existing design and alternative design 2

Parameter	Current design	Alternative design 2
Total assembly time	73.04	44.7
Number of parts in design	10	4
Total assembly cost	1.22	1.78
Design efficiency (%)	17.91	27

- Less automatic assembly equipment
- Less dedicated fabrication tooling
- Less paperwork and fewer drawings

Table 6.3 Analysis results for the existing design and alternative design 3

Parameter	Current design	Alternative design 3
Total assembly time	73.04	27
Number of parts in design	10	4
Total assembly cost	1.22	1.08
Design efficiency (%)	17.91	44

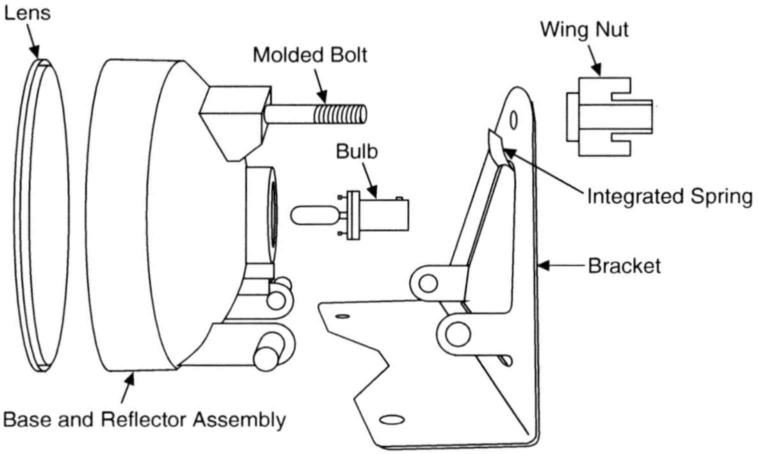

Fig. 6.7 Alternative design 1

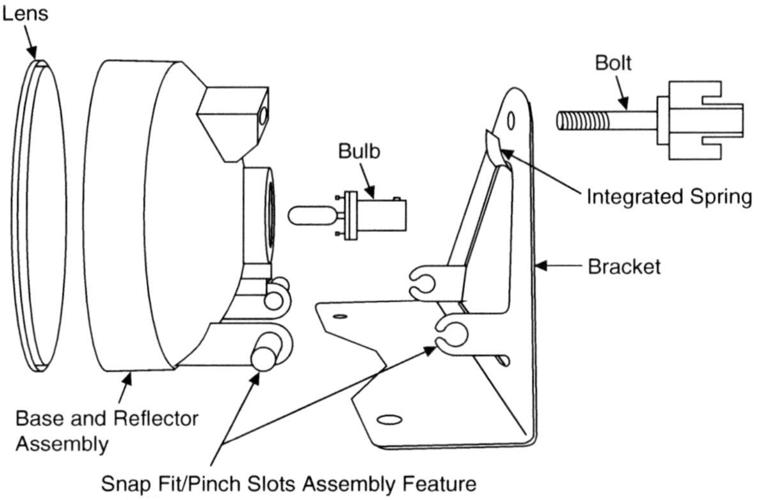

Fig. 6.8 Alternative design 2

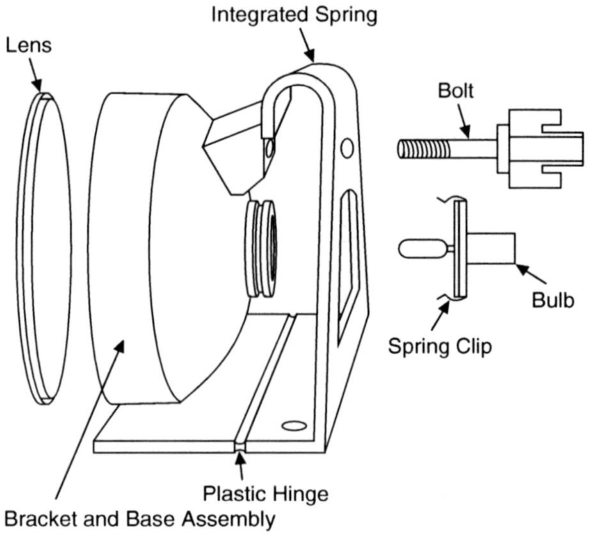

Fig. 6.9 Alternative design 3

6.5 LUCAS Design for Assembly Analysis and Evaluation Method

The Lucas's Design for Assemble method was developed in 1980s. Unlike the B&D methods the Lucas DFA analysis is not based on cost rather on the relative measure of assembling difficulty. This method is carried out in three separate and sequential steps. The parts count reduction is the primary goal of this method. The analysis is carried out in three sequential and separate stages and the results for each analysis are recorded in the separate Assembly Analysis Chart for comparison [4]. An example of the Assembly Analysis Chart is shown in Fig. 6.10.

***Functional Analysis*:** In this step, components are divided into two groups. The components that perform a primary function are considered as essential components, or "A" components. The nonessential components or "B" components perform only secondary functions. These components are mainly used in support of the "A" components and can be eliminated. The initial target for the measure of design efficiency is 60%. If this is not achieved, the initial step to improve the design is by eliminating the nonessential components. "A" and "B" components are selected using the functionality analysis diagram as shown in Fig. 6.19. The Design for Efficiency is measured as:

$$\text{Design efficiency} = 100 \times \text{Essential number of components} / \text{Total number of components.} \qquad (6.1)$$

***Feeding Analysis*:** Issues regarding the handling of components (sub-assembly) is analyzed using this ratio until they are acceptable for assembly. Feeding/handling

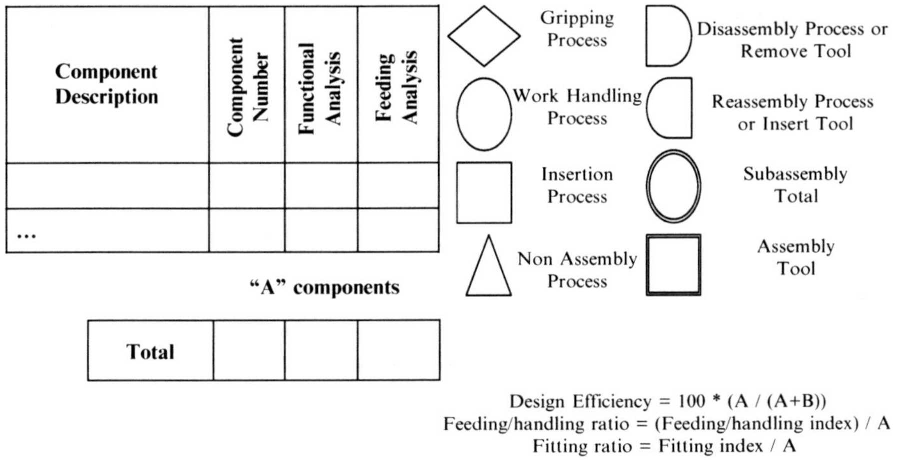

Fig. 6.10 LUCAS DFA assembly analysis chart

Table 6.4 Size and weight analysis

Very small	Requires handling aids	1.5	Select one for *A*
Easy	Requires one hand only	1	
Large and/or heavy	Requires more than one hand or grasping aid	1.5	
Large and/or heavy	Requires more than one person or hoist	2	

index is calculated based on size, weight, handling difficulties, and orientation of a part. The ideal value for this ratio is 2.5. It is calculated as:

$$\text{Feeding / handling ratio} = (\text{Feeding / Handling index}) / \text{Essential number of components} \tag{6.2}$$

***Fitting analysis*:** The fitting analysis is divided into gripping, insertion, and fixing analyses. An index is given to each part based on its required fixturing, resistance force to insertion, and restricted vision during assembly. High values for these measures indicate costly operations. For an acceptable design, the fitting ratio should be 2.5. The fitting index is used to measure the fitting ratio as:

$$\text{Fitting ratio} = \text{Fitting index / Essential number of components} \tag{6.3}$$

Feeding/handling and fitting ratios are used as "*measures of performance*" to indicate the effectiveness of the design. Handling analysis is estimated using the following Tables (Tables 6.4–6.6) [5].

The handling index is the sum of *A*, *B*, *C*, and *D* values. Manual and automated fitting analysis is estimated using the following Tables (Tables 6.7–6.10) [5].

The fitting handling index is the sum of *A*, *B*, *C*, *D*, *E*, and *F* values. Non-assembly process analysis is estimated as shown in Table 6.11 [5].

A sample part analysis using this method is presented in Fig. 6.11. In this example a drain pump is analyzed [4]. Only the redesign analysis is presented here.

Table 6.5 Handling analysis

Delicate	0.4	Select all applies for B ($B = 0$ if delivered automatically)
Flexible	0.6	
Sticky	0.5	
Tangle	0.8	
Severely nested	0.7	
Sharp	0.3	
Untouchable	0.5	
Gripping problem	0.2	

Table 6.6 Orientation analysis

Orientation along the axis of insertion		Select one for C
Symmetrical – no orientation required	0	
Easy to see orientation – end to end	0.1	
Not easy to see orientation – end to end	0.5	
Rotational orientation along the axis of insertion		Select one for D
Rotational symmetry not required	0	
Easy to see rotational symmetry – end to end	0.2	
Not easy to see rotational symmetry – end to end	0.4	

Table 6.7 Part placing and fastening analysis

Part placing analysis		Select all applies for A
Self-holding	$M = 1.0, A = 1.0$	
Holding down required	$M = 2.0, A = 1.2$	
Part fastening analysis		
Self-securing	$M = 1.3, A = 1.1$	
Screwing	$M = 4.0, A = 1.0$	
Riveting	$M = 4.0, A = 1.3$	
Bending	$M = 4.0, A = 1.6$	

M, manual; A, automated

Table 6.8 Direction analysis

Straight line from above	$M = A = 0$	Select one for B
Straight line not from above	$M = 0.1, A = 0.2$	
Not straight line – bending required	$M = 1.6, A = 1.2$	

M, manual; A, automated

Table 6.9 Insertion analysis

Single	$M = A = 0$	Select one for C
Multiple	$M = 0.7, A = 1.2$	
Concurrent	$M = 1.2, A = 1.2$	

M, manual; A, automated

Table 6.10 Access, alignment, and resistance analyses

Restricted access and/ or vision	$M = 1.0, A = 0$	Select one for D
Difficult to align		Select one for E
Yes	$M = 0.7, A = 0.8$	
No	$M = A = 0$	
Resistance to insertion		Select one for F
Yes	$M = 0.6, A = 0.8$	
No	$M = A = 0$	

M, manual; A, automated

Table 6.11 Non-assembly process analysis

Additional screwing	$M = 4.0, A = 1.0$	Select all applies
Mechanical deformation	$M = 4.0, A = 1.0$	
Soldering/welding	$M = 6.0, A = 1.6$	
Adhesive	$M = 5.0, A = 1.2$	
Re-orientation	$M = 1.5, A = 1.5$	
Inspection easy	$M = 1.5, A = 1.0$	
Inspection hard	$M = 7.5, A = 3.0$	
Liquid/gas fill/empty	$M = A = 5$	

Component Description	Component Number	Functional Analysis	Feeding Analysis
Shaft	1	B	1.1
Piston	2	A	1.0
Valve	3-a	A	1.1
Body	4	A	1.1
Base	5	B	1.1
Valve	3-b	A	1.1
		"A" components	
Total	**6**	**4**	**6.5**

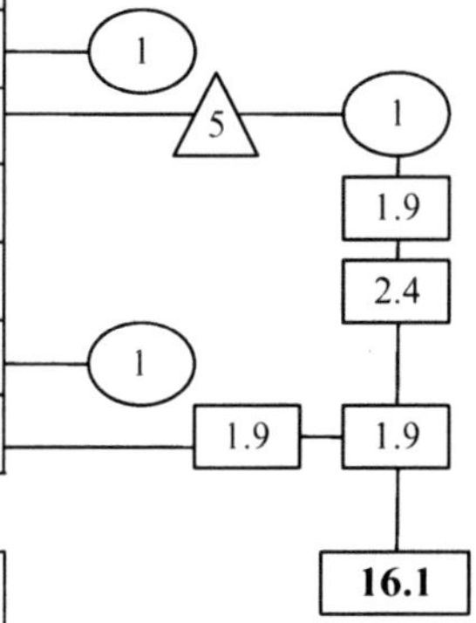

Design Efficiency = 100 * (A / (A+B)) = 4/6 * 100 = 66.7 percent

Feeding/handling ratio = (Feeding/handling index) / A = 6.5 /4 = 1.63

Fitting ratio = Fitting index / A = Grip + Fit + Fixturing/ A = 16.1/4 = 4.0

Fig. 6.11 LUCAS DFA analysis of pump

6.6 Design for Manufacture

DFM is an integrated approach to product and process development. In an integrated approach, the product is developed while its impact on functionality and manufacturability is being considered. DFM utilizes several forms of information and data. These may include product specifications data, alternatives for variant designs, information on the availability and suitability of material, machine and process suitability, production and assembly data, and, finally, costs. Figure 6.12 shows an overview of a DFM environment. In this model the process begins with a series of alternative feasible designs. The design is then interpreted through the parametric file and/or feature-extraction analyzer. The information is then transferred to a knowledge-intensive environment, which is then evaluated using manufacturability rules. Rules are structured based on four major criteria. These are compatibility, complexity, efficiency, and coupling. Compatibility is concerned with the degree of associativity between features, material, and the process. In complexity, issues such as GD&T, level of detail, feature size, and accessibility are evaluated.

Efficiency deals with the process availability, material usage, configurability, and modularity. These first three elements are integrated and optimized through

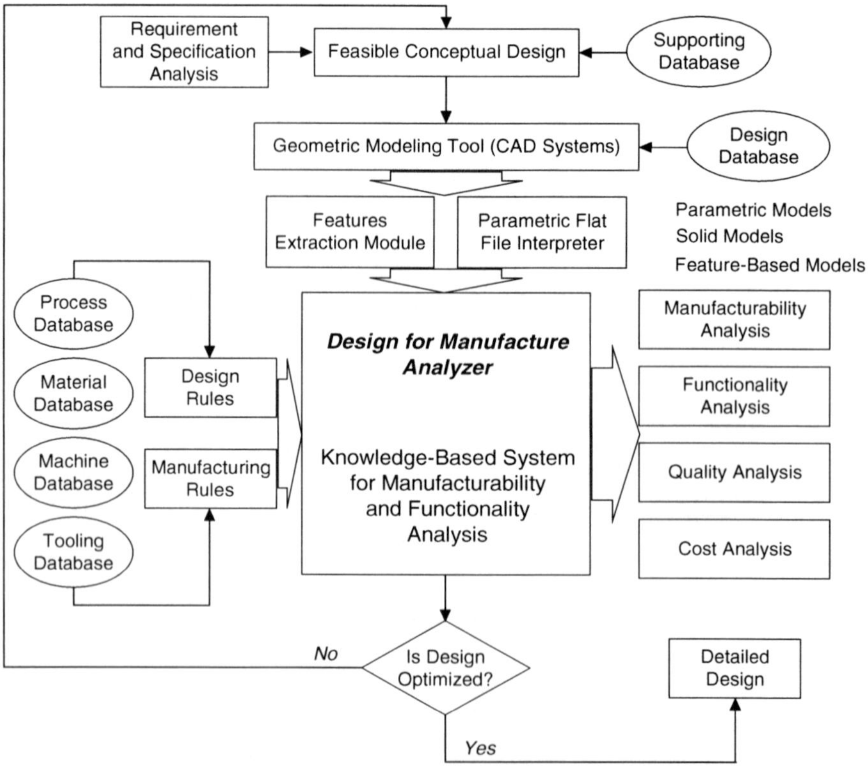

Fig. 6.12 DFM process

coupling. Coupling of features deals with evaluating the impact of the decisions made on both manufacturability and functionality of the design. For example, a decision on the GD&T may improve the design functionality, but it could also increase the cost of production and quality.

6.7 LUCAS Design for Manufacturing Analysis and Evaluation Method

LUCAS DFM method uses the concept of group technology to classify components based on their design features and material characteristics [4]. It estimates an index known as "*manufacturing cost index*" which is used to analyze the suitability of the selected manufacturing processes and operations. This index is calculated as [4]:

$$M_i = R_c \times P_c + M_c \tag{6.4}$$

where M_i is manufacturing cost index, R_c the relative cost, P_c the primary processing cost (net or near net shape), and M_c the total material cost.

The primary processing cost of the component is based on the concept of near net shape. The best primary process is selected based on achieving the entire (net shape) or major design features (near net shape) of the component. This process is selected based on both design characteristics and material, although material has the major impact on this selection. The next step of this process is to incorporate design complexity. This is used to further analyze the relative cost associated with processing of the features that caused these complexities. The R_c is calculated as:

$$R_c = C_c \times C_{mp} \times C_s \times C_t \quad \text{or} \quad C_f \tag{6.5}$$

where C_c is the shape complexity, C_{mp} the material complexity, C_s the minimum section complexity, C_t the tolerance complexity, and C_f the surface finish complexity.

If a net shape process is used then $R_c = 1$. The shape complexity is determined based on general shape classification into rotational, prismatic, and flat wall. This is further categorized into other characteristics. The descriptions of these features are listed in Table 6.12.

Table 6.12 Feature descriptions

Features	Descriptions
Basic	Operation without a change of setting or tooling
Secondary features	Requires additional processing or more complex tooling
Multi-axis	Processed in multiple axis or setup
Non-uniform	Requires the development of more complex process or setup
Complex	Requires dedicated tooling and specialized process
Single axis	Axis along largest dimensions of the component
Through	Through features from one end or side to the other

Tables 6.13–6.15 list the estimates values for shape complexity values using sand casting, forging, sheet metal working, and machining as the primary processes (P_c).

The material suitability is selected by using the material/process compatibility matrix. For example, the material suitability factor for cast iron and aluminum are listed in Table 6.16 using the same processes used above for the complexity measure. Other complexity parameters such as minimum area, tolerance and surface finished are estimated using similar Tables (Tables 6.17–6.19) [4]. The basic (near net shape) processing cost is an average cost (based on the volume of production) using factors such as equipment, direct labors, operations, and tooling costs. The second part of the Mi cost analysis is the total cost of the material. The total material cost is estimated as:

$$M_c = V \times C_{mt} \times W_c \tag{6.6}$$

where V is material volume, C_{mp} the material cost (cost/volume), and W_c the waste factor

This value can be estimated using Tables 6.17–6.19.

Table 6.13 Shape complexity analysis for rotational features and C_c values [4]

		Single/primary axis		Secondary axis		Complex shape
Processes	Features	R1	R2	R3	R4	R5
Sand casting		1.0	1.2	1.3	1.8	3.2
Forging		1.0	2.1	2.3	2.6	3
Sheet metal working		*Process not used*				
Machining		1.0	1.2	2.9	5.3	6.1

Table 6.14 Shape complexity analysis for prismatic features and C_c values [4]

		Single axis		Multiple axis		Complex shape
Processes	Features	P1	P2	P3	P4	P5
Sand casting		1.1	1.2	1.4	1.8	2.6
Forging		1.0	2.2	2.2	2.3	2.7
Sheet metal working		*Process not used*				
Machining		1.0	1.3	2.6	2.6	2.8

Table 6.15 Shape complexity for flat wall features and C_c values [4]

		Single axis	Secondary features		Regular shapes	Complex shape
Processes	Features	F1	F2	F3	F4	F5
Sand casting		2.1	2.3	2.8	3.7	5.0
Forging		1.0	1.2	1.6	2.5	3.6
Sheet metal working		1.0	1.2	1.5	2.2	2.5
Machining		1.0	1.4	3.1	5.4	6.5

Table 6.16 Material suitability and C_{mp} values [4]

Processes Material	Cast iron	Aluminum alloy
Sand casting	1.0	1.0
Forging	*Process not used*	1.0
Sheet metal working	*Process not used*	1.0
Machining	1.2	1.0

Table 6.17 Waste factor analysis for rotational features and W_c values [4]

	Single/primary axis		Secondary axis		Complex shape
Processes Features	R1	R2	R3	R4	R5
Sand casting	1.1	1.1	1.2	1.3	1.4
Forging	1.1	1.1	1.2	1.2	1.3
Sheet metal working	*Process not used*				
Machining	1.6	2.0	2.5	3.0	4.0

Table 6.18 Waste factor analysis for prismatic features and W_c values [4]

	Single/primary axis		Secondary axis		Complex shape
Processes Features	P1	P2	P3	P4	P5
Sand casting	1.1	1.1	1.2	1.3	1.4
Forging	1.1	1.1	1.2	1.2	1.3
Sheet metal working	*Process not used*				
Machining	1.7	2.2	2.8	4.0	6.0

Table 6.19 Waste factor analysis for flat wall features and W_c values [4]

	Single/primary axis		Secondary axis		Complex shape
Processes Features	F1	F2	F3	F4	F5
Sand casting	1.1	1.2	1.3	1.4	1.6
Forging	1.1	1.1	1.1	1.2	1.3
Sheet metal working	1.2	1.2	1.4	1.5	1.6
Machining	1.8	2.4	4.0	6.0	8.0

6.8 Case Study: DFM Analysis Radiator Structure Front-End Support

The purpose of this analysis is to estimate the manufacturability cost of the front-end structure of a radiator support. The goal of this analysis is to (1) increase the performance and the functionality of the radiator support and (2) reduce cost. The customer has identified safety and fuel efficiency requirements are top priority for design of the truck. A base model (benchmark) was used to conduct LUCAS DFM analysis. The final conceptual model of the front-end structure support for the radiator is illustrated in Fig. 6.13. The DFM analysis results are listed in Table 6.20. The total manufacturing index of the base model is $146.15.

Table 6.20 DFM analysis for radiator support structure

Components	Quantity	C_c	C_c code	C_{mp}	Material	C_s	C_t	C_f	C_{tf}	P_c	V (mm^3)	W_c	C_{mt} (\$/mm^3)	M_i
Mg casting	1	3.6	C5	1.2	Zinc die cast	1	1	1	1	0.95	54,208	1.3	0.00139	$102.06
Al FESM spacer	2	1	C1	1	Aluminum alloy	1	1	1	1	0.57	537	1.2	0.00039	$0.82
Al Shotgun spacer	2	1	C1	1	Alumindm alloy	1	1	1	1	0.57	298	1.2	0.0009	$0.71
Al upper spacers	2	1	C1	1	Aluminum ally	1	1	1	1	0.57	298	1.2	0.00039	$0.71
FESM bracket	2	1	C1	1.2	Low carbon steel	1	1	1	1	0.57	31,007	1.2	0.00039	$15.20
Screws	4					Purchased components								
Shotgun bracket	2	1	C1	1.2	Low carbon steel	1	1	1	1	0.57	36,519	1.2	0.00039	$17.77
Rivets	4					Purchased components								
Upper air deflector	1	1	C1	1	Nylon plastic	1	1	1	1	0.95	7,425	1	.000061	$5.48
Push pins	5					Purchased components								
Side air deflector	2	1	C1	1	Rubber mold	1	1	1	1	0.95	4,015	1	0.00061	$3.40
Push pins	4					Purchased components								
Lower air deflector	1	1	C1	1	Rubber mold	1	1	1	1	0.95	6,875	1	0.00061	$5.14
Push pins	3					Purchased components								
Fascia bracket	2	2.2	C4	1.2	Low carbon steel	1.2	1	1	1	0.57	103,356	1.5	0.00039	$62.27
Screws	2					Purchased components								
Hood latch bracket	1	1.2	C2	1.2	Low carbon steel	1	1	1	1	0.57	31,007	1.2	0.00039	$15.33

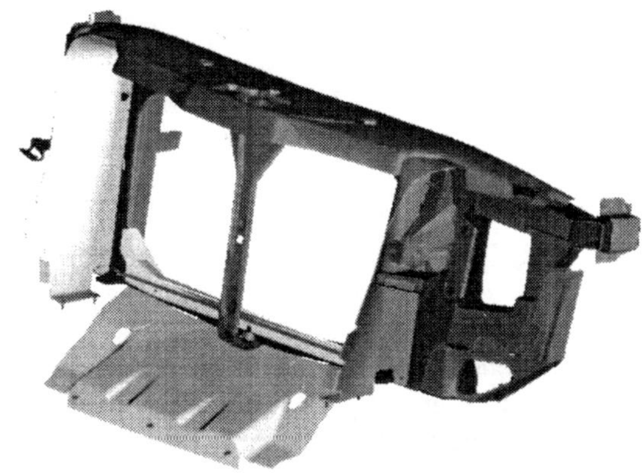

Fig. 6.13 Conceptual CAD model for radiator support structure

6.9 Case Study: Automotive Recliner Mechanism (Based on the Work Done by DiCicco et al 2003)

An automotive seat recliner mechanism was "quickly" designed and released to eliminate a noise condition that was unacceptable to the customer. This case study focuses on finding and selecting an alternative design for the seat recliner subassembly using both DFA and manufacturing methodologies. Competitive products are researched and customer and functional needs identified. Functional analysis and affinity diagrams are developed to organize functional needs in a logical sequence. Subsystem-level specifications are developed using a functional decomposition methodology.

Four design alternatives are selected for the study. A financial analysis, DFA, and design for manufacturing analysis are presented for each design alternative. A final design selection is chosen based on the results of all phases of the product development analysis. Findings show that the best design alternative reduces part count, design complexity, cost, and number of assembly operators. Advantages of the final design, long-term value to the customer, and future research directions are outlined.

6.10 Introduction

6.10.1 Seat Recliner Subsystem History

During design testing of the seat recliner subsystem shown in Fig. 6.14, a "squeak, rattle" condition occurred.

The noisy condition was created by the return spring, which would serpentine, rubbing against the rod in the mechanism when the recliner was adjusted. In other words, when the seatback was adjusted, the spring would wiggle hitting the rod, creating

Fig. 6.14 Seal recliner subsystem

a noisy condition. This condition was unacceptable to the customer. The "buzz, squeak, rattle" condition was caused by the difference between the I.D. of the spring return and the O.D. of the rod. To eliminate this condition a plastic sleeve was added over the rod. The rod sleeve has an O.D. closer to the I.D. of the spring return preventing the spring from winding. The recliner is assembled with the rod built into the mechanism on the assembly line. The mechanism is then unloaded from the line and placed into an off-line machine which then compresses the spring return over the rod and locks it onto the mechanism with the spacer at the end of the rod.

It was later determined that the sleeve needed to be located at the paddle end of the rod. In order to keep the sleeve in a fixed position crushed ribs were added to the ID of the rod sleeve. This created a hard to load condition for the operator. Due to the assembly process, which was assigned before this condition was found; an extra operator was added to the process. The operator loads the rod sleeve onto the rod at an off-line station using a semi-automatic fixture. The rod with the rod sleeve pre-assembled is then transferred to the operator at the assembly line.

6.10.2 Problem Statement

This seat recliner mechanism is a high volume product which is built on two manufacturing lines. The addition of the plastic sleeve on both manufacturing lines has resulted in the need for two new operators. The addition of these operators is a big expense to the company. The scope of this project will be to redesign the spring end of the recliner subassembly including the spacer, spring, rod sleeve, pawl, and pivot pawl as shown in Fig. 6.15.

The seat recliner mechanism has two customers. These customers are defined as internal management and the end user of the product. Internal management needs are concerned with assembly issues: (1) the number of assembly operators, (2) complexity reduction, (3) total cost, (3) product quality, and (4) durability/life of the product. External customer needs are concerned with issues relating to the operation of the mechanism: (1) adjusting the seatback to multiple positions smoothly and with little force, (2) returning the seatback with little to no help, and (3) little or no chuck "felt" to the seat. The objective is to combine or eliminate components and incorporate DFA and design for manufacturing (DFM) principles along with consideration of the total cost of redesigning the seat recliner mechanism.

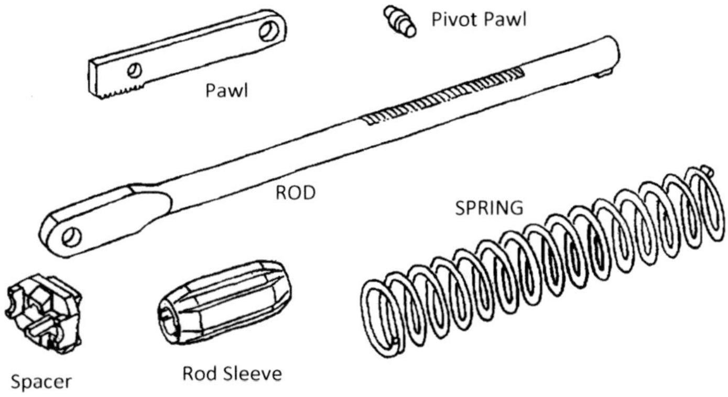

Fig. 6.15 Seat recliner components

6.11 Competitive Benchmarking Study

Competitive benchmarking for the seat recliner subassembly under study is shown in Table 6.21.

The baseline seat recliner subassembly (P221) is a linear single-sided mechanism. Five similar seat recliner subassemblies were studied. These recliners include the W-Dual linear recliner (dual sided), U-152 linear recliner (dual sided), GMT-800 power recliner (single sided), SW-162 linear recliner (dual sided), and P221 ISS recliner (single sided). The metrics used to complete the benchmarking included: *ultimate forward and rearward loads defined by the customer to support the seatback, inner diameter of the return-spring, outer diameter of the rod, and overall length of the rod.* Additional customer needs included: *number of components required for each subassembly system, number of operators required to build the system, ease of assembly based on 1 (easy) to 5 (difficult), and type of recliner – single- or dual sided.*

6.12 Customer Functional Requirements

6.12.1 Customer Needs

Two lists of customer needs, internal and external, were identified based on information collected from company market research studies and competitive bench marketing studies. Internal customer needs are based on the needs of the company management and the needs of the manufacturing plant management. External customer needs are based on the needs of the end user (customer) of the vehicle. Internal customer and external customer needs are shown in Table 6.22.

Table 6.21 Competitive needs/metrics – seat recliner subassembly

Benchmark	Metrics					Needs			
models	Forward load (N)	Rearward load (N)	Spring dia. I.D. (mm)	Rod dia. (mm)	Length of rod (mm)	Component	Required operators	Assimilability	Single- or dual side
Baseline P221	7,364	21,538	27.40	12.635	286.58	6	3	4	S
W-Dual	7,364	21,538	15.30	12.635	224.70	4	2	1	D
U-152	7,364	21,538	14.59	12.635	224.70	4	2	1	D
GMT-800	8,844	18,145	27.40	10.040	303.25	5	2	3	S
SW-164	8,673	22,386	15.30	12.635	224.70	4	2	2	D
P221 55	8,844	1,845	26.00	10.040	303.23	5	2	3	S

Table 6.22 Customer requirements

Internal customer (management and manufacturing plant)	External customer (end user of product)
• Minimize the number of operators	• No noise (buzz, squeak, and rattle)
• Lower manufacturing cost	• Adjust the seatback to multiple positions
• Higher quality	• Adjust smoothly
• Consistent quality	• Adjust with little force
• Reduce complexity	• Seatback with little to no help
• Longer life of the product	• Exhibits little or no chuck felt in the seat
• Minimize # of assembly operations	• Durable – longer life
• Safety (no sharp edges)	• Durable – strength
• Easy to assemble	• Antirust
• Low logistic cost	• Compact size
• Low inspection cost	• Light weight
	• Robust
	• Handling
	• Low product cost
	• Easy of serviceability
	• Requires less maintenance
	• Safe

6.12.2 Functional Requirements

Functional requirements were developed for the purpose of (1) identifying high-order functions for product design performance, (2) identifying links among functions for product design or improvement, and (3) creating alternatives for product design or improvement. Seven categories of functional definitions were identified including: recliner, ergonomics, quietness, durability, safety, light weight, and assembly and manufacturing for this purpose. These seven categories were further defined into 26 different functions. Functional requirements for the seat recliner subassembly are shown in Table 6.23.

6.12.3 FAST Diagram

The Function Analysis System Technique (FAST) diagram is a communication vehicle which uses verb–noun rules. The FAST is used to define, analyze, and understand product functions, how functions relate to one another, and which functions require priority to increase product value. The FAST diagram displays the functions in a logical sequence, prioritizes them, and tests their dependency. The critical path of the seat recliner subassembly contains both the higher-order function and the assumed function. The higher-order function of the seat recliner subassembly is to recline seats. The assumed function is to provide force. The FAST diagram for the seat recliner subassembly was developed using the functional requirements defined in Table 6.22 and is shown in Fig. 6.16.

Table 6.23 Functional requirements list

		Functional requirements	
Categories list	Function definition list	Verb	Noun
1	Recliner	Move	Seatback to multiple positions
		Stop	Movement at one position
		Contain	Return spring
		Contain	Holding fixture (pawl, rod)
		Perform	Robustness
2	Ergonomics	Support	Loads
		Minimize	Applied force
		Feel	Smooth in adjustment
		Replace	Recline easily
		Apply	Grease properly
3	Quietness	Contain	Rod sleeve
		Reduce	Gap between ID and OD
4	Durability	Contain	Oil
		Maintain	Force during the product life cycle
		Maintain	Stainless
		Extend	Product life
5	Safety	Eliminate	Sharp edges/corners
		Restrict	Dangerous substances
6	Light weight	Reduce	Weight Size
		Reduce	
7	Assembly and manufacturing	Reduce	Number of parts
		Reduce	Number of operators
		Reduce	Number of operations
		Transport	Parts economically
		Reduce	Complexity
		Minimize	Rejects and defects

6.12.4 *Affinity Diagram*

The opportunity statement for the seat recliner subassembly was defined by the team as to "provide a safe, easy to operate, multi-position seat recliner mechanism for automotive seats with competitive features and at a price lower than the competition." Functional requirements from Table 6.23 and the FAST diagram shown in Fig. 6.25 are used to generate related, natural, and logical grouping for the affinity diagram as shown in Fig. 6.17.

6.12.5 *Product/Concept Integration*

The results of the decomposition are shown in Table 6.24. From the physical and functional decomposition, system-level specifications are determined and used for design impact analysis. The physical and functional decomposition are based on attachment, proximity, tooth connection/lock, and movement.

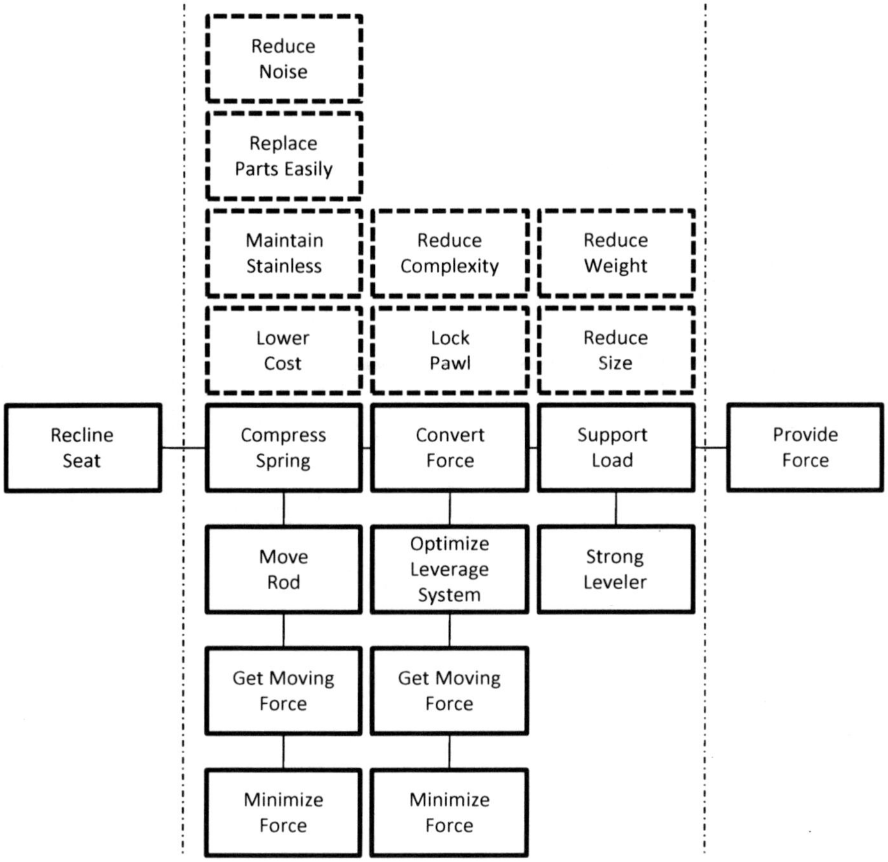

Fig. 6.16 Sample FAST diagram

- Attachment is either removable or non-removable components. The relationship is defined removable if the components could be separated from each other without damage to the structures and no re-work was needed.
- Proximity is defined as either direct-connected or separated. The relationship is labeled direct-connected if components are within the close proximity or attached to one another. If the components were separated in location the relationship is labeled separated.
- Tooth connection/lock is used to categorize relationship if such function is present. The relationship is true, if both components have tooth and locking mechanism.
- Movement, which resulted from the functional decomposition, is defined by the relationship between components as "moves" or "stationary/stable." The relationship is defined "moves" if one of the two components moved in relation to the others.

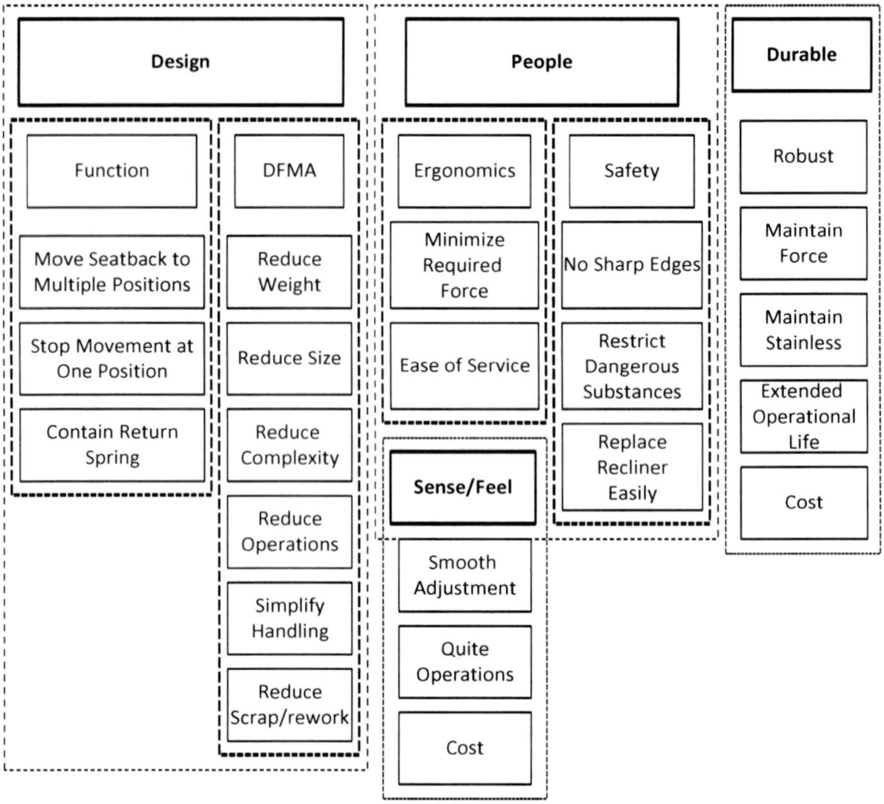

Fig. 6.17 Affinity chart for the seat recliner

The subsystem-level specifications impact on the general functional requirements (GFR) is listed in Table 6.25. The GFR selected for this design are ergonomics, durability and performance for the seat recliner subassembly.

Two independent modules are identified for this design. Module 1 includes the rod, sleeve, spacer, and spring-return, and module 2 includes the pawl and pivot-pawl. The summary of this analysis is listed in Table 6.26.

6.13 Conceptual Designs Analysis and Comparisons

6.13.1 Current Design

Seat recliner subassembly design shown in Fig. 6.18 is the current production design. This design is the result of an attempt to eliminate a noise condition created when the spring hits the rod. A plastic sleeve with crushed ribs was added to prevent the noise condition.

Table 6.24 Decomposition and integration analyses

Components		Components							
		Physical				Functional			
		Attachment		Proximity		Teeth		Movement	
A	B	Removable	Non-removable	Direct	Separated	Lock	Non-lock	Present	Stationary
Rod	Sleeve	1	0	1	0	0	1	0	1
	Spacer	1	0	1	0	0	1	0	1
	Spring-return	1	0	1	0	0	1	1	0
	Pawl	0	1	1	0	1	0	1	0
	Pivot-pawl	0	1	0	1	0	1	0	1
Sleeve	Spacer	1	0	0	1	0	1	0	1
	Spring-return	1	0	1	0	0	1	1	0
	Pawl	0	1	0	1	0	0	0	1
	Pivot paw	0	1	0	1	0	1	0	1
Spacer	Spring-return	1	0	1	0	0	1	0	1
	Pawl	0	1	0	1	0	0	1	0
	Pivot-pawl	0	1	0	1	0	1	0	1
Spring return	Pawl	0	1	0	1	0	0	1	0
	Pivot-pawl	0	1	0	1	0	1	0	1
Pawl	Pivot-pawl	1	0	1	0	0	0	1	0

Table 6.25 GFR vs. SLS

System-level specifications		General functional requirement		
		Ergonomics	Durability	Performance
Physical	Removable	1	0	0
	Non-removable	0	0	0
	Direct connect	0	0	I
	Separated	1	0	0
	Teeth present	0	−1	1
	Not-present	0	0	0
Functional	Moves	0	0	1
	Stationary	0	0	0

Table 6.26 Modularity analysis

	Module 1				Module 2	
	Rod	Sleeve	Spacer	Spring-return	Pawl	Pivot-pawl
Rod		2	2	2	1	1
Sleeve	2		2	2	1	1
Spacer	2	2		2	1	1
Spring-return	2	2	2		1	1
Pawl	1	1	1	1		2
Pivot-pawl	1	1	1	1	2	

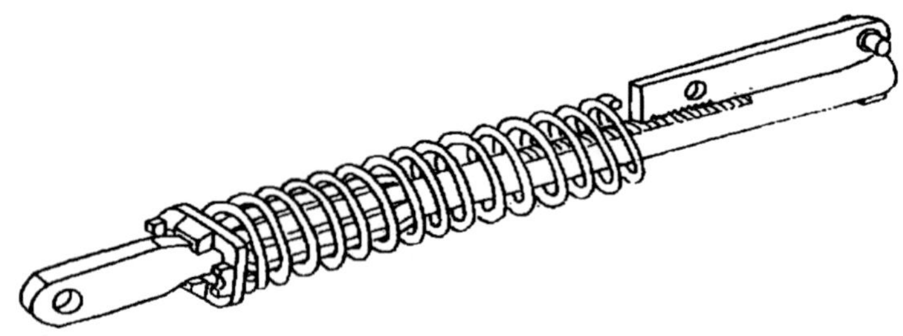

Fig. 6.18 Original design

Results of DFA and DFM analyses of this design are listed in Tables 6.27–6.32. The design efficiency of the seat recliner subassembly for this design is 49.18%. The DFM analysis determines the relative cost of producing each component based on the manufacturing processes used. The processing cost is determined using a basic near net shape processing cost, design-dependent relative cost (R_c), and the cost of the material used (M_c). The seat recliner subassembly material cost (M_c) is \$0.7625, and the manufacturing index cost (M_i) is \$314.95.

6.13.2 Pivot-Pawl Combinations

This design reduces the number of components the assembly operator needs to load during assembly. Since two mechanisms are build on a pallet, this will save the operator time to add two components. This design is illustrated in Fig. 6.19. Results of DFA and DFM analyses of this design are listed in Tables 6.33–6.38. The design efficiency of the seat recliner subassembly for this design is 57.42%. The seat recliner subassembly material cost is \$0.7625, and the manufacturing index cost is 300.48.

6.13.3 Clip Design

Design Option 3 includes the pivot-pawl combination with addition of a new design to the rod-sleeve. Currently in production the rod-sleeve is placed on the rod prior to installation of the return- spring. This assembly process adds an additional operator to the subassembly process. Changing the rod-sleeve to a clip, which is installed after the return-spring has been placed over the rod, will eliminate the need for the additional operator since both components will be installed at the same station. The loading process is also easier than the rod-sleeve. This design is illustrated in Figs. 6.20 and 6.21.

Table 6.27 Design for assembly analysis – Design Option 1

1	2	3	4	5	6	7	8	9			
Part I.D. no.	No. of times the operation is earned out consecutively	2-digit manual handling code	Manual handling hem per part	2-digit manual insertion code	Manual Insertion time per part	Operation time seconds: columns 2 x (4+6)	Operation cost (cents): 0.4 *x* column 7	Figures for estimation of theoretical minimum parts	Automotive seat recline design 1: current design	α	β
1	1	30	2	00	1.5	3.50	1.40	1	Rod	360	360
2	1	20	1.8	30	2	3.80	1.52	0	Spacer	360	180
3	1	20	1.8	31	5	6.80	2.72	1	Rod-sleeve	360	180
4	1	05	1.8	00	1.5	3.30	1.32	1	Return-spring	360	0
5	1	30	2	00	1.5	3.50	1.40	1	Pivot-pawl combination	360	360
6	1	10	1.5	30	2	3.50	1.40	0	Pivot-pawl combination	360	0
						TM 24.4	CM 9.76	NM 4.00	*3xNM /TM*	49.18%	

Table 6.28 Relative costs (R_c) – Design Option 1

	Shape complexity	Material	C_c	C_{mp}	C_s	C_t	R_c
Rod	A4	Steel	5.3	1.4	1.0	1.6	11.87
Spacer	B3	Plastic	1.8	1.0	1.0	1.0	1.8
Rod sleeve	A3	Plastic	1.3	1.0	1.2	1.0	1.56
Return-spring	Purchased						
Pawl	A4	Steel	5.3	1.4	1.6	1.4	16.62
Pivot-pawl	Purchased						

Table 6.29 Ideal process cost (P_c) – Design Option 1

			P_c		
	Material	Process	1.5 ml	2.0 ml	2.5 ml
Rod	Steel	Machine	5.21	5.21	5.21
Spacer	Plastic	Plastic mold	0.85	0.83	0.82
Rod-sleeve	Plastic	Plastic mold	0.85	0.83	0.82
Return-spring	Purchased				
Pawl	Steel	Machine	5.21	5.21	5.21
Pivot-pawl	Purchased				

Table 6.30 Material volume – Design Option 1

	Volume calculations (mm)	V (mm³)
Rod	((12.5/2) × (12.5/2) × 3.14) × 300	36,796.88
Spacer	28 × 35.5 × 17.3	17,196.20
Rod-sleeve	((26.6/2) × (26.6/2) × 3.14 × 55) – ((13.5/2) × (13.5/2) × 3.14 × 55)	22,680.26
Return-spring	Purchased	
Pawl	91 × 12.7 × 4.76	5,501.13
Pivot-pawl	Purchased	

Table 6.31 Material cost (M_c) – Design Option 1

	Shape complexity	Material	V (mm³)	C_{mt}	W_c	M_c
Rod	A4	Steel	36,796.88	0.00039	3.0	43.05234
Spacer	83	Plastic	17,196.20	0.00061	1.1	11.53865
Rod-sleeve	A3	Plastic	22,680.26	0.00061	1.1	15.21845
Return-spring	Purchased					
Pawl	A4	Steel	5,501.13	0.00039	3.0	6.43632
Pivot-pawl	Purchased					

Table 6.32 Manufacturing cost index (M_i) – Design Option 1

	R_c	P_c (qty.: 2 m)	M_c	M_i (=R_c × P_c + M_c)	M_i (qty.: 1.5 ml)	M_i (qty.: 2.5 ml)
Rod	11.87	5.21	43.05234	104.90546	104.91	104.91
Spacer	1.80	0.83	11.53865	13.03265	13.07	13.01
Rod-sleeve	1.56	0.83	15.21845	16.51325	16.54	16.50
Return-spring	Purchased			73.00000	73.00	73.00
Pawl	16.62	5.21	6.43632	93.03069	93.03	93.03
Pivot-pawl	Purchased			14.50000	14.50	14.50

Total 76.25 314.98 315.05 314.95

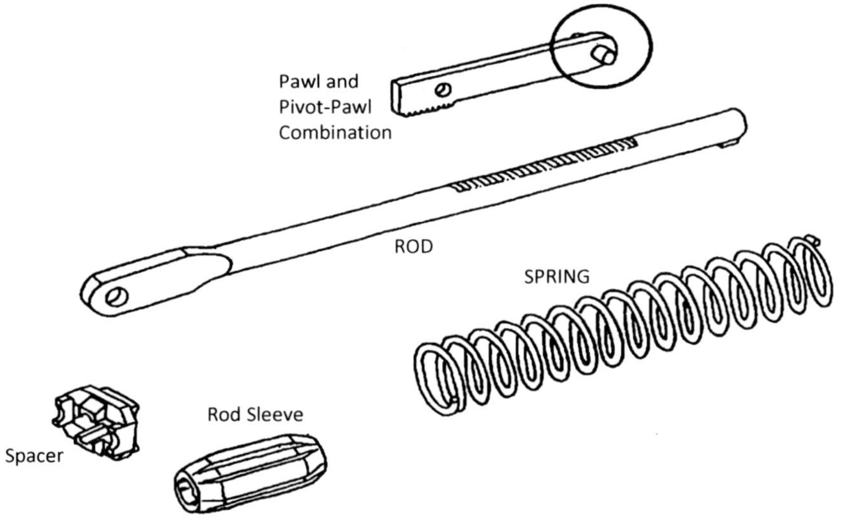

Fig. 6.19 Combined pawl and pivot-pawl design

Results of DFA and DFM analyses of this design are listed in Tables 6.38–6.43. The design efficiency of the seat recliner subassembly for this design is 68.18%. The total material cost is \$0.6329, and the total manufacturing index cost is 287.15.

6.13.4 Latch Design

Design Option 4 combines the spacer with the rod-sleeve into what is called as "Latch" design. This option is shown in Fig. 6.22. This design reduces components. It will also allow the sleeve to be added onto the rod after the return-spring has been installed.

Results of DFA and DFM analyses of this design are listed in Tables 6.43–6.48. The design efficiency of the seat recliner subassembly for this design is 85.11%. The total material cost is \$0.6287, and the total manufacturing index cost is 285.81 Tables 6.49 and 6.50.

6.14 Conclusion and Discussions

The costs estimates for all four design options are shown in Table 6.51. Estimated annual volume is 2,000,000 pieces.

Table 6.33 Design for assembly analysis – Design Option 2

1	2	3	4	5	6	7	8	9			
Part I.D. no.	Number of times the operation is earned out consecutively	2-digit manual handling code	Manual handling hem per part	2-digit manual insertion code	Manual insertion time per part	Operation time seconds: columns 2 x (4+6)	Operation cost (cents): *0.4 x* column 7	Figures for estimation of theoretical minimum parts	Automotive seat recliner design 2: pivot-pawl combination	α	β
1	1	30	2	00	1.5	3.50	1.40	1	Rod	360	360
2	1	20	1.8	30	2	3.80	1.52	0	Spacer	360	180
3	1	20	1.8	31	5	6.80	2.72	1	Rod-sleeve	360	180
4	1	05	1.8	00	1.5	3.30	1.32	1	Return-spring	360	0
5	1	30	2	00	1.5	3.50	1.40	1	Pivot-pawl combination	360	360
						TM 29.90	CM 8.36	NM 4.00	*3xNM /TM*	57.42%	

Table 6.34 Relative costs (R_c) – Design Option 2

	Shape complexity	Material	C_c	C_{mp}	C_s	C_t	R_c
Rod	A4	Steel	5.3	1.4	1.0	1.6	11.87
Spacer	83	Plastic	1.8	1.0	1.0	1.0	1.80
Rod-sleeve	A3	Plastic	1.3	1.0	1.2	1.0	1.56
Return-spring	Purchased						
Pivot-pawl combo	A4	Steel	5.3	1.4	1.6	1.4	16.62

Table 6.35 Ideal process cost (P_c) – Design Option 2

	Material	Process	P_c (Ow: 2,000,000)
Rod	Steel	Machine	5.21
Spacer	Plastic	Plastic mold	0.83
Rod-sleeve	Plastic	Plastic mold	0.83
Return-spring	Purchased		
Pivot-pawl combo	Steel	Machine	5.21

Table 6.36 Material volume – Design Option 2

	Volume calculations (mm)	V (mm^3)
Rod	$((12.5/2)\times(12.5/2)\times 3.14)\times 300$	36,796.88
Spacer	$28\times 35.5\times 17.3$	17,196.20
Rod-sleeve	$((26.6/2)\times(26.6/2)\times 3.14\times 55) - ((3.5/2)\times(13.5/2)\times 3.14\times 55)$	22,680.26
Return-spring	Purchased	
Pivot-pawl combo	$91\times 12.7\times 4.76$	5,501.13

Table 6.37 Material cost (M_c) – Design Option 2

	Shape complexity	Material	V (mm^3)	C_{mt}	W_c	M_c
Rod	A4	Steel	36,796.88	0.00039	3.0	43.05234
Spacer	83	Plastic	17,196.20	0.00061	1.1	11.53865
Rod-sleeve	A3	Plastic	22,680.26	0.00061	1.1	15.21845
Return-spring	Purchased					
Pivot-pawl combo	A4	Steel	5,501.13	0.00039	3.0	6.43632

Table 6.38 Material cost index (M_i) – Design Option 2

	R_c	P_c	M_c	M_i (= $R_c \times P_c + M_c$)
Rod	11.87	5.21	43.05234	104.90546
Spacer	1.80	0.83	11.53865	13.03265
Rod-sleeve	1.56	0.83	15.21845	16.51325
Return-spring	Purchased		0	73.00000
Pivot-pawl Combo	16.62	5.21	6.43632	93.03069

Total 76.25 300.48

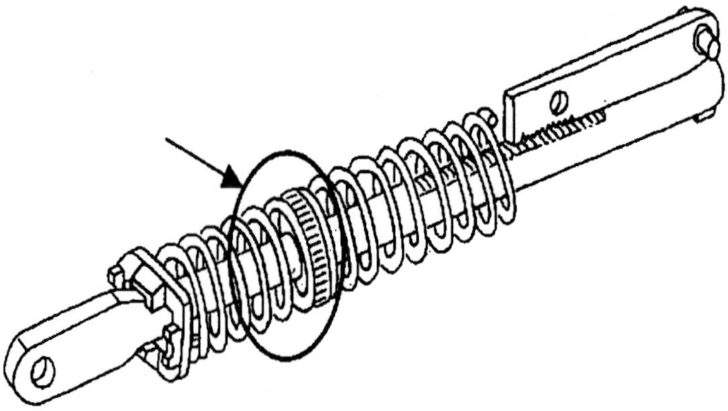

Fig. 6.20 Alternative design 3

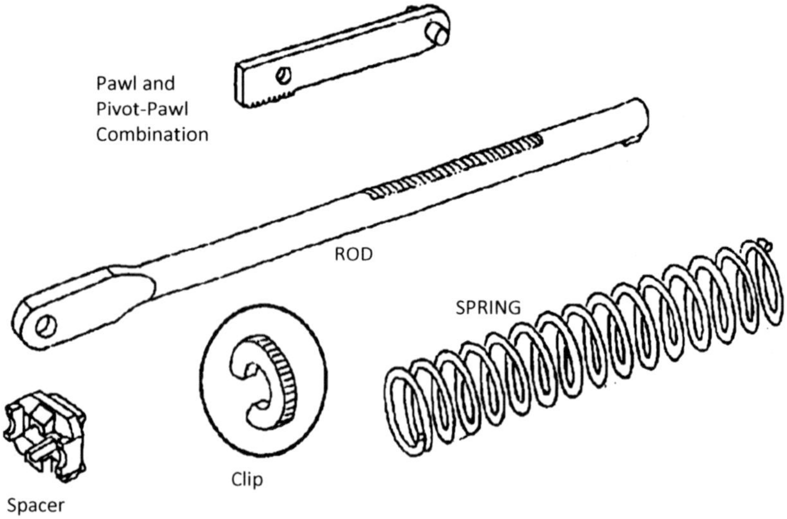

Fig. 6.21 Design components

The final selection for the seat assembly is based on using both qualitative and quantitative methods. Pugh [6] developed a design concept evaluation matrix for multiple design concepts formalizing the qualitative approach. Results of the qualitative Pugh concept selection for the seat recliner subassembly are shown in Table 6.52. The quantitative methods used were benchmarking, customer/functional needs assessment, product/concept integration, similarity matrix, material analysis, and DFA and DFM analyses. The summary results of the DFA and manufacturing is shown below in Table 6.53.

Table 6.39 Design for assembly analysis – Design Option 3

1	2	3	4	5	6	7	8	9			
Part I.D. no.	Number of times the operation is repeated	2-digit manual handling code	Manual handling hem per part	2-digit manual insertion code	Manual insertion time per part	Operation time seconds: columns 2 x (4 + 6)	Operation cost (cents): *0.4 x* column 7	Figures for estimation of theoretical minimum parts	Automotive seat recliner design 3: clip design	α	β
1	1	30	2	00	1.5	3.50	1.40	1	Rod	360	360
2	1	20	1.8	30	2	3.80	1.52	0	Spacer	360	180
3	1	10	1.5	30	5	3.50	1.40	1	Clip	360	0
4	1	05	1.8	00	1.5	3.30	1.32	1	Return-spring	360	0
5	1	30	2	00	1.5	3.50	1.40	1	Pivot-pawl combination	360	360
						TM 17.60	CM 7.04	NM 4.00	*3xNM /TM*	68.18%	

Table 6.40 Relative costs (R_c) – Design Option 3

	Shape complexity	Material	C_c	C_{mp}	C_s	C_t	R_c
Rod	A4	Steel	5.3	1.4	1.0	1.6	11.87
Spacer	B3	Plastic	1.8	1.0	1.0	1.0	1.80
Clip	A2	Plastic	1.1	1.0	1.0	1.0	1.10
Return-spring	Purchased						
Pivot-pawl Combo	A4	Steel	5.3	1.4	1.6	1.4	16.62

Table 6.41 Ideal process cost (P_c) – Design Option 3

	Material	Process	P_c (qty: 2,000,000)
Rod	Steel	Machine	5.21
Spacer	Plastic	Plastic mold	0.83
Clip	Plastic	Plastic mold	0.83
Return-spring	Purchased		
Pivot-pawl combo	Steel	Machine	5.21

Table 6.42 Material volume – Design Option 3

	Volume calculations (mm)	V (mm^3)
Rod	((12.5/2) × (12.5/2) × 3.14) × 300	36,796.88
Spacer	28 × 35.5 × 17.3	17,196.20
Clip	((35.1/2) × (35.1/2) × 3.14 × 4) – ((12.5/2) × (12.5/2) × 3.14 × 4)	3,377.89
Return-spring	Purchased	
Pivot-pawl combo	91 × 12.7 × 4.76	5,501.13

Table 6.43 Material cost (M_c) – Design Option 3

	Shape complexity	Material	V (mm^3)	C_{mt}	W_c	M_c
Rod	A4	Steel	36,796.88	0.00039	3.0	43.05234
Spacer	B3	Plastic	17,196.20	0.00061	1.1	11.53865
Clip	A2	Plastic	3,377.89	0.00061	1.1	2.26656
Return-spring	Purchased					
Pivot-pawl combo	A4	Steel	5,501.13	0.00039	3.0	6.43632

Design is an iterative method. This case study focused on meeting identified customer needs by applying all phases of the product design process including DFA and DFM analyses. Future technology will bring new opportunities in the design of automotive seat recliner mechanisms. Materials will continue to be developed to reduce weight and provide increased damping of noise, vibration, and harshness. Lean manufacturing principles also offer additional opportunities to gain efficiencies in production processes for the seat recliner assembly to eliminate waste such as overproduction, defects (scrap/rework), unnecessary inventory, excessive transportation, and unnecessary tasks.

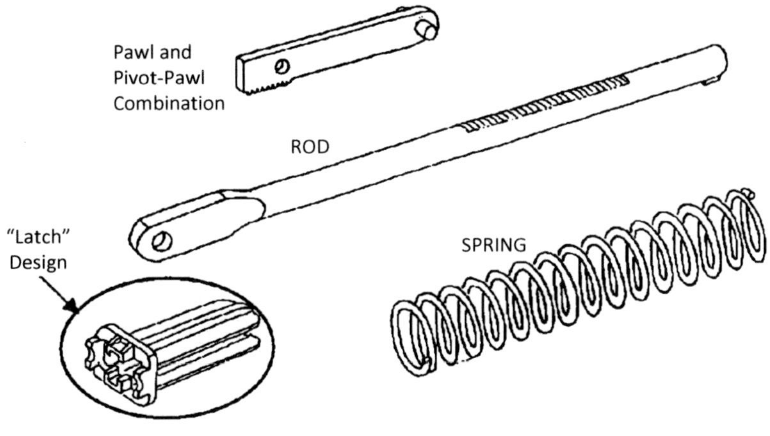

Fig. 6.22 Alternative design 4

Table 6.44 Manufacturing cost index (M_i) – Design Option 3

	R_c	P_c	M_c	M_i (=$R_c \times P_c + M_c$)
Rod	11.87	5.21	43.05234	104.90546
Spacer	1.80	0.83	11.53865	13.03265
Clip	1.10	0.83	2.26656	3.17956
Return-spring	Purchased		0	73.00000
Pivot-pawl combo	16.62	5.21	6.43632	93.03069

Total 63.29 287.15

6.15 Summary

In this chapter, the concepts of DFMA are presented. Product design is the first step in manufacturing and is where the critical decisions are made that will affect the final form and cost of the product. DFMA concentrates on simplifying designs while also evaluating assembly improvements to further enhance the overall design for manufacturability and quality. DFMA is a product development process and improvement methodology that provides a systematic process to achieve improved product design, robustness, and cost reductions through simplifications of the overall design. Finally, many case studies are shown for the purpose of understanding the concepts presented in this chapter.

Table 6.45 Design for assembly analysis – Design Option 4

1	2	3	4	5	6	7	8	9			
Part I.D. no.	Number of times the operation is repeated	2-digit manual handling code	Manual handling hem per part	2-digit manual insertion code	Manual insertion time per part	Operation time seconds: columns 2 x (4+6)	Operation cost (cents): 0.4 *x* column 7	Figures for estimation of theoretical minimum parts	Automotive seat recliner design 4: latch design	α	β
1	1	30	2	00	1.5	3.50	1.40	1	Rod	360	360
2	1	20	1.8	30	2	3.80	1.52	1	Latch	360	180
3	1	05	1.8	00	1.5	3.30	1.32	1	Return-spring	360	0
4	1	30	2	00	1.5	3.50	1.40	1	Pivot-pawl combination	360	360
						TM 14.10	CM 5.64	NM 4.00	*3xNM /TM*	85.11%	

Table 6.46 Relative costs (R_c) – Design Option 4

	Shape complexity	Material	C_c	C_{mp}	C_s	C_t	R_c
Rod	A4	Steel	5.3	1.4	1.0	1.6	11.87
Duckbill	B3	Plastic	1.8	1.0	1.0	1.0	1.80
Return-spring	Purchased						
Pivot-pawl combo	A4	Steel	5.3	1.4	1.6	1.4	16.62

Table 6.47 Ideal process cost (P_c) – Design Option 4

			P_c		
	Material	Process	1.5 ml	2.0 ml	2.5 ml
Rod	Steel	Machine	5.21	5.21	5.21
Duckbill	Plastic	Plastic mold	0.85	0.83	0.82
Return-spring	Purchased				
Pivot-pawl combo	Steel	Machine	5.21	5.21	5.21

Table 6.48 Material volume – Design Option 4

	Volume calculations (mm)	V (mm^3)
Rod	$((12.5/2)\times(12.5/2)\times3.14)\times300$	36,796.88
Duckbill	$(28\times35.5\times17.3)+(50\times27.4\times2)$	19,936.20
Return-spring	Purchased	
Pivot-pawl combo	$91\times12.7\times4.76$	5,501.13

Table 6.49 Material cost (M_c) – Design Option 4

	Shape complexity	Material	V (mm^3)	C_{mt}	W_c	M_c
Rod	A4	Steel	36,796.88	0.00039	3.0	43.05234
Duckbill	B3	Plastic	19,936.20	0.00061	1.1	13.37719
Return-spring	Purchased					
Pivot-pawl combo	A4	Steel	5,501.13	0.00039	3.0	6.43632

Table 6.50 Material cost index (M_i) – Design Option 4

	R_c	P_c (qty.: 2 ml)	M_c	M_i (=$R_c\times$ $P_c + M_c$)	M_i (qty.: 1.5 ml)	M_i (qty.: 2.5ml)
Rod	11.87	5.21	43.05234	104.90546	104.91	104.91
Duckbill	1.80	0.83	13.37719	14.87119	14.91	14.85
Return-spring	Purchased		0	73.00000	73.00	73.00
Pivot-pawl combo	16.62	5.21	6.43632	93.03069	93.03	93.03

Total 62.87 285.81 285.84 285.79

Table 6.51 Cost estimation – four design options

Parameters	Design option 1 current	Design Option 2 pivot-pawl	Design Option 3 clip	Design Option 4 latch
Total assembly costs	0.10	0.08	0.07	0.06
Total manufacturing costs				
Material costs	0.76	0.76	0.63	0.63
Purchasing parts costs	0.88	0.73	0.73	0.73
Manufacture costs	1.51	1.51	1.51	1.50
Total	3.15	3.00	2.87	2.86
Total life cycle costs				
Tooling and equipment	0.06	0.06	0.06	0.05
Overhead	0.15	0.15	0.15	0.15
Others	0.10	0.10	0.10	0.10
Total	0.31	0.31	0.31	0.30
Total competitive issues	0.80	0.80	0.80	0.80
Total estimated unit cost	4.36	4.20	4.05	4.01

Table 6.52 Pugh "qualitative" concept selection matrix

		Concepts							
		Design Option 1 (current)		Design Option 2 (pivot-pawl combo)		Design Option 3 (clip design)		Design Option 4 (latch design)	
Selection criteria assembly	Weight (%)	Rating	Weighted score	Rating	Weighted score	Rating	Weighted score	Rating	Weighted score
Ease of assembly	20	2	0.40	3	0.60	4	0.80	5	1.00
Ease of manufacture	20	3	0.60	3	0.60	4	0.80	4	0.80
Ease of maintenance	10	3	0.30	3	0.30	4	0.40	4	0.40
Durability	15	4	0.60	4	0.60	2	0.30	3	0.45
Quietness	10	4	0.40	4	0.40	3	0.30	3	0.30
Cost	25	3	0.75	3	0.75	4	1.00	4	1.00
Total score:			3.05		3.25		3.60		3.95
Rank:			4		3		2		1
Continue?			No		No		No		Develop

6.16 Review Questions

2.1. Describe in details DFMA Methodology.
2.2. Explain Boothroyd–Dewhurst method for manual assembly analysis.

Table 6.53 "Quantitative" DFM and DFA results – all options

Parameter	Design Option 1 current	Design Option 2 pivot-pawl	Design Option 3 clip	Best Design Option 4 latch
Design for manufacturing (DFM) summary				
Material cost	47.15	47.15	39.30	38.87
Manufacturing cost index	54.07	54.07	46.22	44.88
Design for assembly (DFA) summary				
Total assembly time	30.00	26.50	23.20	19.70
No. of parts in design	6	5	5	4
Total assembly cost	12.00	10.60	9.28	7.88
Design efficiency (%)	40.00	45.28	51.72	60.91

2.3. What are the steps of design for manual assembly?

2.4. Select any product in the market and do all the required analysis that you learned in this chapter.

References

1. Lefever DD, Wood KL (1996) Design for assembly techniques in reverse engineering and redesign. In: Proceedings of the ASME design and engineering technical conference, Irvin, California
2. Boothroyd G, Dewhurst P, Knight W (1994) Product design for manufacture and assembly, M. Dekker, New York
3. Boothroyd G (1991) Assembly automation and product design. Dekker Publishing, New York
4. Whitney D (2004) Mechanical assembly. Oxford University Press, New York
5. Munro Associates Inc. (1989) Design for manufacture training manual. Munro Associates Inc.
6. Pugh S (1993) Total design: integrated methods for successful product engineering. Addison-Wesley Publishing Co., Wokingham, England; Reading, Massachusetts
7. Akiyama K (1991) Function analysis: systematic improvement of quality and performance. Dillon AP (translator), Productivity Press, Cambridge, Massachusetts
8. Amirouche F (1993) Computer aided design and manufacturing. Prentice Hall, Englewood Cliffs, New Jersey
9. Anderberg MR (1973) Cluster analysis for applications. Academic Press, New York

Part II
CAD/CAM and Features-Based Technologies

Chapter 7
Computer-Based Design and Manufacturing

Abstract This chapter provides an overview of computer-aided design and manufacturing (CAD/CAM), the most important reasons of using CAD systems in the manufacturing environment, computer integrated manufacturing (CIM), the implementation of the automation in the production organization, the role of CAD/CAM systems in the manufacturing facility, the CAM cycle in a feature based design environment, and the different types of features.

Keywords CAD • CAM • CIM

7.1 Introduction

Product design could be considered as one of the critical functions in the design life cycle. The quality of the product design (i.e., how well the design department does its job) is the single most important factor in determining the commercial success and societal value of a product. If the design is poor, no matter how well it is manufactured, the product is very likely doomed to contribute little to the wealth and well-being of the firm that produced it. If the design is good, there is still the question of whether the product can be produced at sufficiently low cost to contribute to the company's profits and success. It is important to know that a very significant portion of the cost of any product is determined by its design [11].

Design and manufacturing are two major components in the production engineering process. A product design that cannot be realized through manufacturing processes is a poor design. On the other hand, manufacturing processes cannot be effective without a thoughtful design and plan [5].

Traditionally, design and manufacturing are treated as two separate stages and usually managed by two different groups of people. The design group often does not anticipate the manufacturing implications of the decisions [16]. After the detail design is completed, it is passed to the manufacturing group as annotated engineering drawings. The manufacturing group has to determine the methods of how to produce parts based on its interpretation of the design group's drawings, which may

A.K. Kamrani and E.A. Nasr, *Engineering Design and Rapid Prototyping*,
DOI 10.1007/978-0-387-95863-7_7,

not be the same as the designer's intent. It usually takes a few passes back and forth between the two groups in order to reach a satisfactory design [24].

This sequential approach is a slow and a costly process. Therefore, to reduce the time and cost, it is important to achieve a good integration of design and manufacturing that provides a common language to interact and to anticipate design changes and coordinates them with the manufacturing processes [6]. The goal of the proposed methodology with the integration of computer-aided design (CAD) and computer-aided manufacturing (CAM) is to support the design, modification, analysis, and manufacture of parts with prismatic shapes automatically and efficiently within a unified frame.

7.1.1 Computer-Aided Design

Computer-aided design can be defined as the technology used to prepare technical drawings and plans used to build everything from manufactured products such as toys, toasters, industrial machinery, and spacecraft to structures such as houses, office buildings, and oil and gas pipelines. CAD drawings provide visual procedures; show the technical details of the products and structures; and specify dimensions, materials, and procedures. CAD systems create and store drawings electronically that can then be viewed, printed, or programmed directly into automated manufacturing systems. Personnel who produce technical drawings with CAD still function as drafters and need the knowledge of traditional drafters, in addition to their CAD skills [21].

7.1.2 Computer-Aided Manufacturing

Computer-aided manufacturing is the use of computer-based software tools that assist engineers and machinists in manufacturing or prototyping product components. Its primary purpose is to create a faster production process and components with more precise dimensions and material consistency, which in some cases, uses only the required amount of raw material (thus minimizing waste), while simultaneously reducing energy consumption. CAM is a programming tool that makes it possible to manufacture physical models using CAD programs. CAM creates real-life versions of components designed within a software package. CAM was first used in 1971 for car body design and tooling [33].

7.2 Computer-Aided Design and Computer-Aided Manufacturing Integration

The idea of using computer technologies in design and manufacturing is traced back to the late 1950s and early 1960s, when numerical control (NC) machines and computer graphics were first introduced [8]. Early applications were mainly in computer-assisted NC part programming and computer-aided drafting [10].

Computer-aided design is defined as any design activity that involves the effective use of the computer to create, modify, analyze, optimize, and document an engineering design. CAD is most commonly associated with the use of an interactive computer graphics system, referred to as a CAD system. On the other hand, CAM is defined as the effective use of computer technology in manufacturing management planning and control. CAM is most closely associated with functions in manufacturing engineering, such as process planning and NC part programming. Other applications of CAM include inspection, material handling, robotics, assembly, etc. [3, 4, 28]. The term CAD/CAM system is also used if it supports manufacturing as well as design applications.

With improved computer power and better understanding of its usages of design and manufacturing, CAD and CAM systems [12, 25, 32] were used to allow interactive design of geometric models and direct or indirect use of the model data in manufacturing processes. There are several good reasons for using a CAD system [11]:

1. ***To increase the productivity of the designer***: To help the designer to conceptualize the product and its components. Therefore, this helps reduce the time required by the designer to create, analyze, and document the design.
2. ***To improve the quality of the design***: Use of a CAD system with appropriate hardware and software capabilities permits the designer to perform a comprehensive engineering analysis and to involve a larger number and variety of design alternatives. The quality of the resulting design is thereby enhanced.
3. ***To improve design documentation***: The final graphical output of a CAD system results in better documentation of the design than what is practical with manual drafting. The engineering drawings are superior, and there is more standardization among the drawings, fewer drafting errors, and greater legibility.
4. ***To create a manufacturing database***: During the process of documentation, the product design (geometric specification of the product, dimensions of the components, materials specifications, bill of materials, etc.) and much of the required data for the manufacture of the product are also created.

7.3 Computer-Integrated Manufacturing

Computer-integrated manufacturing (CIM) in engineering is a method of manufacturing in which the entire production process is controlled by computer. The traditional separated process methods are joined through a computer by CIM. This integration allows that the processes exchange information with each other and they are able to initiate actions. Through this integration, the manufacturing can be faster and with fewer errors, although the main advantage is the ability to create automated manufacturing processes. Typically, CIM relies on closed-loop control processes based on real-time input from sensors. It incorporates a flexible design and manufacturing capabilities.

To further increase the level of automation in production processes, CIM was introduced [24, 34]. CIM includes not only the engineering function from CAD/

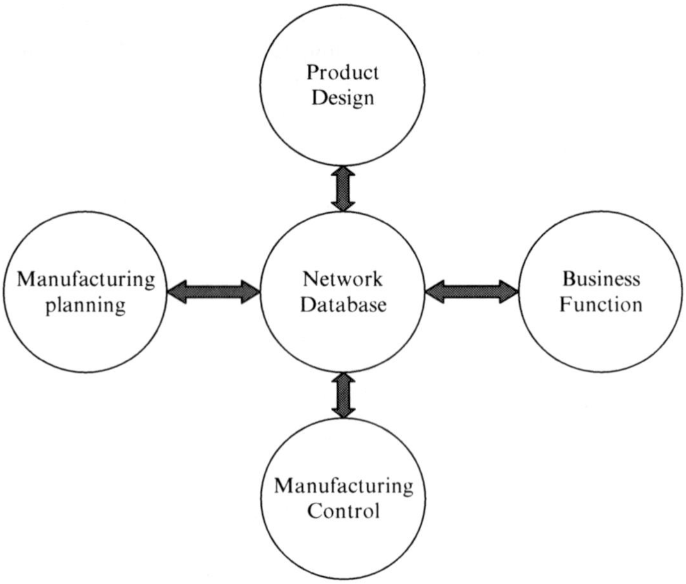

Fig. 7.1 A typical model of CIM implementation

CAM, but also the business aspect of a product, such as marketing, accounting, management, etc. The CIM concept includes all of the operations related to the production incorporated in an integrated computer system to assist, augment, and automate the processes. The same information must be shared throughout the entire life cycle of a product. Existing CAD/CAM systems are made to work together through format conversion. Internally, each application works independently and uses its own data representation. A typical model of CIM implementation is a star-shaped structure with a network database as the central node (Fig. 7.1).

One problem of this implementation is that each of the existing subsystems usually uses a different data format or needs some information that is not available from other subsystems. Data format conversion between each subsystem and the central database is then required. Sometimes, manual augmentation of the product design is applied to obtain extra information required for a particular application. Standard data formats, such as Initial Graphics Exchange Specifications (IGES) and Data Interchange Format (DXF), are used to store models in the database [1, 17].

For example, the CAD design may contain only points, lines, and arcs in the model data while the CAM system needs to know the width/length/depth of a slot that is not available from the part design explicitly. Such information has to be available in a format that can be used by a CAM system. This heterogeneous approach can easily integrate various existing components. It does, however, build barriers against a full implementation of the CIM concept [18].

7.3.1 The Role of CAD/CAM in Manufacturing

The CAD/CAM is concerned with engineering functions in both design and manufacturing. The engineering activities in design are product design, engineering analysis, and documentation of the design. On the other hand, process planning, NC part programming, and other activities associated with CAM represent engineering activities in manufacturing. The CAD/CAM systems developed during the 1970s and early 1980s were designed primarily to address these types of engineering problems. In addition, CAM has evolved to include many other functions in manufacturing, such as material requirements planning, production scheduling, computer production monitoring, and computer process control.

The successful link between CAD and CAM will be defined by the efficient methods that can process a decomposed object from the CAD database and reconstruct the entities for downstream applications [19]. Features provide access to related production-process and resource methods. Thus, features have a higher level semantically than the primitive elements used in the ordinary CAD systems. Features are expected to form a basis for linking CAD with downstream manufacturing applications such as process planning and for organizing databases for reusing design data. The efficiency and speed of algorithms to do this task have been the focus of many researchers. The algorithms should be such that they can be applied to a wide range of complexities of solid parts and use minimal data and processing time to arrive at the desired manufacturing planning and execution [12]. Numerous CAD systems on the market today necessitate that the computer programs be flexible and portable on different systems.

It should also be noted that CAD/CAM requires an integration of design and manufacturing activities by means of computer systems. The method of manufacturing a product is a direct function of its design. With conventional procedures, engineering drawings were prepared by design draftsmen and later used by manufacturing engineers to develop the process plan. The activities involved in designing the product were separated from the activities associated with process planning. Fundamentally, a two-step procedure was employed. This was time consuming and involved duplication of effort by design and manufacturing personnel [16].

A CAM system in general should have the ability to manipulate the geometric data from a product model and other information about the machining process, such as selection of cutting tools, machining sequence, and strategy. Most traditional cutter path generation methods implemented in CAD/CAM systems use surface, wireframe, or solid models. User interaction is often required to ensure that features are accessible, no collisions occur, and the part is not over or undercut [23]. Using the feature-based approach, the agent that provides the interpretation, or more explicitly the "feature interpreter," decomposes the part into a group of manufacturing standard features that are suitable for machining operations.

With the advance of CAD/CAM technology, it is possible to establish a direct link between product design and manufacturing engineering. The goal of CAD/

CAM is not only to automate certain phases of design and certain phases of manufacturing, but also to automate the transition from design to manufacturing. In the ideal CAD/CAM system, it is possible to take the design specification of the product as it resides in the CAD database and convert it into a process plan for making the product, with the conversion done automatically by the CAD/CAM system [15].

A large portion of the processing might be accomplished on a numerically controlled machine tool. As part of the process plan, the NC part program is generated automatically by CAD/CAM. The CAD/CAM system downloads the NC program directly to the machine tool by means of a telecommunications network. Hence, under this arrangement, product design, NC programming, and physical production are all implemented by computer [28]. Figure 7.2 shows a typical CAM cycle in a feature-based environment. Steps 1–4 detail the feature extraction procedure in

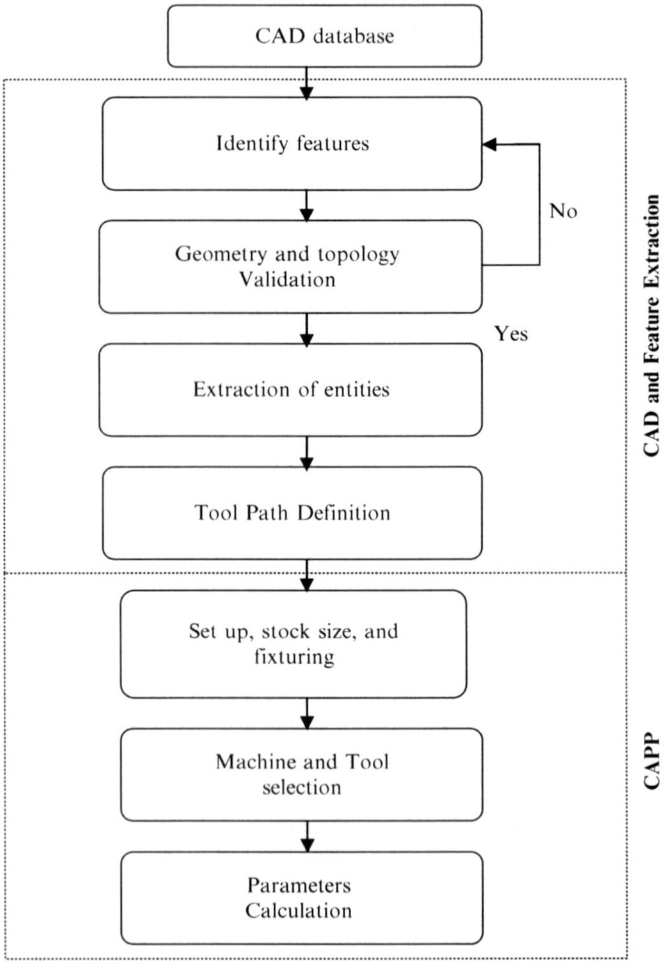

Fig. 7.2 The CAM cycle in a feature-based design environment

which the geometric information of the part design has to transfer to some format that can be used by Computer aided process planning (CAPP) system, while steps 5–7 relate to CAPP in which a generated process plan can be developed that contains all the information required to manufacture the part (setup method, raw material, fixturing methods, cutting tools, machines, cutting parameters) and finally step 8 and 9 to generate NC code.

7.4 Flexible Manufacturing Systems

A flexible manufacturing system (FMS) is a manufacturing system in which there is some amount of flexibility that allows the system to react in the case of changes, whether predicted or unpredicted. This flexibility is generally considered to fall into two categories, which both contain numerous subcategories. The first category, machine flexibility, covers the system's ability to be changed to produce new product types, and ability to change the order of operations executed on a part. The second category is called routing flexibility, which consists of the ability to use multiple machines to perform the same operation on a part, as well as the system's ability to absorb large-scale changes, such as in volume, capacity, or capability. Most FMS systems comprise of three main systems. The work machines that are often automated CNC machines are connected by a material handling system to optimize parts flow and the central control computer that controls material movements and machine flow. The main advantage of an FMS is its high flexibility in managing manufacturing resources like time and effort in order to manufacture a new product. The best application of an FMS is found in the production of small sets of products like those from a mass production [30].

7.5 Concurrent Engineering

The concurrent engineering method is a new design management system, but has had the chance to mature in recent years to become a well-defined systems approach toward optimizing engineering design cycles. For this reason, concurrent engineering has gained much attention from industrial environment and has been applied in a multitude of companies, organizations, and universities, such as the aerospace industry.

The basic principle for concurrent engineering revolves around two concepts. The first is the idea that all elements of a product's life cycle, from functionality, producibility, assembly, testability, maintenance issues, environmental impact, and finally disposal and recycling, should be taken into careful consideration in the early design phases. The second concept is that the preceding design activities should all be occurring at the same time, or concurrently. In general, the main objective of the concurrent engineering process is increasing productivity and product quality. This philosophy is a key to the success of concurrent engineering because it allows for errors and redesigns to be discovered early in the design process when the project is still in a more abstract and possibly digital realm. By locating and fixing these issues early, the design team can avoid what often become costly errors as the project moves to more complicated computational models and eventually into the physical realm [26].

As stated before, part of the design process is to ensure that the entire product's life cycle is taken into consideration. This includes establishing user requirements, propagating early conceptual designs, running computational models, creating physical prototypes, and eventually manufacturing the product. Included in the process is taking into full account funding, work force capability and time, subject areas that are extremely important factors in the success of a concurrent engineering system. As before, the extensive use of forward planning allows for unforeseen design problems to be caught early so that the basic conceptual design can be altered before actual physical production commences. The amount of money that can be saved by doing this correctly has proven to be significant and is generally the deciding factor for companies moving to a concurrent design framework [23].

One of the most essential reasons for the huge achievement of concurrent engineering is that by definition it redefines the basic design process structure that was common place for decades. This was a structure based on a sequential design flow (Waterfall Model). Concurrent engineering significantly modifies this outdated method and instead opts to use what has been termed an iterative or integrated development method. The difference between these two methods is that the "Waterfall" method moves in a completely linear fashion by starting with user requirements and sequentially moving forward to design, implementation, and additional steps until you have a finished product. The problem here is that the design system does not look backwards or forwards from the step it is on to fix possible problems. In the case that something does go wrong, the design usually must be scrapped or heavily altered. Conversely, the iterative design process is more cyclic in that, as mentioned before, all aspects of the life cycle of the product are taken into account, allowing for a more evolutionary approach to design.

7.6 Feature-Based Technologies

A graphical model can capture geometric shapes that form the bulk of the information needed in an engineering design [25, 28]. Three-dimensional models can be directly used in visualization, saving much effort in prototype fabrication and making it easier to integrate with other manufacturing functions [29]. In general, the geometric data for the design can be represented in the following ways: *Wireframe Representation*, *Boundary Representation* (*B-rep*), or *Constructive Solid Geometry* (*CGS*).

These feature representation methods are discussed in detail in Chap. 8. After a product's geometric model is constructed, the geometric data need to be transfered into a format that can be used to generate the required manufacturing processes. This conversion of the geometric data is called feature recognition or extraction. Several approaches have been developed for feature recognition [2, 7, 9, 13, 14, 22, 27].

7.6.1 Types of Features

The feature-based approach to the application of CAD is an intelligent form of design representation in which the design is expressed in terms of some high level definition that has direct relevance in various downstream manufacturing activities such as process planning [7]. Features are defined in a different way based on their specific applications as shown in Fig. 7.3. Therefore, it is difficult to determine an accurate and specific definition of a feature. For example, in the manufacturing domain, a feature-based design representation is in terms of holes, slots, pockets, steps, and chamfers. Traditional CAD represents the design in terms of either two-dimensional entities (lines, arcs, and circles), or three-dimensional entities (wire-frame, surfaces, and solids). On the other hand, these representations have little significance to the process planner who thinks of manufacturing features such as holes, slots, and pockets. Feature information permits the process planner to determine the machining tools and manufacturing processes required to machine the designed objects. Making a decision for the types of configurations a designer wants to be featured can be difficult and depend on the application at hand. For example, Fig. 7.4 shows a simple designed object in which designers may think about different features in a different way than manufacturing engineers do. In this figure, for the designer, this shape can be represented using a single slot bisected by the rib "A," whereas for a manufacturing engineer, this shape can be represented using two pockets, "B" and "C."

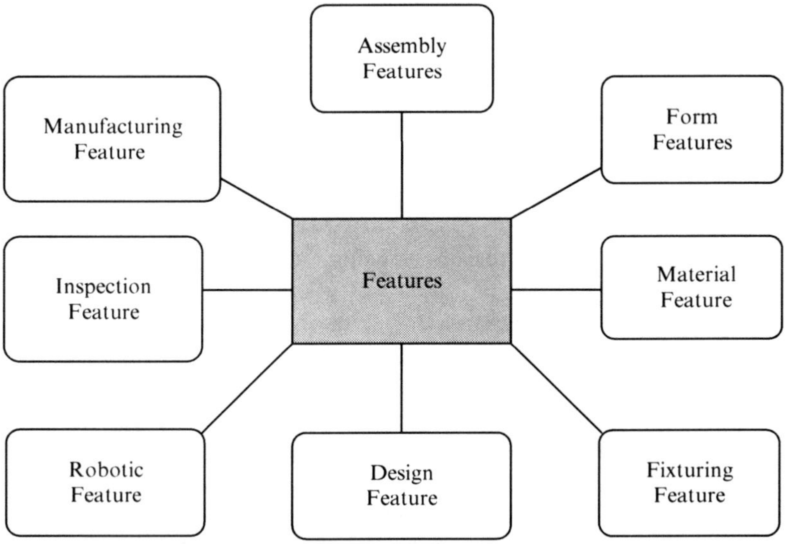

Fig. 7.3 Types of features

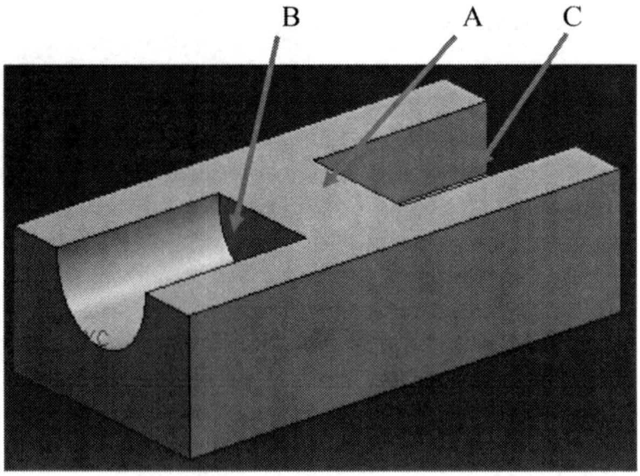

Fig. 7.4 A simple designed object

7.7 Summary

This chapter provides an overview of CAD/CAM, the most important reasons of using CAD systems in the manufacturing environment, CIM, the implementation of the automation in the production organization, the role of CAD/CAM systems in the manufacturing facility, the CAM cycle in a feature-based design environment, and the different types of features.

7.8 Review Questions

7.1. Define CAM.
7.2. What is CAD?
7.3. List the most important reasons of using CAD systems in the manufacturing environment.
7.4. Define CIM and its components. Explain how it can be implemented in the automation of the production organization.
7.5. What is the main obstacle facing the successful implementation of CIM?
7.6. Discuss the role of CAD/CAM systems in the manufacturing facility.
7.7. Why does features model play an important role in downstream applications?
7.8. Describe briefly the CAM cycle in a feature-based design environment.
7.9. What are the different types of features?
7.10. What is the difference between manufacturing features and design features? Explain your answer using drawings.

References

1. Ahmad N, Haque AFM (2001) Manufacturing feature recognition of parts using DXF Files. In: Fourth international conference on mechanical engineering, Dhaka, Bangladesh 1(1):111–115.
2. Alan C, Shou L (1998) A volume decomposition approach to process planning for prismatic parts with depression and protrusion design features. Int J Comput Integr Manuf 11(6):548–563.
3. Allada V, Anand S (1996) Machine understanding of manufacturing features. Int J Prod Res 34(7):1791–1820.
4. Bartholomew O, Nnaji BO, Liu HC (1990) Feature reasoning for automatic robotic assembly and machining in polyhedral representation. Int J Prod Res 2(3):517–540.
5. Bedworth DD, Henderson MR, Wolfe PM (1991) Computer integrated design and manufacturing. McGraw-Hill, New York.
6. Boyer KK (1999) Evolutionary patterns of flexible automation and performance: a longitudinal study. Manag Sci 45(6):824–842.
7. Case K, Harun WA (2000) Feature-based representation for manufacturing planning. Int J Prod Res 38(17):4285–4300.
8. Czerwinski A, Keats J, Leibar K, Seilar A (1993) CAD/CAM data exchange application. Adv Manuf Technol 10(1):306–310.
9. Dong J, Parsaei HR (1994) An automated process planning system. Intell Automat Soft Comput: Trend Res 1(1):57–62.
10. Duan W, Zhou J, Lai K (1993) FSMT: a feature solid modeling tool for feature-based design and manufacturing. Comput-Aided Des J 25(1):29–38.
11. Groover MP (2001) Automation, production systems, and computer-integrated manufacturing. Prentice-Hall, Englewood Cliffs.
12. Jones R, Mitchell S, Newman S (1993) Feature-based systems for the design and manufacture of sculptured products. Int J Prod Res 31(6):863–885.
13. Jung H, Inho H (1999) Manufacturable feature recognition and its integration with process planning. In: Fifth symposium on solid modeling, Ann Arbor, 108–118.
14. Kayacan MC, Celik SA (2003) Process planning system for prismatic parts. Integr Manuf Syst 14(2):75–86.
15. Lee K (1999) Principles of CAD/Cam/CAE Systems. Addison-Wesley, Reading.
16. Lionel R, Otto S, Henri P (2003) Process planning as an integration of knowledge in the detailed design phase. Int J Comput Integr Manuf 16(1):25–37.
17. Linardakis S, Mileham AR (1992) A CAD interpreter for prismatic components. Adv Manuf Technol 8(1):21–25.
18. Linardakis S, Mileham AR (1993) Manufacturing feature identification for prismatic components from CAD DXF Files. Adv Manuf Technol 9(1):37–41.
19. Lin AC, Lin S, Lin Y, Cheng SB (1997) Extraction of manufacturing features from a feature-based design model. Int J Prod Res 35(12):3249–3288.
20. Mari A, Kobu B (2003) Implementation of computer-integrated manufacturing in small and medium enterprises. Ind Commercial Train 35(4):151–157.
21. McGarry T, McHugh B, Buis A, McKay G (2008) Evaluation of the effect of shape on a contemporary CAD system. Prosthet Orthot Int 32(2):145–154.
22. Nagaraj HS, Gurumoorthy B (2001) Automatic extraction of machining primitives with respect to preformed stock for process planning. J Manuf Syst 20(3):210–222.
23. Executing Nategh MJ (2009) Concurrent engineering planning on the basis of forward and backward effects of manufacturing processes. Int J Prod Res 47(18):5147–5161.
24. Pham D, Gologu C (2002) A computer aided process planning system for concurrent engineering. Proc Instn Mech Engrs, part B, J Eng Manuf 215(1):1177–1131.
25. Qiang J, Michael M (1997) Machine interpretation of CAD data of manufacturing applications. ACM Comput Surv 24(3):264–311.

26. Sapuan SM, Osman MR, Nukman Y (2006) State of the art of the concurrent engineering technique in the automotive industry. J Eng Des 17(2):143–157.
27. Sashikumar V, Milind S, Vinay K (2001) A graph-based framework for feature recognition. In: Appearing in ACM symposium on solid modeling and applications, Ann Arbor, 194–205.
28. Somashekar S, Michael W (1995) An overview of automatic feature recognition techniques for computer-aided process planning. Comput Ind 26(1):1–21.
29. Turkay D, Huseyin F (2002) Design for manufacturing system for elimination of critical feature interactions on prismatic parts. J Eng Des 13(2):141–157.
30. Venkata Rao R (2007) Evaluating flexible manufacturing systems using a combined multiple attribute decision making method. Int J Prod Res 46(7):1975–1989.
31. Waldner JB (1992) CIM – principles of computer integrated manufacturing. Wiley, Chichester.
32. Wang HP, Wysk RA (1998) Computer aided manufacturing, 2nd ed. Prentice-Hall, Upper Saddle River, NJ.
33. Xie SQ, Xu X (2006) A STEP-compliant process planning system for sheet metal parts. Int J Comput Integr Manuf 19(6):627–638.
34. Zhao Z, Ghosh SK, Link D (1990) Recognition of machined surfaces for manufacturing based on wireframe models. J Mater Process Technol 24(1):137–145.

Chapter 8
Feature Representations

Abstract This chapter presents discussions related to feature representation methodologies. It includes the definition of features, wireframe modeling, surface modeling, boundary representation (B-rep), constructive solid geometry (CGS), and definition of interacting features. These methods are used to facilitate the feature recognition process.

Keywords Feature representation • B-rep • CGS • Interacting features

8.1 Feature Definitions

The traditional separation of design and manufacturing activities in modern industry is widely recognized as a major contributor to product development costs. Feature-based modeling has proven to be an effective and time-saving approach for product design. Features provide a good way to handle the design at a higher level of description than engineering drawings or traditional CAD descriptions that characterize an object in terms of mathematical surfaces or volumes [1]. Traditionally, parts were represented at a low level using wireframe representation, B-rep, and CSG representation. Manufacturing information needed for higher level applications, such as process planning, could not be extracted directly from these models. The term "feature" has different meanings in different environments depending on the specific domain. For example, in design it refers to a web or a notch section, etc., while in manufacturing it refers to slots, holes, and pockets, while in inspection it is used as a datum or reference on a designed object [2]. In general, classification of features is totally application dependent. It is very difficult to make an application-independent classification of features. Based on the existing literature, a classification of features according to different perspectives is given in Table 8.1. There are many definitions in the literature for the term "feature" [3–7], some of which are as follows:

1. "A feature is any entity used in reasoning about the design, engineering, or manufacturing of a product" [8].

A.K. Kamrani and E.A. Nasr, *Engineering Design and Rapid Prototyping*,
DOI 10.1007/978-0-387-95863-7_8, © Springer Science+Business Media, LLC 2010

Table 8.1 Classification of manufacturing features according to different perspectives

Aspect	Types of features	Characteristics
1. Information support requirement	Form features	Size, shape
	Precision features	Tolerance, surface finish
	Material features	Material type, material grades, heat treatment, etc.
2. Applicability of the features in different phases of manufacturing	Design features	Center line, radius, function of the features
	Process planning features	Starting surface, manufacturing method
	Assembly features	Line, radius, tolerance
3. Configuration of the feature	Elementary features	Hole, step, etc.
	Compound features	Stepped hole
	Functional features	Screw thread
4. Profile of the features	Prismatic features	Slot, pocket, etc.
	Cylindrical features	Hole, fillet, arch, etc.

2. "A geometric form or entity whose presence or dimensions are required to perform at least one CIM function and whose availability as a primitive permits the design process to occur" [9] [19].
3. "A region of interest on the surface of a part" [9].
4. "A geometric form or entity that is used in reasoning in one or more designs or manufacturing activities" [10].
5. "A parametric shape associated with such attributes as its intrinsic geometric parameters – length, width, and depth – as well as position, orientation, geometric tolerances, material properties, and references to other features" [11].

Although many different definitions have been given, the basic of these definitions is that features represent the engineering meaning of the geometry of a part, assembly, or other manufacturing activity. In this book, features are considered as generic shapes with which engineers associate certain attributes and knowledge that are useful in reasoning about the product definitions.

8.2 Features in Manufacturing

The advantage of feature-based modeling in manufacturing applications is based on associating design feature with manufacturing process models. For example, the process model for a machining process would indicate the process resources, such as machines, tools, fixtures, and auxiliary materials; process kinematics, such as tool access direction; process constraints, such as interference and spindle power; process parameters, such as feeds and speeds, and other information, such as time and cost [12]. There is a need for a good link for features, process models, and resource models that leads to the manufacturing knowledge repository organization

[13]. Operatively, the processes potentially suited for the features of a part can be retrieved to construct a manufacturing plan for the designed object. Strategically, the data repository supports adding new products, processes, and resources to reflect results of new product design and manufacturing process development [14]. Some important manufacturing applications are described in the following subsections.

8.2.1 Process Planning

Process planning involves determining the information required for manufacturing a part. Most research on process planning includes machining operations, although a few works include other areas, such as injection-molded parts, stamped and die cast parts, and sheet metal parts. The two primary approaches for automated process planning (*variant and generative planning*) are described in the following subsections.

8.2.1.1 Variant Process Planning

In variant process planning (VPP) as shown in Fig. 8.1, the manufacturing engineer uses group technology coding scheme to map a design (G) into an alphanumeric

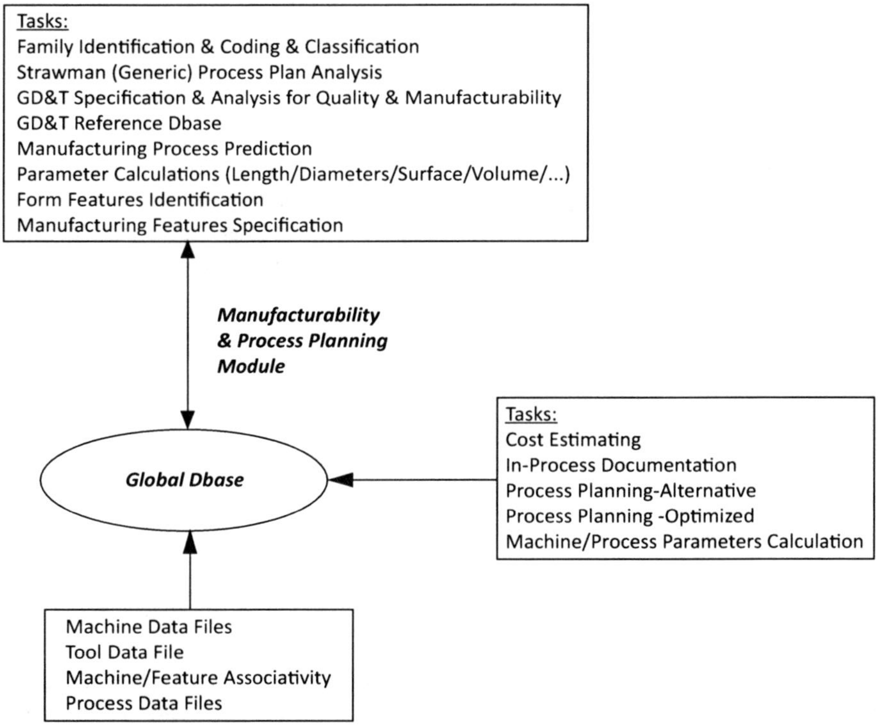

Fig. 8.1 Variant process planning

code. Then, this code can be used as an index to search the process plan database to retrieve a process plan N′ for a design (G′) similar to (G) and then adapts N′ manually to produce a process plan (N) for (G). In VPP methodology, the key contribution of features is to facilitate group technology coding on the basis of the explicitly recorded relationships between the feature configurations. In this approach, a computer is used to help the designer set and adapt the process plan [15]. Several commercial VPP systems deliver significant benefits but have drawbacks. For example, if the process plan N′ uses an obsolete database for manufacturing processes, these processes propagate to (N) unless the process engineer makes the suitable modifications.

8.2.1.2 Generative Process Planning

In the generative process planning (GPP) approach, the planning system seeks to synthesize the process plan (N) directly. For machine-designed objects, the distinctive approach is to perform the planning on the basis of a feature-by-feature methodology by retrieving candidate processes from the manufacturing knowledge repository, selecting the practical processes on the basis of geometric and manufacturing information of the designed objects, and merging the selected processes in a proper sequence.

A GPP engineer would alter industrial practice of the designed objects because of the manufacturability constraints. As a result, much research has been developed on GPP approaches [16–18], and a number of experimental systems have been developed for various characteristics of process planning. On the other hand, GPP systems have proved quite difficult in real manufacturing environments. These difficulties can be attributed to the interaction among various characteristics of the problem, such as workpiece fixturing, process selection, and process sequencing. Consequently, most GPP systems work only in some domains that have restrictions, and few generative systems have achieved significant in the real manufacturing market.

Even in the absence of comprehensive solutions to the entire process planning problem, GPP techniques can be useful in design for manufacturing in which the designer attempts to take into account considerations of manufacturability during the design phase. For instance, generating and evaluating operation plans for a designed object can give feedback to designers about possible manufacturability problems with the designed object or propose changes to the designed object to improve its manufacturability [19].

8.2.2 Assembly Planning

Since most manufacturing products are composed of assemblies of individual components, assembling these components is a key point in manufacturing products. In light of this fact, it is odd that assembly and assembly planning receive far less research attention than individual component manufacturing. Assembly planning can be roughly divided into three phases [20]:

1. ***Selection of assembly method***: identifying the one most suitable method for the product while accounting for the type of assembly system to be used.
2. ***Assembly sequence planning***: generating a sequence of assembly operations (placing each component in its final position in the assembly) to be used to implement an assembly task in a given assembly system.
3. ***Assembly operations planning***: emphasizing the details of individual assembly steps, such as access directions, mating movements, and application of fasteners.

Many assembly process problems can be effectively handled in the early design phase, such as poor quality, inefficiency, and high cost. These problems have yielded the *design for assembly approach* in which the most economical assembly process is selected during design and the product is modified to the chosen method [21].

In the robotic assembly process, an operation planning involves designing a valid sequence of rigid motions, translations, and rotations that bring a component into its final position in the assembly. Various assembly tools should be designed, including grippers, jigs, fixtures, pallets, and component feeders. Assembly cells must be laid out while monitoring the physical restrictions of robot motions.

Features are potentially attractive for many of these tasks. A feature model can record the physical arrangement of components in an assembly as well as attribute information on the physical fit between linked components. These data are fundamental for the necessary assembly operations and for ways to sequence them.

8.2.3 Inspection Planning

A popular manufacturing development is error free *six-sigma* manufacturing [22], which seeks to ensure that bad parts are almost never produced. Part inspection is vital for this goal not only to discard bad parts but also to provide closed loop control on the quality of the products being produced. Consequently, inspection planning is treated as an integral part of manufacturing planning. Several research problems remain unresolved in inspection planning. To decide which problem to concentrate on, it should focus on the product characteristics with the greatest influence on a product's performance. At the same time, it should avoid the cost of inspecting characteristics with little significance. Inspection planning is deeply involved with such product modeling issues as representations of dimensions and tolerances, assembly relationships, and finally with functions and behaviors of physical configurations [23].

8.3 Geometric Data Format

From a manual drawing to computer-aided design (CAD), the geometric data representation also evolved from simple points and lines to 3D and solid data format structure. Geometric modeling can be defined as a group of methods that are used to define the shape and other geometric characteristics of an object. Basically, the methods of geometric modeling are a synthesis of techniques from many fields such as analytic and descriptive geometry, topology, set theory, numerical analysis, vector calculus, matrix methods, and so on [24]. Geometric modeling is a necessary element of CAD and has had rapid development due to enormously increased computing power technology; otherwise, it would not be possible to analyze models of any practical importance. Geometric modeling systems should represent and manipulate 3D objects in an unambiguous way. The main functions of the geometric modeling system are [25] as follows:

1. Easy and interactive generation of new solid object.
2. Easy modification and manipulation of the solid object using geometric operators such as rotation, translation, Boolean operation, etc.
3. Easy analysis of the properties of solid objects.
4. Effective and easy way of displaying the objects.

Currently, geometric modeling is required in the field of computer-aided manufacturing (CAM) as well as CAD. A designer should be able to provide a complete and unambiguous model so that the manufacturing engineers can create a process plan and tool path of the numerical control machine, and perform inspection of the final product. This automatic linkage between CAD and CAM is impossible if the geometric model is incomplete and ambiguous. For manufacturing applications, geometric modeling should be able to represent and manipulate 3D solid objects in a nonambiguous way [26].

There are three different types of geometric modeling techniques for representing 3D objects: *wireframe*, *surface*, and *solid modeling* [27].

8.3.1 Wireframe Modeling

The result of the reverse engineering process is typically a wireframe model that is ambiguous and lacks solidity. This representation does not define the surface information nor does it give the volume of the various regions of the part. Because only points, lines, and curves are used, the model cannot determine the interior or exterior shapes of a surface. Therefore, it is difficult to use this representation in automatic manufacturing feature recognition.

Although the wireframe representation is not considered the major format for feature representation, for three main reasons it is still popular in design models. First, it requires a less expensive facility that is good for small businesses where it is still the fundamental representation [28]. Second, this method of modeling needs less computational effort and it is extensively used in computer graphics such as the area of display [7]. Third, some models need to be interpreted from a paper drawing or physical part by optical equipment. The information received from the optical procedure can only form the wireframe data and such a procedure is one of the important steps in the reengineering area [29].

8.3.2 *Surface Modeling*

Surface modeling represents a 3D object by describing part surfaces, but the interior properties are not defined in this model. This model can generate a picture equivalent to those of solid modeling by facilitating hidden line removal and shading effects [30]. Although this model can automatically estimate volume property from the surface definitions, the mass property and other important attributes for manufacturing cannot be extracted from this model because of undefined interior attributes. Therefore, the section cut of an object reveals only the edge definitions of the cut surface.

The surface model can represent a more accurate and sophisticated surface definition than an equivalent solid model. Hence, this model is useful for the detection of the NC tool path [31]. Basically, there are three types of surfaces that can be generated from this model and each type of surface fits each particular domain of the designing work [32]:

1. Ruled or extruded surfaces
2. Surfaces of revolution
3. Sculptured surfaces

The extruded surfaces are formed by skinning two 2D closed curves or extruding one 2D closed curve. This technique is useful for piping design or airplane wing design. The surface of evolution can be created by rotating the 2D curve about an axis. For example, this method is useful in creating such a symmetrical object as a cylindrical part. The sculptured surface is used to create more complex surfaces such as a ship's hull or an automobile's fender. There are many ways to create 3D surfaces such as polygon meshes and parametric cubic patches. Even though the algorithm of polygon meshes is simple, the major disadvantage is that this model represents a curve approximately by using polygon meshes. As a result, the exact definition of the curve is impossible. Parametric cubic patches identify the coordinates of points on a curved surface using three parametric equations. Each bicubic patch is constructed from the corresponding curve, and there are three typical types: *Ferguson*'s, *Bezier*, and the *B-spline methods* [33].

8.3.2.1 Ferguson's Curve

Ferguson's curve is defined by the two end points and the two tangents at the end points as shown in Fig. 8.2.

The Ferguson's curve is defined by the following equation [34]:

$$r(t) = \begin{bmatrix} P_0 & P_1 & T_0 & T_1 \end{bmatrix} \begin{bmatrix} 2 & -3 & 0 & 1 \\ -2 & 3 & 0 & 0 \\ 1 & -2 & 1 & 0 \\ 1 & -1 & 0 & 0 \end{bmatrix} \begin{bmatrix} t^3 \\ t^2 \\ t \\ 1 \end{bmatrix} \quad 0 \le t \le 1 \tag{8.1}$$

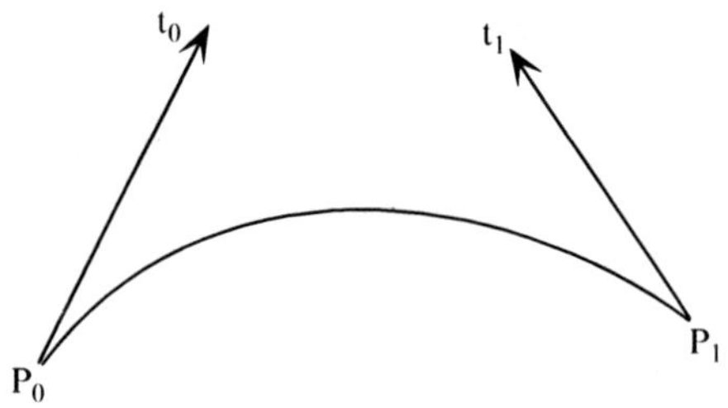

Fig. 8.2 A Ferguson's curve segment

The end points and the tangents at the end points are explicitly defined. Consequently, it can be easily used to design continuous and smooth curve segments. However, it is difficult to relate the tangent vectors to the overall shape of the curves.

8.3.2.2 Bezier's Curve

One solution to the problem of efficiently saving and redrawing a smooth curve was independently developed in the late 1960s by two French automobile engineers, Pierre Bezier, who worked for Renault automobile company and P. de Casteljau, who worked for Citroen. Originally, the solutions were considered industrial secrets, but Bezier's work was eventually published first. The curves that result using Bezier's method are called Bezier curves.

The method of Bezier curves allows us to efficiently store information about smooth shapes and to quickly stretch, rotate, and distort these shapes. Bezier curves are now commonly used in CAD work and in most computer drawing programs. They are also used to specify the shapes of letters of the alphabet in different fonts. By using this method, a computer and a laser printer can have many different fonts in many different sizes available without using a large amount of memory. The Bezier's curve $B(t)$ defined for the four points [33] P_0, P_1, P_2, and P_3 is shown in Fig. 8.3 and is represented as:

$$B(t) = \begin{bmatrix} P_3 & P_2 & P_1 & P_0 \end{bmatrix} \begin{bmatrix} 1 & 0 & 0 & 0 \\ -3 & 3 & 0 & 0 \\ 3 & -6 & 3 & 0 \\ -1 & 3 & -3 & 1 \end{bmatrix} \begin{bmatrix} t^3 \\ t^2 \\ t \\ 1 \end{bmatrix} \quad 0 \le t \le 1 \qquad (8.2)$$

where the four points P_0, P_1, P_2, and P_3 are called *control points* for the Bezier's curve.

Properties of Bezier Curves

Bezier curves have a number of properties that make them particularly useful for design work, and some of them are stated below:

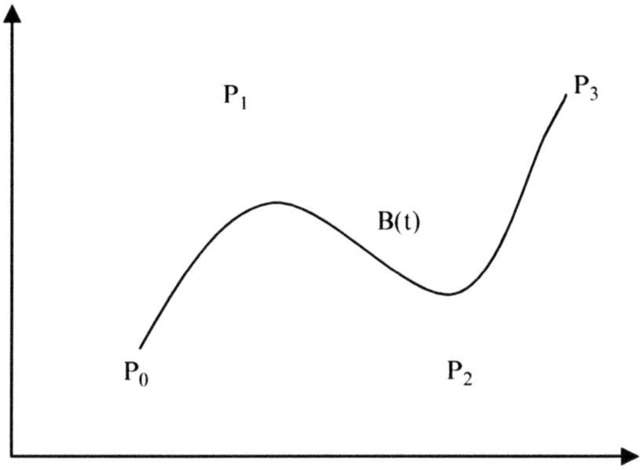

Fig. 8.3 Bezier's curve $B(t)$

1. $B(0) = P_0$ and $B(1) = P_3$. So the Bezier curve goes through the points P_0 and P_3. This property guarantees that $B(t)$ goes through specified points. If two Bezier curves need to be fit together, it is important that the value at the end of one curve matches the starting value of the next curve. This property guarantees the values of control points of the Bezier curves at their endpoints can be controlled by choosing appropriate values for the control points P_0 and P_3.
2. $B(t)$ is a cubic polynomial. This is an important property because it guarantees that $B(t)$ is continuous and differentiable at each point so its graph is connected and smooth at each point. It also guarantees that the graph of $B(t)$ does not distort too much between control points.
3. $B'(0)$ = slope of the line segment from P_0 to P_1: $B'(1)$ = slope of the line segment from P_2 to P_3. This is an important property because it means it is impossible to match the ending slope of one curve with the starting slope of the next curve to result in a smooth connection.
4. For $0 \le t \le 1$ the graph of $B(t)$ is in the region whose corners are the control points. Visually, property (4) means that if a rubber band is put around the four control points P_0, P_1, P_2, and P_3 as shown in Fig. 8.4, then the graph of $B(t)$ will be inside the rubber banded region. This is an important property of Bezier curves because it guarantees that the graph of $B(t)$ does not get too far from the four control points.

B-Spline Curve

B-spline curve (nonuniform) is a general case of Bezier's curve. It is usually defined in a Cox–deBoor recursive function [33].

$$r(t) = \sum_{i=0}^{L} N_i^n(t) p_i \quad t_i \le t \le t_{i+1},$$
$$N_i^n(t) = \frac{(t - t_i)}{(t_{i+n-1} - t_i)} N_i^{n-1}(t) + \frac{(t_{i+n} - t)}{(t_{i+n} - t_{i+1})} N_{i+1}^{n-1}(t), \tag{8.3}$$

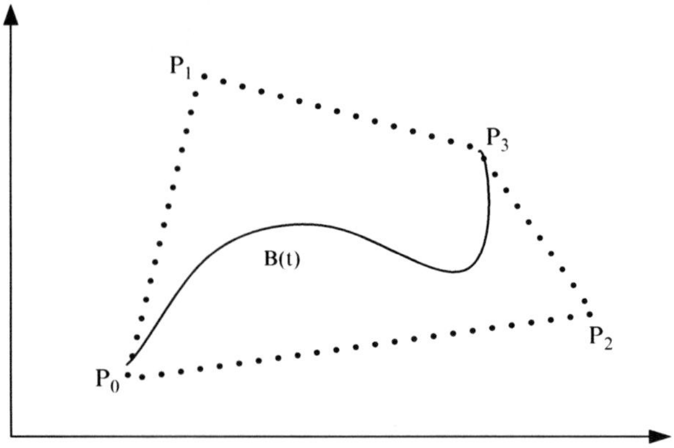

Fig. 8.4 Property 4 of Bezier's curve

where

$$N_i^1(t) = \begin{cases} 1, & t \in [t_i, t_{i+1}] \text{ and } \quad t_i < t_{i+1} \\ 0, & \text{otherwise} \end{cases},$$

and L is the number of control points and n is the degree of the curve.

The Bezier and B-spline methods have many common advantages. One is that control points can be adjusted in a predictable way making them ideal for use in an interactive CAD environment. Another advantage is that local control of the curve shape is possible. In this case, the Bezier method has a disadvantage because if the order of polynomial is increased by adding more control points when more control of shape is needed, the order of polynomial does not change in the B-spline method.

8.3.3 *Solid Modeling*

Solid modeling is the most comprehensive model from the manufacturing aspect. This model overcomes the drawbacks of both wireframe and surface models by providing a full solid definition of a 3D object. This model recognizes the solid object as a volumetric description, and it can also handle both surface and edge definitions of an object. Therefore, the mass property, volume property, surface property, and other important engineering properties can be derived in this model [24]. In this chapter, *solid modeling* is used to represent the designed object.

8.3.3.1 History and Overview

The origins of solid modeling are traced back to the key technological inventions of the 1950s – computer graphics and NC machining. These spawned the developments of computer-based geometric systems to aid in the description of the object's geometry, which is the main activity in design and manufacture of mechanical parts. This resulted research into the development of CAD and CAM. Preliminary systems used electronic drafting and wireframe models to represent the shape of 3D objects. Subsequent systems developed in the 1960s used polygonal and surface-based models which were utilized for a variety of applications in aerospace, marine, and automotive industries. Until the 1970s, these models were used in a broad manner. They were merely a collection of lower dimensional entities (polygons, surfaces, lines, curves, and points) put together in an unstructured manner to represent a real object. The developments in CAD/CAM led to the crucial questions about the uniqueness and the validity of these models, issues that until then were unimportant from the point of view of computer graphics and its applications. In the late 1970s, these issues were resolved by the Production Automation Project at the University of Rochester, where the term "solid modeling" was coined [35]. This group developed new mathematical models for representing solids and identified the relevant properties of an informationally complete representation. They also identified the mathematical operations that could be used to manipulate these models. After that several other models have been proposed along with different representation schemes [7]. In the 1980s, several solid modeling systems were developed and used in the commercial CAD/CAM world, including the automobile, aerospace, and manufacturing industries. Moreover, many advanced CAD/CAM applications of solid modeling have emerged such as feature- and constraint-based modeling, automatic mesh generation for finite element analysis, assembly planning including interference checking, higher dimensional modeling for robotics and collision avoidance, tolerance modeling, automation of process planning tasks, etc. Currently, solid modeling techniques have gained importance in the industry and are also actively pursued as a research field in academic institutions. Figure 8.5 provides a summary of solid modeling history technology evolution.

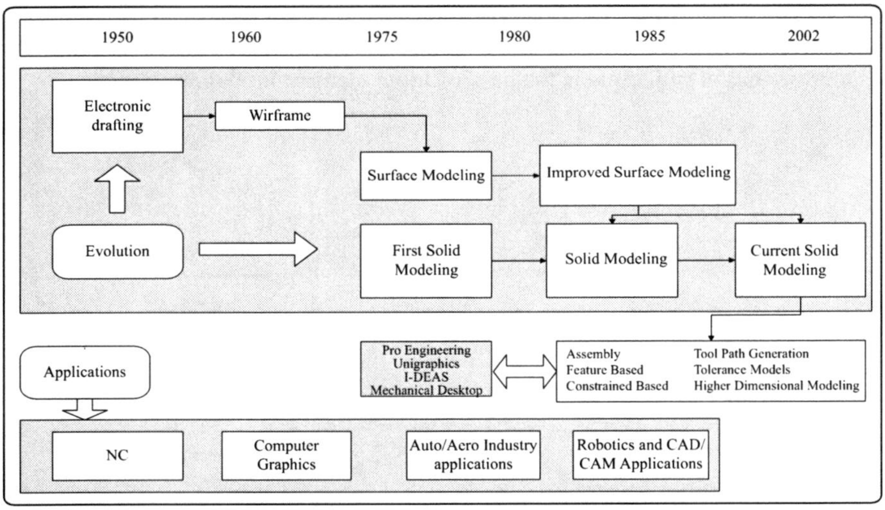

Fig. 8.5 Solid modeling technology evolution

To summarize, solid modeling provides a framework to model and represents an object's shape in the computer, and to perform operations. In addition, a group of application-independent geometric tools and algorithms is provided that can be used to query/analyze the model to obtain unambiguous results. These tools can be used or combined with other application-specific tools to perform the required task. The issues related to data structures and geometric algorithms, and their efficiency, reliability, and robustness also form an important aspect of solid modeling.

The academic effort in solid modeling utilizes several disciplines in many applications, including algebraic geometry and topology, differential geometry and topology, combinatorial topology, computer science, and numerical analysis. Another well established and closely related field is *computer-aided geometric design*, which concentrates on developing techniques for freeform surface design used to model curves and surfaces [31, 36].

8.3.3.2 Types of Solid Modeling

There are six basic types of solid modeling representations [33, 37] as shown in Figs. 8.6–8.12.

1. ***Boundary Representation***: The object is represented by the means of the bounded faces that enclose it. Each face is represented by its bounding edges and vertices.
2. ***Constructive Solid Geometry***: The object is represented by an ordered tree of the Boolean operations. Nonterminal nodes represent Boolean operators such as regularized union, intersection, or difference. Terminal nodes represent a primitive solid and transformation data.
3. ***Cell Decomposition***: The object is represented as the sum or union of a set of cells into which it is divided. The disjoint cells can be of any shape and size. This representation technique is the base of finite element modeling.

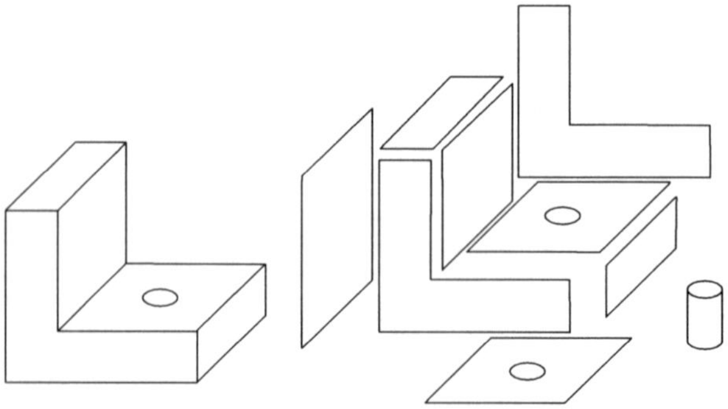

Fig. 8.6 B-Rep solid modeling representation

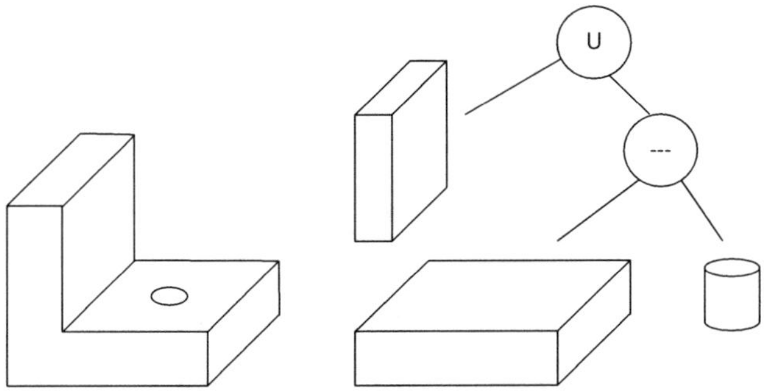

Fig. 8.7 CSG solid modeling representation

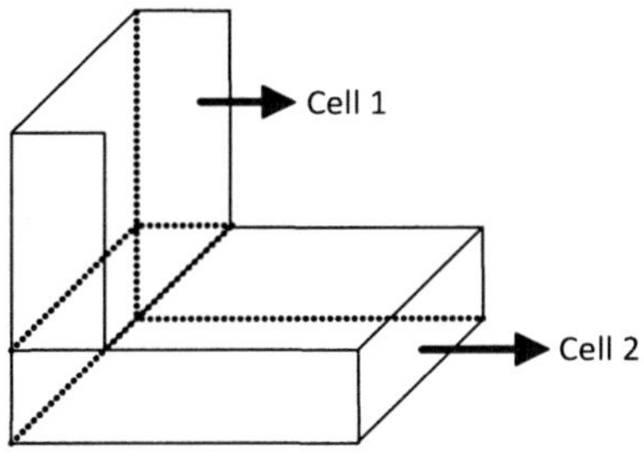

Fig. 8.8 CD solid modeling representation

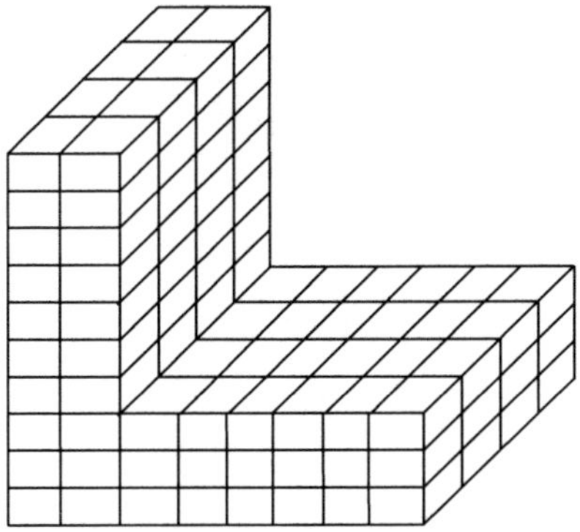

Fig. 8.9 SR solid modeling representation

4. ***Spatial Occupancy Enumeration***: The object is represented by a list of the cubical disjoint spatial cells that it occupies. This is the special case of the cell decomposition where the shape of the cells is cubical.
5. ***Primitive Instancing***: The object is represented by a set of solid primitives such as cuboid, cylinder, cone, etc. Each primitive is usually defined parametrically and located in space [38].

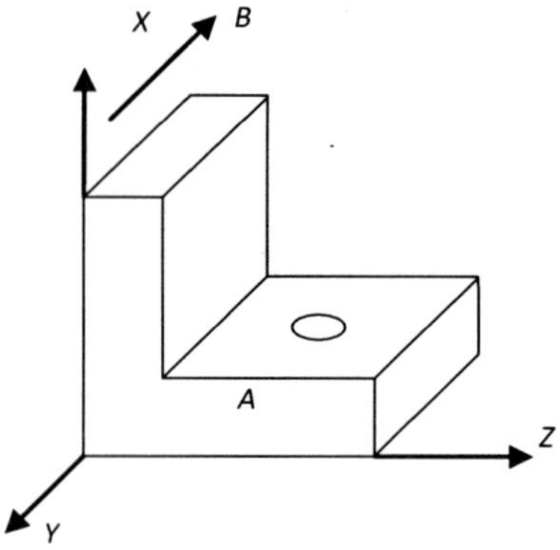

Fig. 8.10 Sweeping form of solid modeling representation

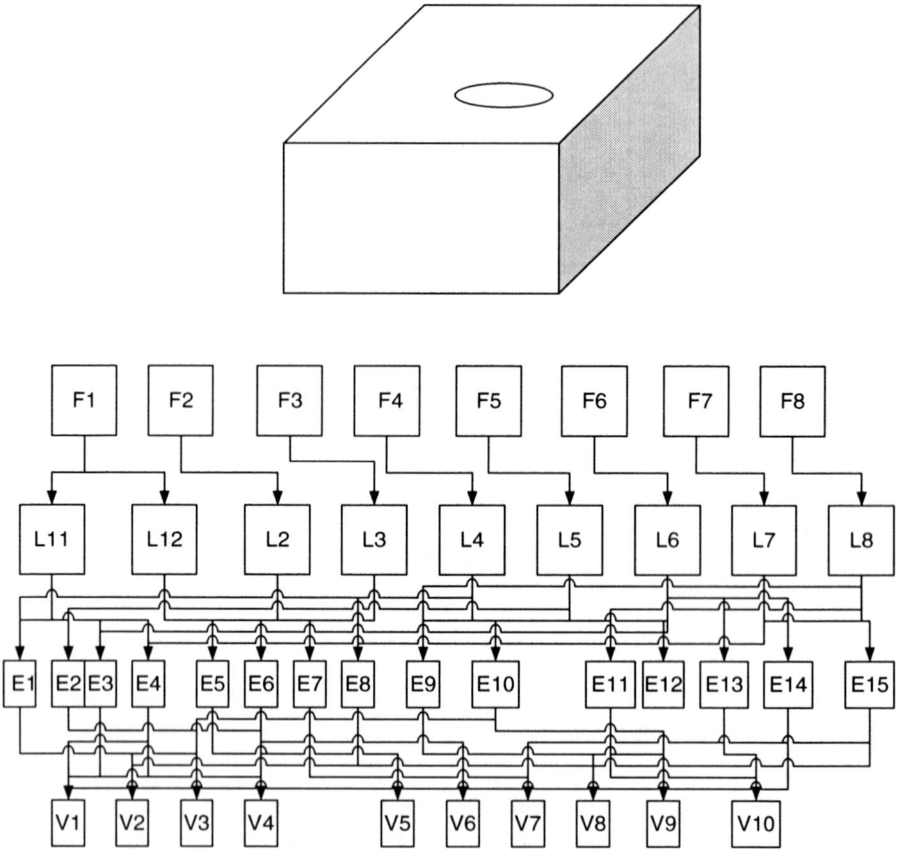

Fig. 8.11 A boundary representation [33]

6. ***Sweeping***: The object is represented by moving a curve or a surface along some paths. This method is useful to model constant cross-sectional parts and symmetrical parts [39].

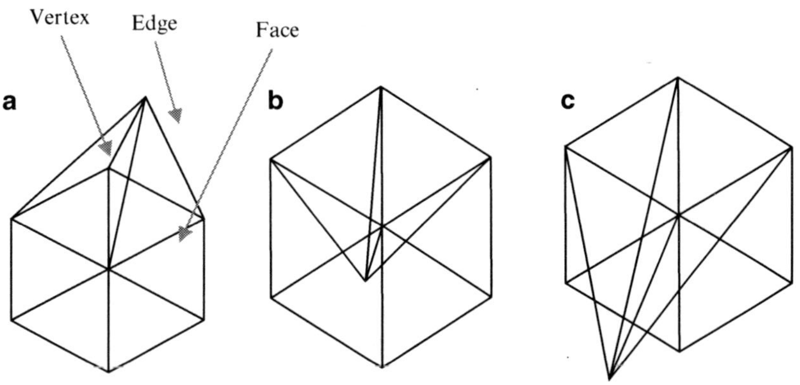

Fig. 8.12 Valid and invalid objects

8.4 Boundary Representation

Boundary representation is one of the solid modeling methods that are extensively used to create a solid model of a physical object and also used in a geometric data model [6]. B-rep describes the geometry of an object in terms of its boundaries, namely the vertices, edges, and surfaces which represent entities of two dimensions, one dimension, and zero dimension, respectively [40]. In order to represent a solid object by its surfaces, the orientation of each surface must be defined as the interior or exterior of the object. Typically, the inside is the material part and the outside is the void space. In a B-rep, a face must satisfy the following conditions:

1. A finite number of faces define the boundary of the solid.
2. The face of a solid is a subset of the solid's boundary.
3. The union of all the faces defines the boundary.
4. The face itself is a limited region or subset of the more extensive surface.
5. A face must have a finite area and is dimensionally homogeneous.

A solid is bounded by orientable surfaces, and one can then define a solid by a set of faces. Consequently, the enclosed volume of the object can be completely defined. Figure 8.11 shows the topology of the solid model which presents the object as a set of faces. Each face is bounded by edges, and each edge is bounded by vertices. To separate the points that are inside or outside of the object, the direction of the surface normal is used to attach the face. This information is typically encoded by numbering the edges in a sequence such that the right-hand rule defines the vector that points outward from the object.

Geometrical data include coordinates of the vertices and transformation (translation and rotation) and metric information, such as distances, angles, area, and volumes. Much of the research in this area uses B-rep to study the machining features [41–46]. However, in order to use B-rep, the topology relations should be stated between each set of surfaces in order to identify the surfaces in order to identify the shape and volume of the object. Parts can be classified as either *polyhedral* or *curved* objects.

A polyhedral object is presented by planar faces connected with straight edges which, in turn, are connected at vertices. A B-rep model is not limited to a planar surface. Different types of surfaces geometries can be described by different B-rep models which approximate curved surfaces as a combination of planar surfaces.

8.4.1 Euler's Formula

The validity of B-rep model can be checked by using Euler's formula [34]. Each edge in a topological valid B-rep is always adjacent to exactly two faces and is terminated by two vertices. Euler's formula states that for any topological valid solid, the number of vertices in addition to the number of faces is equal to the number of edges plus two.

$$V + F = E + 2 \tag{8.4}$$

where V is the number of vertices; F, number of faces; and E is the number of edges.

Topological correctness (all necessary geometric elements are presented and connected properly) does not guarantee a valid solid. If the geometry is altered, the topological model no longer represents the actual shape of the designed part. The geometric and topological definitions of the model may become contradictory, which results in an invalid object. Therefore, Euler's formula does not always give a valid object. For example, Fig. 8.11 shows the geometrical and topological definitions which are conflicting and therefore cause an invalid object. Figure 8.12a, b show the surface modification that result in a valid object. On the other hand, Fig. 8.12c shows a surface modification that results in an invalid object.

The result when two B-rep blocks are combined is shown in Fig. 8.13. Originally, there are eight vertices on each block and two are inside the combined object. Six new

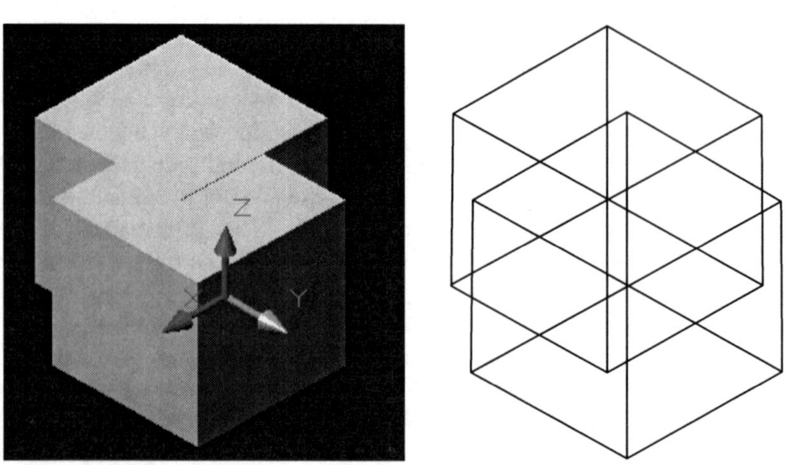

Fig. 8.13 Two cubes combined

vertices are created from the combined object. The total of vertices (V) is now 20. The basic two blocks have 24 edges, and all exist with the object. Six more edges are created by the union operation. The total number of edges (E) is 30 (24 + 6) and the total number of faces is 12. Therefore, after applying Euler's formula (20 + 12 = 30 + 2), it can be concluded that the resulting object has the right number of topological elements.

8.5 Constructive Solid Geometry

Another popular data format in solid modeling is CSG. The CSG is a modeling method that defines the complex solids as compositions of simple solid primitives [35]. In CAD systems, primitives can be either from a fixed set of predefined primitives built in the CAD system or defined primitives created specially by the users [40]. For example, the standard solid primitives shown in Fig. 8.14 are block, cylinder, sphere,

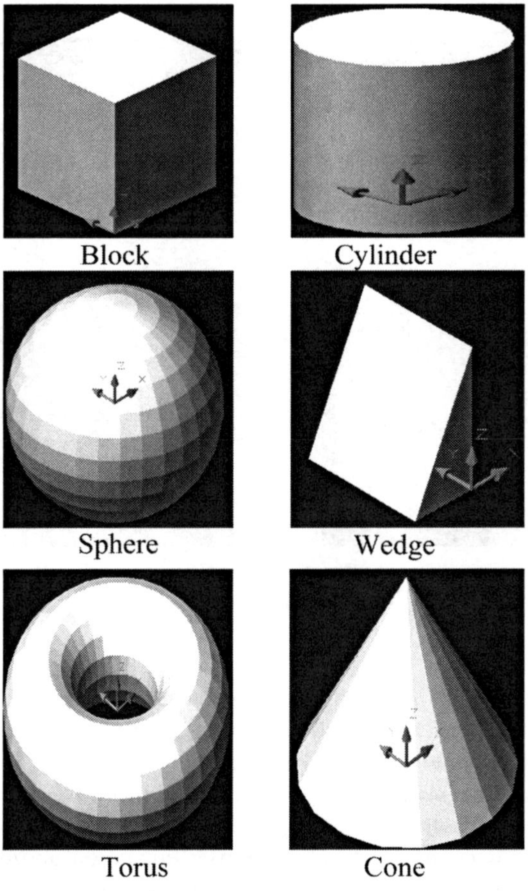

Fig. 8.14 Solid primitives

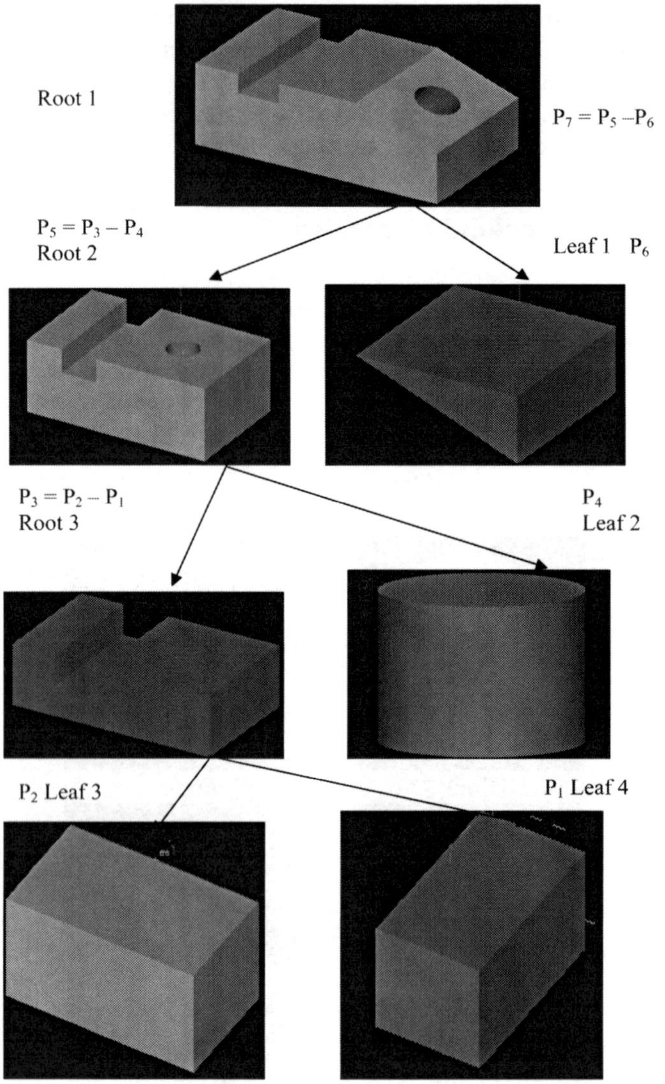

Fig. 8.15 Constructive solid geometry tree

torus, wedge, and cone. In the CSG model representation, the part design is represented by an ordered binary tree. The ordered binary tree consists of nodes, where these nodes are either terminal or nonterminal [47] as shown in Fig. 8.15.

Terminal nodes (leaf) can be either primitive or primitive with transformations. Nonterminal nodes (roots) are either Boolean operators (union U, difference –, and intersection ∩) or solid object motions (translation and/or rotation). Each nonterminal node represents a combination of two solid primitives in the tree [37].

8.6 Advantages and Disadvantages of Constructive Solid Geometry and Boundary Representation

Most solid modeling systems have adopted either CSG or B-rep as internal representations of solids. In this section, the advantages and disadvantages [4, 34] are summarized in Table 8.2. Two distinct approaches are used to incorporate features into the part modeling. These are *feature recognition* and *feature-based design*.

8.7 Feature Recognition and Design

Feature recognition involves the identification and grouping of feature entities from a geometric model. Such "postdefinition" of features can be done interactively or automatically. Usually, identified entities (i.e., the recognized features) are extracted from the model and additional engineering information such as tolerances and non-geometric attributes are then associated with the feature entities [8]. Figure 8.16 shows the procedure of feature recognition.

Table 8.2 Advantages and disadvantages of CSG and B-rep

Representation type	CSG	B-rep
Advantages	1. The data file of CSG is concise	1. The information is complete especially for adjacent topology relations
	2. CSG guarantees, automatically, that objects created by CSG are valid	2. Considerable number of methodologies developed for engineering analyses is based on B-Rep
	3. CSG is more user friendly	3. B-rep has the most refined geometric information compared to the others
	4. Algorithms for converting CSG into B-Rep have been developed	
Disadvantages	1. CSG database contains information about a solid in an unevaluated form	1. B-rep needs further feature extraction procedures to extract features from its face-edge-vertex database
	2. The validity of a feature of an object cannot be assessed without evaluating the entire tree	2. B-rep model does not provide any explicit information spatial constraints between features
	3. The tree is not unique for the same part design	3. The data structure of B-rep is complex compared with that of CSG

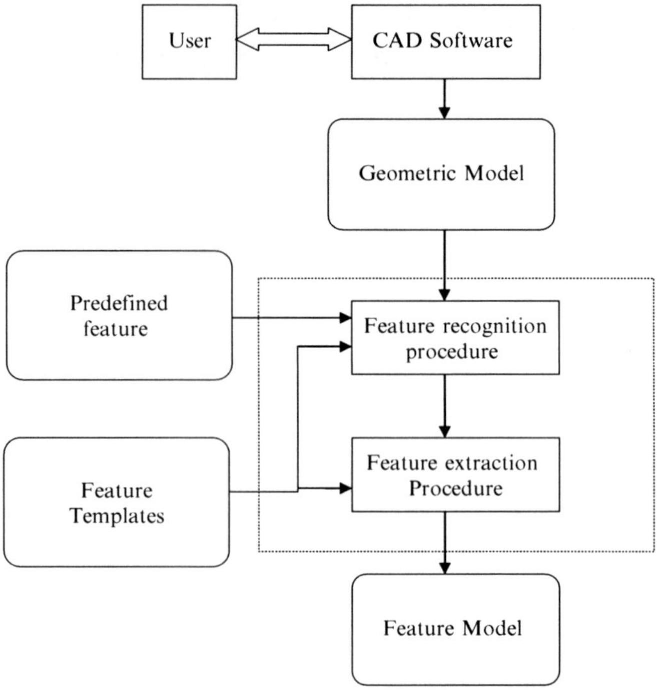

Fig. 8.16 Feature recognition procedure

8.7.1 Feature-Based Design

Design by features, or the so-called feature-based design (FBD), uses a library of 2D or 3D features as design primitives on the product modeling level. The use of features provides a more natural interface between design and solid model [18]. In mechanical design, for example, a designer can work directly with high-level entities such as a pocket rather than associating low-level entities in which vertices and edges form a pocket. Features allow the capability of providing additional information useful for process planning [48]. Since features reflect specific manufacturing processes, they assure the parts can be produced [17, 49, 50]. Figure 8.17 shows the design by feature procedure.

The integration of design by feature and feature recognition is applicable for design and manufacturing integration. A user can create a context-dependent representation that is specific and informative enough to satisfy the requirements of an application. While in the recognition process, the user builds the geometric model of a part that is successively mapped into the feature-based model [51]. Feature recognition is also used to directly recognize features from a CAD model of a part. It is easy to derive a geometric feature from the higher level entities of a part model, in which each feature is associated with a component of a solid model. On the other

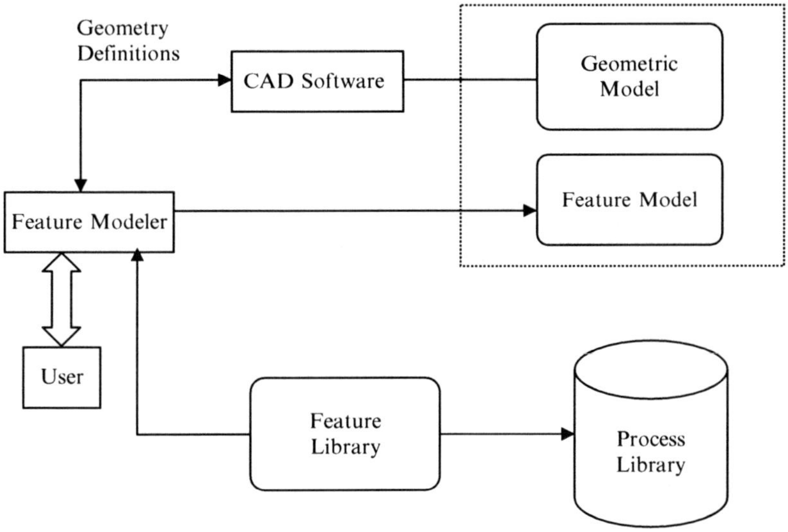

Fig. 8.17 Feature-based design

hand, to transform the geometric model into a feature representation is more difficult. There are many methods that are used to extract features from a CAD model. These methods are discussed in detail in the next chapter with literature review.

8.7.2 Feature Interactions

Interacting features are complex feature subparts on the part that cannot be recognized as any isolated standard feature. Shah and Mantyla [52] defined feature interactions as "feature interactions are intersections of feature boundaries with those of other features such that either the shape or the semantics of a feature are altered from the standard or generic definition."

An important distinction must be made between intersecting and interacting features. Interacting features are defined in the context of design by features in which design features are used to build the model [53]. The addition (or subtraction) of a feature to the part can result in the creation of several new features due to interaction between the features. On the other hand, intersecting features are present on conventional representations and not generated by feature operations [18].

Algorithms are needed to extract both intersecting and interacting features, but operating on different input representations. Feature interactions also take on a different meaning in process planning. If the set of machining features is not machined in the correct sequence, one feature can interact with another feature already on the part, thus generating an incorrect part.

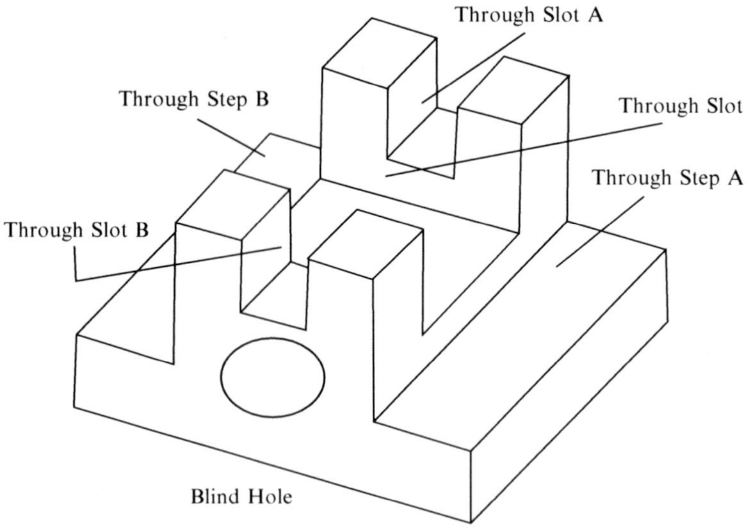

Fig. 8.18 Interacting features

Table 8.3 Possible feature interactions

Feature	Type
Simple	Blind hole
Interacting	Through step A interacting through slot
	Through step B interacting through slot
	Through slot A interacting through slot
	Through slot B interacting through slot

The feature interaction problem can be treated as a problem of feature relationships. There is no existing feature recognition system that could recognize all types of 3D solid primitive features and their various interacting combinations [54, 55]. The remainder of this section defines the problem of interacting features with an example. Interacting features are very common on manufacturing designed objects. As shown in Fig. 8.18, there are six features on the raw material of the workpiece, and the interaction between the features is listed in Table 8.3.

8.8 Summary

In this chapter, feature definition, wireframe modeling, surface modeling, B-rep, CGS, and definition of interacting features were discussed. Techniques and approaches that are used to extract and recognize the features are discussed in Chap. 9.

8.9 Review Questions

8.1. Define the term "feature."
8.2. Classify the manufacturing features.
8.3. What are the advantages of feature-based modeling in manufacturing applications?
8.4. Describe briefly the definition of process planning. Discuss its primary approaches.
8.5. Construct a diagram showing how a feature-based manufacturing knowledge repository is implemented in the process planning.
8.6. Distinguish between VPP and GPP.
8.7. What are the phases of assembly planning?
8.8. Define geometric modeling.
8.9. What are the main functions of geometric modeling?
8.10. Discuss the three different types of geometric modeling techniques.
8.11. What is the disadvantage of using wireframe model in automatic manufacturing feature recognition?
8.12. What are the differences between a surface model and a solid model?
8.13. Discuss the three different types of surface modeling.
8.14. Describe the Bezier curve and list its properties.
8.15. Why is solid modeling considered the most comprehensive model from the aspect of manufacturing?
8.16. Explain the basic types of solid modeling representations.
8.17. Describe briefly B-rep.
8.18. Construct the B-rep for the following component:

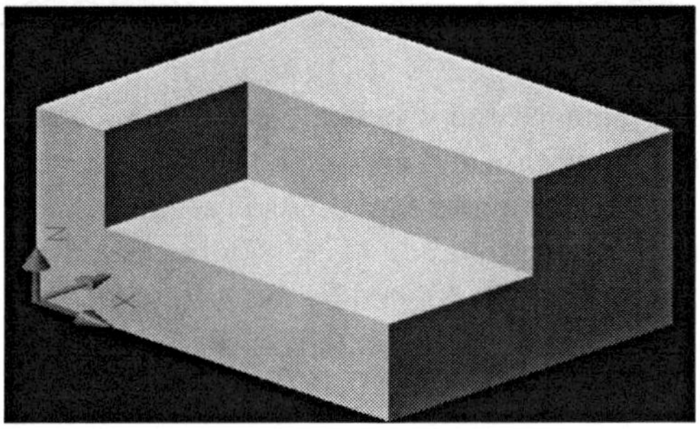

8.19. By using Euler's formula, is the following design valid?

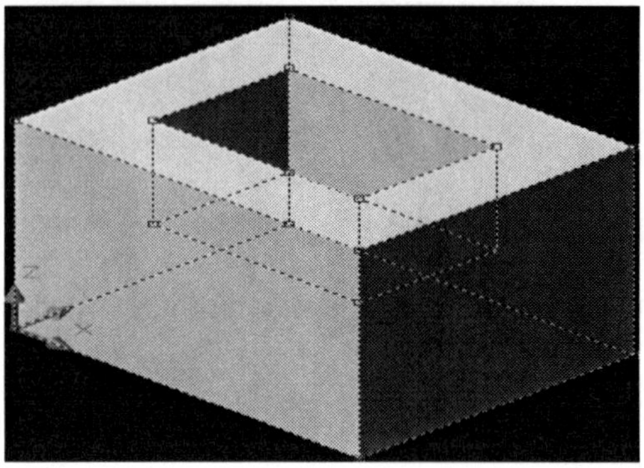

8.20. Explain CSG.

8.21. Construct a CSG tree for the following component:

8.22. What are the advantages and disadvantages of CSG and B-rep?

8.23. Define feature recognition and explain its procedure.

8.24. Define design by features and explain its procedure.

8.25. What are feature interactions?

8.26. The Ferguson curve can be defined by the following data:
Two points: $P_0 = (0, 0, 0)$, $P_1 = (3, 0, 0)$
Two tangents: $T_0 = [1\ 3\ 0]$, $T_1 = [1\ {-3}\ 0]$
Develop an equation representing the curve.

8.27. The Bezier curve can be defined by the following data:
$P_0 = (2, 0, 1)$, $P_1 = (3, 0, 4)$
$P_2 = (7, 0, 4)$, $P_3 = (11, 0, 1)$
Develop an equation representing the curve.

8.28. The B-spline curve can be defined by the following data:

$L = 5$ $n = 3$ $t_0 = 0$ $t_1 = 1$ $t_2 = 3$
$t_3 = 4$ $t_4 = 5$ $t_5 = 6$ $t_6 = 8$
$P_0 = (0, 0)$ $P_1 = (2, 2)$ $P_2 = (2, 1)$ $P_3 = (4, 0)$
$P_4 = (6, 1)$ $P_5 = (5, 2)$

Develop an equation representing the curve.

References

1. Merat FL, Radack G (1992) Automatic inspection planning within a feature based CAD system. Robot Comput Integr Manuf 9(1):61–69.
2. Chandra RD, Ghosh K (1999) Feature-based modeling and neural networks-based CAPP for integrated manufacturing. Int J Comput Integr Manuf 12(1):61–74.
3. Case K, Harun WA (2000) Feature-based representation for manufacturing planning. Int J Prod Res 38(17):4285–4300.
4. Chiyokura H (1988) Solid modeling with designbase: theory and implementation. Addison-Wesley, Reading.
5. Santosh K, Shanker K, Lal GK (2003) A generative process planning system for cold extrusion. Int J Prod Res 41(2):269–296.
6. Sheu LC, Lin JT (1993) Representation scheme for defining and operating from features. Comput Aided Des 25(6):33–347.
7. Somashekar S, Michael W (1995) An overview of automatic feature recognition techniques for computer-aided process planning. Comput Ind 26(1):1–21.
8. Sreevalsan PC, Shah JJ (1992) Unification of form feature definition methods. Intelligent Computer Aided Design, Browns DC, Waldron M, Yoshikawa H eds 83–106.
9. Pratt MJ, Wilson PR (1985) Requirements for support of form features in a solid modeling system. Final report, CAM-I Report R-85-ASPP.-01.
10. Dixon J, Cunningham J (1988) Design with features: the origin of features. In: Proc ASME computer in engineering conference, San Francisco.
11. Mantyla M, Dana N, Jami S (1996) Challenges in features-based manufacturing research. Commun ACM 39(2):77–85.
12. Ahmad N, Haque AFM, Hasin AA (2001) Current trend in computer aided process planning. In: Proceedings of the 7th annual paper meeting and 2nd international conference of the Institution on Engineering, Dhaka, Bangladesh, 1(10):81–92.
13. Madurai SS, Lin L (1992) Rule-based automatic part feature extraction and recognition from CAD Data. Comput Ind Eng 22(1):49–62.
14. Kao C-Y, Kumara SRT, Kasturi R (1995) Extraction of 3D object features from CAD boundary representation using super relation graph method. IEEE Trans Patt Anal Mach Intell 17(12):1228–1233.
15. Nasser B, Hamid P, Leep HR (1996) A prototype solid modeling based automated process planning system. Comput Eng 31(1/2):169–172.

16. Alan C, Shou L (1998) A volume decomposition approach to process planning for prismatic parts with depression and protrusion design features. Int J Comput Integr Manuf 11(6):548–563.
17. Hsin-Chi C, Wen FLU (2002) Machining process planning of prismatic parts using case-based reasoning and past process knowledge. Appl Artif Intell 16(1):303–331.
18. Wen FL, Ronak M (1997) Feature-based design in an integrated CAD/CAM system for design for manufacturability of machining prismatic parts. Concurr Prod Des Environ Consc Manuf 5(1):95–112.
19. Pham D, Gologu C (2002) A computer aided process planning system for concurrent engineering. Proc Instn Mech Engrs, part B, J Eng Manuf 215(1):1177–1131.
20. Groover MP (2001) Automation, production systems, and computer-integrated manufacturing. Prentice-Hall, Englewood Cliffs.
21. Beno B (2003) Manufacturing design, production, automation and integration. Marcel Dekker, New York.
22. Goh TN, Xie M (2004) Improving on the six sigma paradigm. TQM Mag 16(4):235–240.
23. Bartholomew O, Nnaji BO, Liu HC (1990) Feature reasoning for automatic robotic assembly and machining in polyhedral representation. Int J Prod Res 2(3):517–540.
24. Hoffmann C, Rossignac J (1996) A road map to solid modeling. IEEE Trans Vis Comput Graph 2(1):45–54.
25. Navazo L, Ayala D, Brune P (1986) A geometric modeler based on the exact octtree representation of polyhedra. Comput Graph Forum 5:91–104.
26. Yong Y, Tang R (2000) Historical procedures and G-DSG method based manufacturing planning. Chin J Aeronaut 13(2):123–128.
27. McMahon C, Jimmie B (1998) CAD/CAM: principles, practice and manufacturing management. Addison-Wesley, Reading.
28. Vosniakos G (1998) Conversion of wireframe to ACIS solid models for 2½-D engineering components. Int J Adv Manuf Technol 14(3):199–209.
29. Lin AC, Lin S, Lin Y, Cheng SB (1997) Extraction of manufacturing features from a feature-based design model. Int J Prod Res 35(12):3249–3288.
30. Kobbelt LP, Botsch M, Schwanecke U, Seidel H (2001) Feature sensitive surface extraction from volume data. In: Proceedings of the ACM siggraph conference on computer graphics, 57–66.
31. Sundaarajan V, Wright PK (2004) Volumetric feature recognition for machining components with freeform surfaces. Comput Aided Des 36(1):11–25.
32. Chasen SH (1986) Principles of geometric modeling. CIM Technol 15–18.
33. Wang HP, Wysk RA (1998) Computer aided manufacturing, 2nd ed. Prentice-Hall, Upper Saddle River.
34. Zeid I (1991) CAD/CAM theory and practice. McGraw-Hill, New York.
35. Reqicha A, Voelcker H (1982) Solid modeling: a historical summary and contemporary assessment. IEEE Comput Graph Appl 2(2):9–24.
36. Natekar D, Zhang X, Subbarayan G (2004) Constructive solid analysis: a hierarchal, geometry-based meshless analysis procedure for integrated design and analysis. Comput Aided Des 36(5):473–486.
37. Mantyla M (1988) Introduction to solid modeling. Computer Science Press, Rockville.
38. Requicha AAG (1980) Representations of rigid solids: theory, methods, and systems. Comput Surv 12(4):437–464.
39. Taylor D (1992) Computer aided design. Addison-Wesley, Readings.
40. Chang T-C (1990) Expert process planning for manufacturing. Addison-Wesley, Reading.
41. Gossard DC, Zuffante RP, Sakurai H (1988) Representing dimensions, tolerances and features in MCAE system. IEEE Comput Graph Appl 51–59.
42. Henderson MR, Anderson DC (1984) Computer recognition and extraction of form feature: a CAD/CAM link. Comput Ind 5(4):329–339.
43. Kayacan MC, Celik SA (2003) Process planning system for prismatic parts. Integr Manuf Syst 14(2):75–86.

44. Nagaraj HS, Gurumoorthy B (2002) Machinable volume extraction for automatic process planning. IIE Trans 34(1):393–410.
45. Roy U, Liu CR (1988) Feature-based representational scheme of a solid modeler for providing dimension and tolerancing information. Robot Comput-Integr Manuf 4(3/4):335–354.
46. Wang N, Ozsoy TM (1991) A scheme to represent features, dimensions, and tolerances in geometric modeling. J Manuf Syst 10(3):233–240.
47. Waco DL, Kim YC (1994) Geometric reasoning for machining features using convex decomposition. Comput Aided Des 26(6):477–489.
48. Shpitalni M (1990) CSG representation as a basis for extraction of machining features. CIRP Annu Manuf Technol 40(1):157–160.
49. Aslan E, Seker U, Alpdemir N (1999) Data extraction from CAD model for rotational parts to be machined at turning centers. Turk J Eng Environ Sci 23(5):339–347.
50. Rong KL, Taur BW, Shyur HJ (1991) A two-stage feature-based design system. Int J Prod Res 29(1):133–154.
51. Martino TD, Falcidieno B, Giannini F, Hassinger S, Ovtcharova J (1994) Feature based modeling by integrated design and recognitions approaches. Comput Aided Des 26(8):646–653.
52. Shah JJ, Mantyla M (1995) Parametric and feature-based CAD/CAM-concepts, techniques, and applications. Wiley, New York.
53. Tseng YJ, Joshi SB (1998) Recognizing of interacting rotational and prismatic machining features from 3D mill-turn parts. Int J Prod Res 36(11):3147–3165.
54. Allada V, Anand S (1996) Machine understanding of manufacturing features. Int J Prod Res 34(7):1791–1820.
55. Huikange K, Nandakumar M, Shah J (2002) CAD/CAM integration using machining features. Int J Comput Integr Manuf 15(4):296–318.

Chapter 9
Feature Extraction Techniques

Abstract This chapter presents a brief review of the previous work on the related topics of feature representation and recognitions. The first section describes previous research efforts in the area of feature representation. Previous research in the area of feature recognition is described in the second section. In third section, a methodology for feature analysis and extraction of prismatic parts for CAM applications is developed and presented. This approach aims to achieve the integration between CAD and CAM.

Keywords Feature representation • Feature extraction • Integration

9.1 Feature Representation

The recognized features and their relationships, which are the result of the feature recognition process, are used to recreate the designed object. A feature has a higher level of abstraction than a geometric solid primitive in the traditional solid model [56]. Manufacturing information, which is related to a specific application activity, can be incorporated with the model. Features not only represent the form but should also contain information on its functions and interrelationship with other features [50].

Feature representations by a boundary representation (B-rep) and a constructive solid geometry (CSG) are the most broadly used representation methods, because CSG and B-rep are more commonly used and can specify the actual spatial occupancy of an object. The CSG of the solid model is specified with a set of Boolean operations and a set of 3D solid primitives. On the other hand, the B-rep of a solid model contains information about faces, edges, and vertices of a surface model and at the same time includes topological information that defines the relationship between the faces, edges, and vertices. In the next subsections, a literature survey in the area of feature representations is presented.

A.K. Kamrani and E.A. Nasr, *Engineering Design and Rapid Prototyping*,
DOI 10.1007/978-0-387-95863-7_9, © Springer Science+Business Media, LLC 2010

9.1.1 Feature Representation by B-rep

Henderson and Anderson [17] used production rules in the PROLOG format structure to identify part features. A 3D solid model in B-rep is converted into PROLOG facts. The pattern matching mechanism of PROLOG was used to search the part's facts and try to satisfy the feature rules. Because this system can include only a limited number of features, it cannot recognize all possible machining features.

Kang and Nnaji [22] developed a feature representation and classification model for an automatic process planning system. The representational scheme is based on B-rep. The part feature attributes are defined by the geometry of the workpiece and the function that they serve. Based on this, a generalized feature definition is presented and schemes for classification and representation of features in the domains of mechanical assembly and sheet metal fabrication are developed and presented.

Srikantappa and Crawford [52] developed a feature representation using a vocabulary of structural and geometric primitives. The structural primitives are defined as important partial areas of features and the geometric primitives are defined as various spatial configurations of structural primitives or symmetry axes of features. The representation includes a set of structural and geometric primitives, along with the methodology for modeling features with interacting and interfeature relationships. Semantic networks implemented as graphs were used for representation of intermediate geometry. Nodes of the graph represent features and the spatial relationships are represented as labeled arcs between nodes.

Liu et al. [28] developed an automatic feature extraction system for prismatic parts. The parts were generated by B-rep. In this system, the automatic feature extraction system used an object-oriented data structure (OODS) that was designed to allow modeling and manipulating of manufacturing parts. The system consisted of three algorithms that translate the data in a CAD file into manufacturing features that could be used in CAPP. The algorithms were the concave edge test, the oblique convex feature extraction algorithm, and the complex convex minimal enclosing box feature extraction procedure. The system also generated a classification scheme that identifies features that were related to manufacturing. However, the features in this system were very limited and the rotational parts were not considered.

Tseng and Joshi [54] developed a method for feature recognition of mill-turned parts. B-rep was used to create rotational and prismatic components. This method was based on machining volume generation approach to recognize and classify features. The feature volumes were generated by sweeping boundary faces along a direction determined by the type of machining operations. In this approach, first, the part was segmented into several rotational machining zones. Next, the prismatic features were recognized on the basis of intermediate rotational shapes using volume decomposition and maximal volume sweeping and reconstruction. Classification of prismatic features was done by using the face adjacency relationships, while classification of rotational parts was performed by using profile edge patterns.

Aslan et al. [5] developed a feature extraction module only for rotational parts to be machined at turning centers. In this research, the DXF file format was used to extract 2D, which was represented by B-rep, for rotational parts. This extraction module was a part of an expert system (ES) called ASALUS. ASALUS was designed to manage the life cycle of rotational parts from design all the way to production by performing process planning using a generative approach and applying postprocessing for two different CNC lathes. Prismatic features and the intersecting features were not involved in this module.

Nagaraj and Gurumoothy [34] described an algorithm to extract machinable volumes that need to be removed from a stock. This algorithm can handle both prismatic and cylindrical components by using B-rep model. The machinable volumes can be used to automate process planning and NC tool path generation. The algorithm identified the cavity volumes in the part with respect to outermost faces in the part and filled them with the appropriate primitive volume to obtain the stock from which the part can be realized.

Kayacan and Celik [25] developed a feature recognition system for process planning for prismatic parts. This system was achieved with the B-rep modeling method to give vectoral direction knowledge and adjacent relationships of surface using the STEP standard interface program. The feature recognition system was prepared on IPM compatible PC by using C++ programming language.

Fu et al. [14] developed a feature extraction approach that had the ability to be compatible with different CAD systems through STEP and IGES formats. Their feature classification was related to product design and manufacturing. In this book, multilevel feature taxonomy was proposed based on the feature geometry and topological characteristics. The lowest level features in the feature taxonomy were equivalent to the features defined in STEP AP224. The other levels of features were defined such that the entire feature identification approach could be elaborated and implemented. The approach and the detailed algorithms were implemented in the CAS.CADE object library environment in B-rep. The features were restricted to simple form features but the interaction between them was not involved.

9.1.2 Feature Representation by CSG

Requicha and Chan [38] developed a system for representing surface features in a solid modeler based on CSG and for associating tolerances and other attributes with such features. Their approach treats tolerances as properties or attributes of object features that are part of the object's surface or topological boundary. They developed a graph structure called the variation graph (V graph) to represent these features and attributes.

Shah and Roger [46] developed an integrated system for form features, precision features, and material features. The solid representation of the form features is stored as a feature-producing volume (CSG tree) and Boolean operators. An object-oriented programming approach is used to represent the feature descriptions. The feature relationship graph is created at the top level of the model, where both the adjacency and the parent–child dependency of the form features are stored.

Perng et al. [36] developed a method that converts a pant's CSG tree representation into its equivalent destructive solid geometry (DSG) tree representation, and the types of machinable features are then identified from the DSG tree. A DSG tree can be defined as a special kind of CSG tree in which all geometric operations are of a difference type. The disadvantage of this method is that the CSG-rep model does not uniquely represent a given shape, causing nonstandard representation of a shape possibly producing erroneous process plans.

Sheu [49] developed a computer-integrated manufacturing system for rotational parts. The CSG representation was used to construct the solid geometry of the part. The parametric design and feature-based solid model were used to specify the manufacturing information required to the proposed system. In this system, the boundary of a solid model could be transferred directly into the line and arc profiles. The boundary profiles were the output to a drawing with the DXF format. The part model was created by using the cylinder, the cone, the convex arc, and the concave arc as primitives. This system had the ability to convert the wireframe part model into a CSG representation by the feature recognition approach. Only a limited number of features could be recognized by this system. Prismatic components and their interactions were not considered.

9.1.3 Feature Representation by B-rep and CSG (Hybrid Method)

Roy and Liu [40] presented the framework of an integrated design and manufacturing system: a hybrid CSG and B-rep coupled with a structured faces adjacency graph representation. The main idea behind the representational scheme is to capture topological knowledge and geometrical relationships among the lower level entities and to incorporate the object attributes and their semantics.

Gossard et al. [16] presented a method for explicitly representing dimension tolerances and geometric features in solid models. They created a hybrid CSG/B-rep structure using the relative-position operator (RPO). The RPO is used to represent dimension and tolerance, explicitly. The scheme can support dimension driven geometry and other engineering applications.

Nnaji and Liu [6] described a system that extracts features relevant to assembly and machining from a CSG-based CAD system with the ability to transform a CSG structure into a B-rep. The system transforms CAD information into PROLOG rule format structure. The PROLOG database was used to determine concave edges, convex edges, concave faces, and convex faces. In finding groups of concave faces, a concave face is added to the group if it shares a concave edge with one of the group's members. A depth first search among all the faces of a pan is used to identify all of its concave face groups. Each concave face group is assumed to form a concave feature of the part. Production rules were used to differentiate between the machining features. The recognizable features are also limited, and feature interaction is not considered.

Wang and Ozsoy [61] presented a system for representing features, dimensions, and tolerances in geometric models. The representational scheme is created by using a combination of CSG and B-rep. Features are classified as either form features or primitive features. A generic feature library is established for the parametric representation of the form features. Form features are actually built of primitive features. Form features and primitive features are both considered in the hybrid representation scheme. Dimensions and tolerances are represented by using a constraining operator that describes the dimensional relationship within an object.

Rozenfeld and Kerry [39] tried to present a solution for automated process planning for parametric parts, which at the same time meets the planning requirements of any other parts and products in a real industrial environment. The parts were presented in their CAPP system by a parametric CAD module. In this representation, the parts were represented by parameters such as the ones described through features and their attributes. It was not clear which type of features that CAPP system can be developed.

9.2 Feature Recognition Techniques

Feature recognition algorithms are heavily dependent on the type of data provided by the geometric modeler of the CAD system. The CAD modeler determines the method by which the user can create a design drawing and the data structure for storing the CAD model in the computer memory. The data structure is determined by the types of information to be stored in the CAD database. Many researchers attempted to extract features from the 3D CAD model of a part model [2]. The feature recognition algorithms can be classified by their approaches to the problems as follows:

1. The syntactic pattern recognition approach
2. Logic-based approach
3. Graph-based approach
4. ES/artificial intelligence approach
5. Volume decomposition and composition approach
6. 3D feature recognition from 2D features approach

9.2.1 The Syntactic Pattern Recognition Approach

The syntactic pattern recognition approach uses the semantic primitives for part analysis. The structure information is used to create a description and a classification of the part. Features can be identified by parsing the feature against the object descriptions in the grammar.

In pattern recognition approach, the symmetric part designs are used mainly for feature recognition [53]. The features of the part design are represented in this approach by some semantic (pattern) primitives as shown in Fig. 9.1a. These semantic (pattern) primitives are written in a string form in Fig. 9.1c. A set of grammars and rules is used to define a particular character string. A parser is used to decide

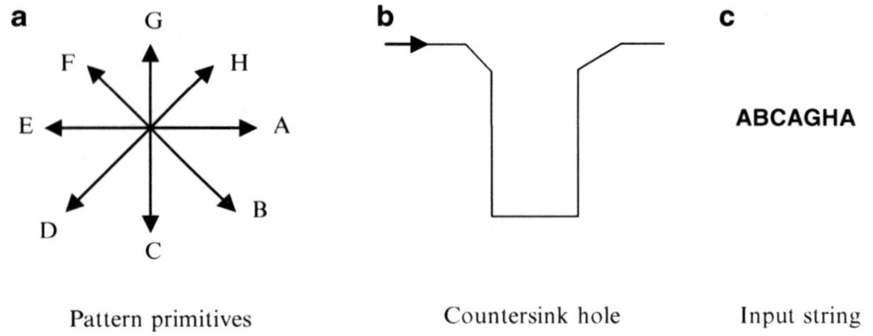

Fig. 9.1 Syntactic pattern recognition approach

if the input character string is true or not. The main components of the syntactic pattern recognition method are an *input string*, a *pattern grammar*, and a *parser*. The input character string can be represented by a combined set of pattern primitives [7]. Each feature's edge can be replaced by a directed pattern primitive. The sequences of the input string primitives are the same as the feature edge directions. For example, the input string **ABCAGHA** represents the direction of edges that form a countersink hole shown in Fig. 9.1b.

The pattern grammar is a set of production rules. The production rules are used to represent the available various types of the input strings that can be recognized by the system. An example of a production rule for a countersink hole shown in Fig. 9.1b is as follows:

IF **the input sting is ABCAGHA**
THEN **the feature is a countersink hole**

A parser can be defined as a process of determining if an string of a model can be generated by the pattern grammar or not. This means a hierarchical analysis of a sentence that is sometimes called syntax analysis. The parser searches for the elements of the input string and tries to match it with the appropriate production rule.

Staley et al. [53] developed a system to recognize holes from a 3D solid model. A string representation was used as an input cross-section of the part. Li and Bedworth [26] developed a system to detect rotation from an initial graphics exchange specification (IGES) file for an automated process planning system. A pattern string is used as a 2D representation associated with the part geometry data. The feature algorithm is performed on the string of patterns to match predefined features.

9.2.2 The Logic-Based Approach

The logic rules are used with the B-rep and the CSG representation modeling approaches for feature recognition. In the B-rep model, the part is described in its lower level such as faces, edges, and vertices [20]. A higher level is needed for accomplishing part recognition where this higher level is called feature.

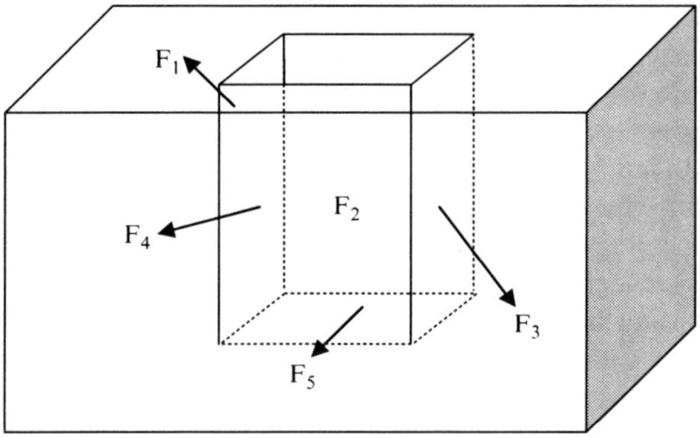

Fig. 9.2 A pocket feature

Each feature can be recognized by a separate rule. Features are recognized one by one in sequence until the whole set of features of the solid model is achieved. For example, to recognize a pocket feature that is shown in Fig. 9.2, the following rule can be used, where adjacency of faces means the existence of common edges:

IF **face F_1 is adjacent to face F_2 and**
face F_2 is adjacent to face F_3 and
face F_3 is adjacent to face F_4 and
face F_5 is adjacent to faces F_1 F_2 F_3 F_4 and
angle between F_1 and F_2 is <180 (concave),
and
angle between F_3 and F_4 is <180 (concave)
THEN **faces F_1, F_2, F_3, F_4, and F_5 form a pocket feature**

In the CSG representation model, the part design is represented in terms of a solid model. This solid model consists of solid primitives (block, cylinder, sphere, torus, wedge, and cone) where the solid primitive (feature) is considered as a higher level of representation [35, 41]. For example, if the Boolean operation between two solid primitives is subtraction and the subtracted solid primitive is a cylinder, then the recognized feature is a hole according to the dimensions and position of the cylinder. The following rule can be implemented for hole recognition:

IF **the Boolean operation is Subtraction,**
and the dimensions of the subtracted solid
primitive are less than the solid model
and the subtracted solid primitive is a cylinder
THEN **the feature is a hole**

Munns et al. [33] developed a methodology to extract the features based on rule-based approach. The CSG representation was used to represent the objects. In this approach, a methodology was developed to examine the spatial position relationships. These relationships were classified into two categories: relationships among geometric elements (two vertices, a vertex, and an edge) and relationships among primitive objects (two blocks, a block, and a cylinder). The extracted features were restricted to a limited number of features such as slot and step. The interaction between the features was not discussed. Furthermore, the example that was demonstrated to explain the approach was related to the construction (a layout of a building) which was not compatible with the manufacturing domain.

Sharma and Gao [48] developed a process planning system using the international standard for the exchange of the product model data (STEP). The developed system can generate process plans without any user interaction. The system consisted of decision logic stored in an external database that makes it generic and compatible with any manufacturing applications. The system also accepted a feature tree generated using the feature model editor which had been developed as an add-on to the system. The system handled only single-piece machined (prismatic) parts.

9.2.3 Graph-Based Approach

Topological relationships between geometric entities of a CAD model can be represented using a graph approach that simplifies the analysis of the geometric entities for feature recognition. Simple heuristics can be applied to the graph to identify features. In this approach, the B-rep of a part is translated into a graph whose nodes and arcs represent the information of a part. Feature recognition is achieved by finding a subgraph from the shape of a designed object where the subgraph is isomorphic to the feature graph.

As shown in Fig. 9.3, the geometrical information of the solid object can be represented with respect to its faces, edges, or vertices. In this model, if an object is represented by its faces, each face of that object can be described by its neighbor faces, its boundary edges, and its boundary vertices as shown in Fig. 9.4. The graph-based model can also be represented in the form of an adjacency matrix as

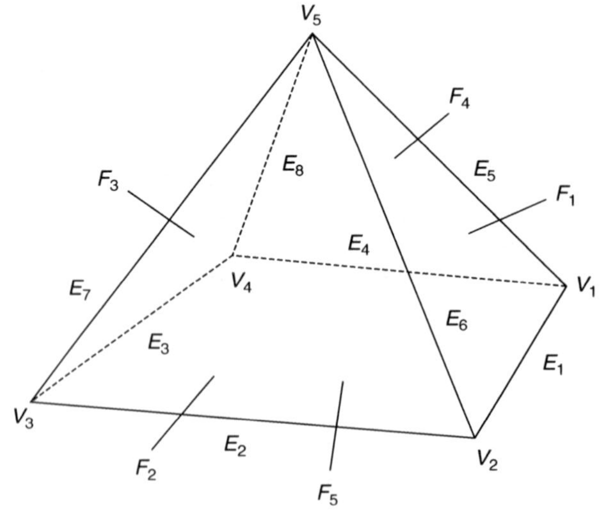

Fig. 9.3 A prism object

	Vertices
	*V*1
Neighbor Vertices	*V*2, *V*5, *V*4
Intersecting Edges	*E*1, *E*4, *E*5
Intersecting faces	*F*1, *F*4, *F*5

	Edges
	*E*1
Neighbor Edges	*E*2, *E*4, *E*5, *E*6
Boundary Vertices	*V*1, *V*2
Intersecting Faces	*F*1, *F*5

	Faces
	*F*1
Boundary Edges	*E*1, *E*4, *E*5
Boundary Vertices	*V*1, *V*2, *V*5
Neighbor faces	*F*2, *F*3, *F*4, *F*5

Fig. 9.4 Representation of vertices, edges, and faces

Fig. 9.5 Adjacency matrix of faces

shown in Fig. 9.5. The adjacency matrix is a binary matrix whose elements are 0s and 1s, where 0 represents no connectivity between faces, edges, and vertices, and 1 represents an existing connectivity.

Floriani [13] proposed the use of a generalized edge-face graph for the recognition of features such as protrusions, depressions, through holes, and handles from a B-rep model. The algorithm was used to extract subgraph components according to the connectivity of the face-edge graph.

Joshi and Chang [20] used an attributed adjacency graph (AAG) for recognizing the machined features of a part form in B-rep. An AAG is defined as a graph in which nodes correspond to the part's faces and arcs correspond to the corresponding part edges. Each face of the feature is represented by a node, and every pair of nodes represents adjacent faces linked by an arc. The arcs of the AAG graph are labeled by binary digits (0s and 1s). The attribute values attached to the arcs are determined by the convexity of the edge. If the faces sharing an edge form a concave angle, the attribute is set to 1. If two adjacent faces form a convex angle, the arc is labeled 1, and if they form a concave angle, the arc is labeled 0 as shown in Fig. 9.6. The feature primitives are invoked one by one and an exhaustive search for the presence of the feature is performed. The search is guided by a heuristic method to increase efficiency. Despite a very fast implementation, it is limited in the number of features that it can recognize.

Gavankar and Henderson [15] proposed an edge-face graph algorithm for the extraction of protrusions and depressions from a B-rep model. Chuang and Henderson [9] developed a method using subgraph isomorphism to recognize shape features and feature-based decomposition of a solid from a B-rep.

Zhao et al. [67] presented a methodology and algorithm of recognizing machined surfaces from a wireframe model. Some problems such as ambiguity and inadequacy

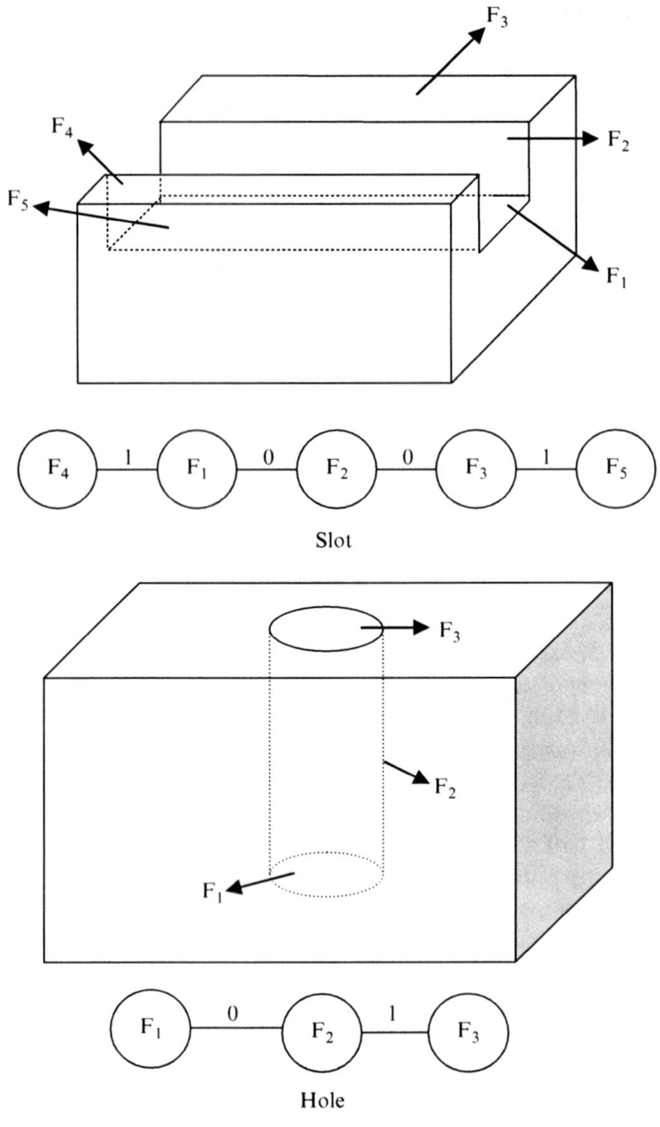

Fig. 9.6 The AAG graph of a slot and a hole

of wireframe models in feature recognition had been solved. The 2D graph was used to generate closed boundaries from wireframe models. The method was powerful especially for canonical parts in which all vertices are in degree three. The elimination of the vertices with degree greater than three needed many more computations because of the difficulties of determining each edge of the facets. Graph theory was employed in a way that could easily remove the ambiguities and would facilitate data modification.

Kao and Kumara [23] proposed the super relation graph (SRG) method to recognize machining features such as slots, holes, and pockets. The nodes of the SRG represent the faces in a depression; the links represent either super-concavity or face-to-face relationships that are generated from a set of new directions of relationships between two faces. The method uses pattern recognition, ANNs, and computational geometry techniques for extracting shape features.

Tzong and Bartholomew [55] developed a classification scheme for mechanical assembly and sheet metal fabrication. Euler's formula and graph theory were used to achieve a coherent and quick way of classifying features. Throughout the development, the role of unique and valid geometric data was emphasized. The developed classification schemes were used for feature recognition as well as feature-based design systems for mechanical assembly and sheet metal fabrication.

Sashikumar et al. [44] developed a feature recognition system for recognizing user defined features (UDF). The feature recognizer used a graph-based approach to represent and recognize features. An attributed facing adjacency graph consisting of topological and geometric attributes was used to represent UDFs. The feature recognition stage involved finding similar subgraphs in the part graph. This technique used a rich set of attributes to recognize a wide range of features such as pockets with variable numbers of side faces.

The major problem with the graph matching approach is the lack of ability to recognize the intersecting feature. When the intersecting features occur, the full patterns of the feature may not appear in the graph representation. Therefore, the pattern matching may fail.

9.2.4 Expert System Approach

An ES is a computer program designed to emulate the problem solving ability of a human expert and consists of two main elements: (1) *production knowledge* and (2) an *inference engine* as shown in Fig. 9.7. The knowledge and experience of an expert are converted to a set of facts and rules used to solve the problem. In the field of feature recognition, the ES is used to determine how to manufacture a solid object.

Production knowledge includes the *procedural knowledge* and the *declarative knowledge* [29]. Procedural knowledge involves the production rules related to generating the process plan from the part design. The production rules are usually expressed in the form of condition–action pairs, if this condition occurs, then do this action, and otherwise do that action [62].

The production rules may include the feature sequence rules, raw material selection rules, tolerance rules, operation selection rules, machine tool selection rules, cutting tool selection rules, cutting parameters rules, and finishing operation rules. Usually, the sequences of firing the production rules are the same as the feature sequences [20]. The feature sequences are predetermined by the user or by using recognition software.

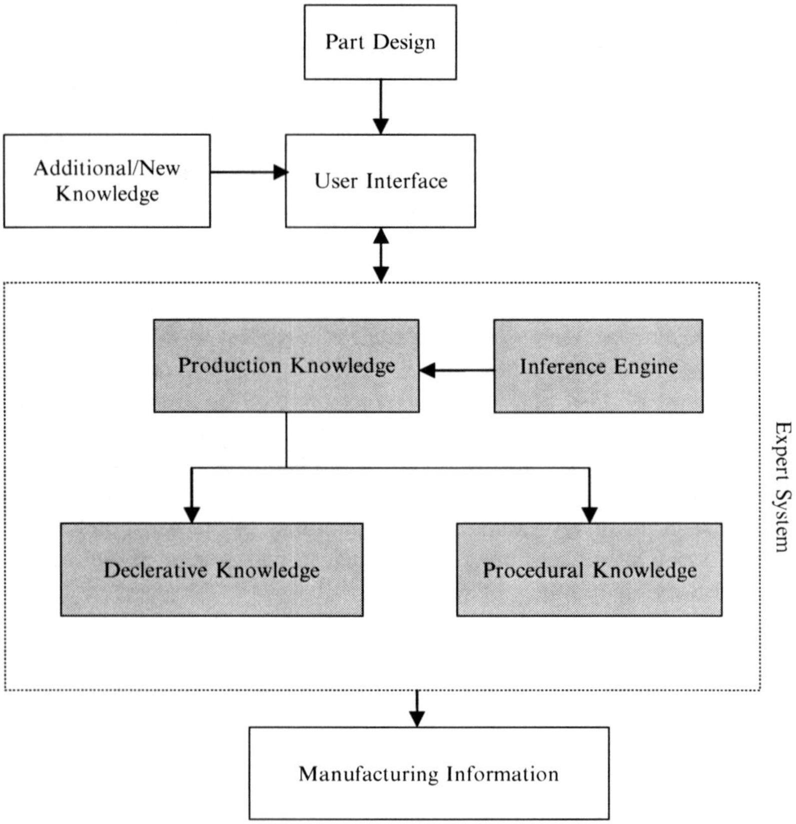

Fig. 9.7 Expert system structure

An example for the feature sequencing rules, the design of the part shown in Fig. 9.8 is a rectangular block that includes two features, a rectangular slot and a circular hole. The process planner could suggest producing this workpiece first on the drilling machine and then on the drilling machine. The first step is to open the slot on the milling machine by using a slot tool. The next step is to drill the hole on the drilling machine. The following rule is an example for representing a slot, prompting the user to enter the required tolerances for machining this slot, and executing the slot milling rules:

IF	**the feature is a slot**
and	**the slot dimensions are: *X, Y*, and *Z***
THEN	**Input the slot tolerances**
	Upper length tolerance
	Lower length tolerance
	Upper width tolerance
	Lower width tolerance
	Upper thickness tolerance

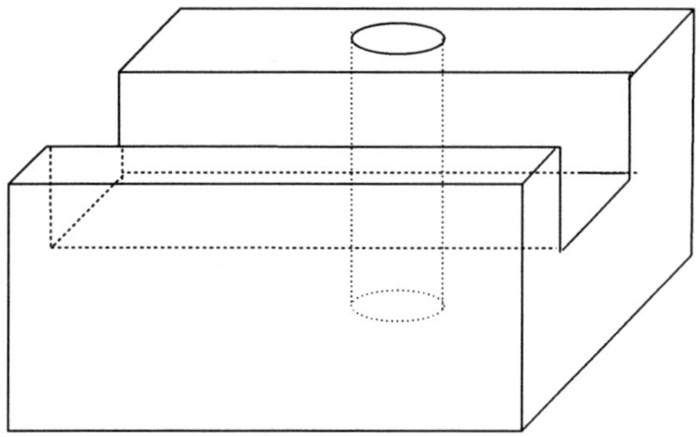

Fig. 9.8 Component with slot and hole features

Lower thickness tolerance
and execute the slot milling rules.

The procedural knowledge could also include rules relevant to the raw materials, their properties, and the machining conditions for these raw materials. Properties could include hardness, thermal conductivity, electrical conductivity, etc. Machining conditions could include allowable cutting speeds and feed rates for certain cutting tool types.

The declarative knowledge involves the production facts. These facts may include feature facts, machine tool capacity facts, machining operation facts, and raw material facts [20, 62]. The feature facts can be defined as the available part design features (e.g., shape features including slot, hole, key, thread, pocket, etc.) which could be machined by the system. The following rule represents the available feature facts in an expert process planning:

Features:
Rectangular slot
Cylindrical hole
Thread
Key
Pocket
Shaft
Step shaft

The second part of the expert process planning system is the inference engine. The inference engine can be defined as the control system of the executed facts and rules that is applied in order to solve a specific problem [62]. Such control system should incorporate a searching algorithm.

Choi et al. [8] developed a method for recognizing the holes, slots, and pockets in a 3D solid model. The definition of regional shape patterns and the recognizer are written as rules in Pascal. Rule-based systems can deal with general 3D shape patterns, but the interaction between the features was not handled in this research.

Henderson and Anderson [17] developed an ES to extract form features such as holes and slots. The system consists of three modules: (a) feature recognition, (b) feature extraction, and (c) feature graph construction. A PROLOG format was used to develop a regional shape pattern recognizer. Patterns are described as rules in PROLOG. A 3D solid model in a B-rep format is converted to facts in PROLOG. Then patterns are recognized with PROLOG's matching mechanism.

Wang [59] presented a process planning system containing a machined surface identifier, a knowledge base, a pattern directed inference engine with concurrent rule firing capability, a machining data selection module, an automated NC part program generator, and an effective relational database manager. A process plan is automatically generated from a part designed using a wireframe CAD modeler. The research contributions include (1) machined surface identification, (2) consistency checking of geometric tolerance, (3) formalization of process planning knowledge, (4) machine learning, (5) creation of alternative plans, (6) addition of new processes, (7) formalization of machining data selection knowledge, and (8) sequential and parallel rule firing mechanism. The limitations are (1) that alternative processes are generated primarily for the internal surfaces of a part and (2) the present research focuses on generating alternative operations. In this CAD module, 2D representation was used for identification. If the part is symmetric, it can be done. However, if some features are not symmetric, these features would not be recognized.

Henderson and Chang [18] developed an automatic process planning system with a feature extraction system called feature recognizer and process planner (FRAPP). The input to FRAPP can be either a CSG or a B-rep model. The feature recognition process consists of recognition, extraction, analysis, computation, and compilation.

Schulte et al. [45] developed a CAPP system that orders the features with the outermost, highest elevation features, and then works toward the innermost, lowest elevation features. The rule base resolves ambiguities in the ordering of the process plan fragments. An example of this is when there is more than one feature at the same elevation and equidistant from the center of the part.

Madurai and Lin [30] developed a system that can automatically extract and recognize part features directly from a CAD database. Part geometric and topological data in IGES format are read by a feature extraction data compactor. Then geometry to feature translator captures manufacturing features in its decision logic expressed as production rules written in LISP. The 2D representation is used in this system. This system was limited to use in rotational parts. It cannot recognize flat surface features that can be generated by processes such as milling.

Devireddy and Ghosh [10] developed a methodology for integrating the design and planning aspects of manufacturing. This methodology was based on the concepts of feature-based modeling (FBM) and artificial neural networks (ANNs). The FBM was developed to give information about both form (shape, size) and precision (tolerance, surface finish) features for the designed rotational components. The FBM shell was programmed in Visual Basic and was operated by Windows 95. The FBM was able to present the designed components in B-rep model.

Ping and Chang [37] developed an artificial intelligent system for dynamic computer-aided process planning. Their system integrated fuzzy logic and ANNs to perform the dynamic recognition and adaptive-learning tasks of the workpieces and process planning. Also, they adopted the idea of an important (critical) feature concept for evaluating the suitability of existing process plans for the incoming product design. In addition, this system integrated the variant and generative CAPP by using an ES module. Thus, the system adaptively decides whether to use the variant or generative procedures.

Kayacan and Celik [25] developed a process planning system for prismatic parts by means of an ES. This system automatically selected operation type and sequence of operations that were defined by using the databases achieved from the feature recognition module. The feature recognition process was achieved with B-rep modeling method to give vectorel direction knowledge and adjacent relationships of surfaces using the STEP standard interface program. The feature recognition system was prepared on an IPM-compatible PC via C++ programming language.

9.2.5 Volume Decomposition and Composition Approach

In this approach, geometry decomposition algorithms operate more directly on 3D volumes that apply Boolean operations during the construction of the CSG model. The objectives of volume decomposition are to decompose a volume to be machined, identify material to be removed from a base stock, and break down this volume into nonoverlapping units corresponding to distinct machining operations. There are two different approaches: convex hull decomposition and volume decomposition. A brief review of the work in this area is presented below.

Woo [64] developed a computer-aided recognition system based on volumetric designs. This system transforms 3D volumetric designs of mechanical components into numerical control descriptions. The concept of cavity was introduced as a geometric configuration. A cavity can be created by using either ADD or REMOVE operations on primitive volumes. The drawback was the limited number of original primitives that the system includes in its domain.

Wang [60] developed the concept of backward growing by reversing the process of machining operations. To determine the cavity shape, the method looks for a set of concavely connected faces, oblique faces, or circular faces that are machined faces. Because each machined face is the result of the machining operation on the part, these faces can be the candidate of the growing base faces. The method uses a set of concave faces of a depression feature to create the feature volume in order to decompose the delta volume as well as a protrusion feature into the basic feature.

Ferreira and Hinduja [12] proposed a convex hull based feature recognition method for 2.5-D components created from a B-rep solid model. This method involved the determination of the convex hull of a component's faces. The edges within the convex hull define the faces that form a feature. Edge inner loops and concave edges are detected because they determine the faces that describe a feature.

The drawback of Ferreira and Hinduja's method is that it cannot recognize some convex features because the rules for merging two or more features are narrow in domain.

Vandenbrande and Requicha [57, 58] developed a system that uses the delta volume to find volumetric features that meet the conditions of manufacturability and correspond to operations typically occurring in a three-axis machining center. The method used a hint feature corresponding to characteristic patterns in the boundary of the object. A feature is completed by first growing or extending it along feature-specific directions. Then the interaction between the extended feature, the stock, and the part was analyzed.

Tseng and Joshi [54] developed a system that used a B-rep as its input and generated multiple sets of features in two stages: volume decomposition and volume composition. Their method started by determining the raw material and identifying the delta volume of the part. The method decomposes the total delta volume of the part to basic removable blocks through the delta volume. Then the volumes are reconstructed by reconnecting small blocks. By using feature definitions, the feature volume can be matched with the predefined feature pattern and classified. There are many ways to build the feature from the blocks. Some basic blocks can be connected with more than one adjacent basic removable block depending on the direction of the connection.

Shah et al. [47] developed a design by feature system that can understand design features from the manufacturing aspect and autonomously generate feasible machining process sequences for undocumented features. The delta volume is obtained by subtracting the part from the stock material. The method decomposes the delta volume of the part into minimum convex cells by half space partitioning at every concave edge. This composition method is called minimum convex decomposition by half space partitioning.

Dong and Parasaei [11] developed an automated process planning prototype system. The prototype system was based on the removable volume. The part and blank were built using SDRC solid modeling I-DEAS. The overall removable volume was generated graphically by comparing the finished part and blank. General machining features were extracted. Process plans were created based on the recognized feature information and production knowledge for prismatic shapes only.

Kao et al. [24] developed the SRG method to extract prismatic features from the CAD B-rep of a machined part. The SRG is constructed by finding the faces in the cavity and grouping them into interacting feature face sets to reduce the search space. The method used is generate-verification for the feature recognition system. Hypotheses are generated by matching SRGs and are verified based on the validity of a feature volume. The final output of the feature recognition is a decomposition of the cavity volume into a set of primitive feature volumes.

Sakurai [42] developed a generic form feature recognition called "spatial decomposition and composition." A feature is defined as feature volume and surface. The method searches for the cavity volume that a user can interactively modify. The delta volume of the part is decomposed into minimal convex cells and then composed back to the volumes. The volumes will be checked if they are feature volumes.

Sakurai and Dave [43] developed the decomposition method which decomposes a delta volume into maximal volumes. A maximal volume is generated by extending the faces or half space of the delta volume. The composition step adds cells adjacent to each other. Combinations of small cells form a maximal volume by examining the relationship among the small cells. Then graph pattern matching is used to classify the volume into a machining feature.

Alan and Lin [3] developed a volume decomposition approach for building an interface that bridges the gap between a 3D CAD model and an automated process planning system. The volume decomposition approach partitions the whole machining volume, which needs to be removed from the raw material to produce the finished part, into some machinable features. Once the machinable features of a part are found, the machining of the part can be viewed as material removal for a series of machinable features. The tasks involved in this approach were extraction of machining volume, decomposition of the machining volume, and determination of procedures required for cutting out the machinable features. Each one of these tasks had its own algorithm.

Jung and Han [21] tried to develop an algorithm to integrate the feature recognition and process planning systems. The objective of their work was to integrate the following: feature recognition for manufacturability, setup minimization, feature dependency construction, and generation of an optimal feature based machining sequence. The feature recognition system was based on volume decomposition and hint-based reasoning approaches. This system was called integrated incremental feature finder (IF^2). The IF^2 system was able to recognize holes, slots, and pockets.

Yong et al. [66] developed a feature recognition approach using alternative sum of volumes with partitioning decomposition. The objective of this approach was to systematically obtain geometry-based relations between machining features that were recognized from the part boundary. The precedence relations between features were achieved by combining the face dependency information and the machining process information. The precedence relations for a given feature decomposition were then recognized as a set of precedence trees. Each precedence tree for a given feature decomposition represented a different set of the features.

The volume decomposition and composition approach show major problems in both the decomposition and composition steps. By extending the faces or half space of the delta volume, the cell decomposition can create a large number of cells. The possible combinations of the cell for composing the predefined features volume grow exponentially with the number of cells.

9.2.6 3D Feature Recognition from a 2D Feature Approach

Meeran and Pratt [32] used 2D entities in a DXF format with no particular order and no relationship in their connectivity. The process of searching and sorting the entities is divided into three groups according to each of three orthographic views. The approach intends to recognize the common machining features of a prismatic part having planar or cylinder surfaces in terms of 2D profiles that are projected in the three orthographic views of the drawing. The approach starts by recognizing the isolated feature and then by providing the library of patterns. The recognition is essentially based on pattern matching.

Liu et al. [27] proposed 3D feature recognition by reconstructing from 2D CAD data. The process is begun by extracting the 2D geometric entities from the IGES file. The divide and conquer strategy is used to extract the vertex-edge data from each 2D orthographic view (front, side, top). When the set of 2D geometric entities of the three views is extracted, a set of production rules is developed to recognize each 2D entity: square, rectangle, triangle, or circular loop. This information is used to facilitate the process of 3D recognition in the next step. The structure of the feature model is constructed by a form feature adjacency graph (FFAG). The recognized form features were used to reconstruct the 3D part as a composition of hierarchies of FFAG. The FFAG can be easily transformed into a CSG/DSG structure, which is readily integrated with the downstream CAPP/CAM systems. The features were limited to prismatic features only without analyzing the interaction between features

Sommerville et al. [51] developed an approach for recognizing features called viewer centered. The approach had the ability to recognize DP features and assembly features. The approach consisted of five steps as follows:

Step 1. Establish Viewpoints: A viewpoint V_p is established outside the boundary of the object.
Step 2. Ray Firing: The object is swept by a cone of rays fired from the viewpoint V_p.
Step 3. Hint Generation: Each ray fired is processed to produce a list of feature hints.
Step 4. Merge Hints: The hint lists generated are merged by combining the hints generated by the same face pairs.
Step 5. Feature Completion: From each possible feature hint a sphere ray is fired from the feature kernel. Figure 9.9 shows a cross-section of a mechanical component assembled from two parts.

However, this approach had the ability to recognize a limited number of features and it did not determine the dimensions of the recognized features, in spite of the authors claim that this approach has the ability to recognize assembled features.

Allada and Anand [4] developed a recognition methodology by recognizing 2D features from polyhedral faces of a part model first. Then they inferred a 3D feature in the 2D features identified. To recognize 2D features, the nonconvex faces of the part model have to be identified. Next, rule- and graph-based approaches are used to derive the 3D feature from the 2D recognized feature. The edges that link to the 2D feature are adjacent to two faces which are identified as the primary face and the secondary face. The face that is associated with nonconvex faces in 2D recognition is considered as the primary face. The other face that is perpendicular is considered as the secondary face. The rules for each type of feature consist of the

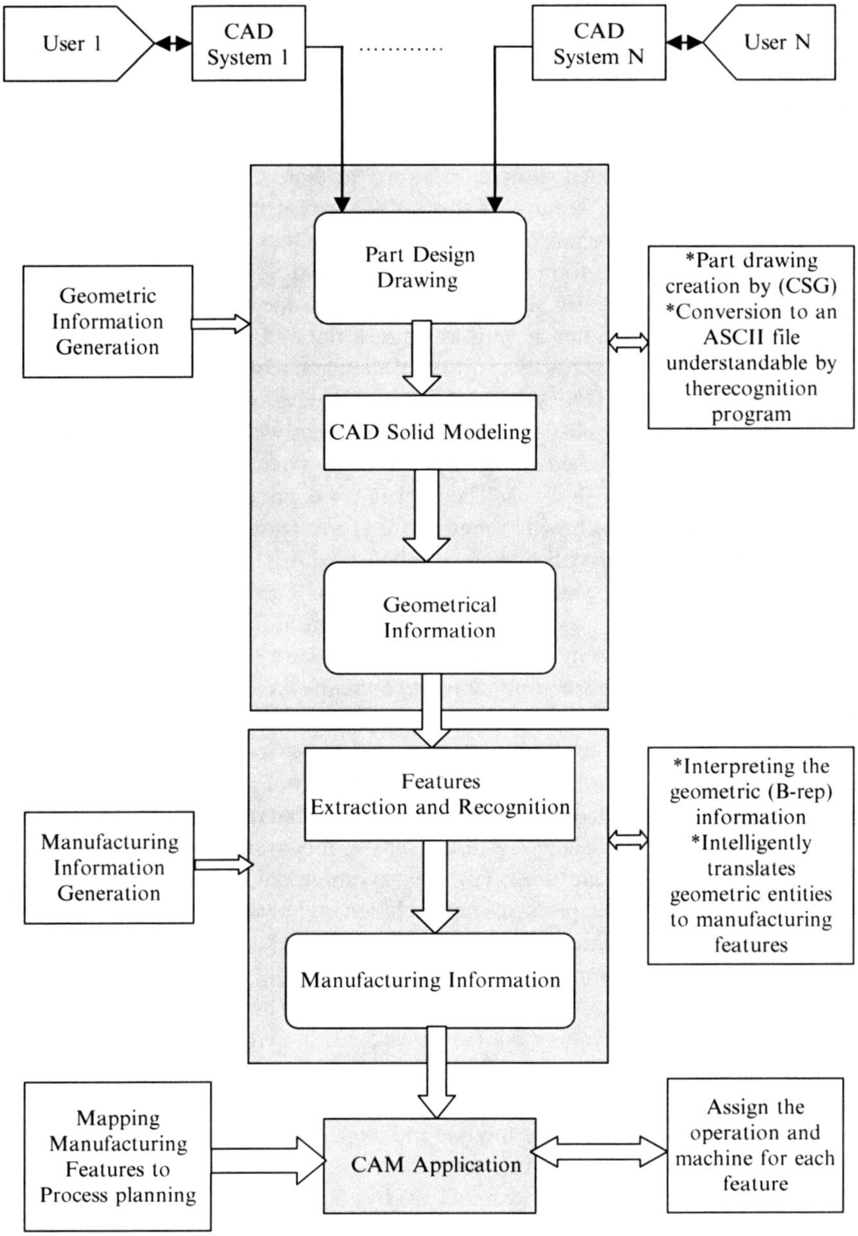

Fig. 9.9 Structure of the proposed methodology

adjacency relationship of the secondary faces of the edges that are associated with the recognized 2D feature.

Ahmad and Haque [1] developed a feature recognition system for rotational components using a DXF file. In this approach, work geometric information of rotational parts is translated into manufacturing information through a data interchange format (DXF). A feature recognition algorithm was used to recognize different features of the part from its DXF file, where geometric information of the part was stored after respective DXF codes. Finally, using the data extracted from DXF file, each feature of the part was recognized. The parts are symmetrical and represented in two dimensions.

Mansour [31] developed simple software to link CAD packages and CNC machine tools. The main feature of this software was its ability to automatically generate part programs for machining sculptured surfaces. To achieve this objective, IGES files of simple or free-form surfaces were exploited. The data extracted from IGES files were used to graphically simulate the tool path due to center and tip of a ball nose tool and a filleted cutter as well as tip of a flat end mill cutter. Depending on the size of the created surfaces, various dimensions of cutting tools should be selected by the user. AUTOCAD and AUTOCAD development systems were used to convert trimmed surfaces to parametric spline surfaces with IGES entity number 114. This entity was closely studied and an algorithm was developed to convert this entity to a large number of straight lines. The package was not developed to handle other IGES entities that provide a wide range of free-form surfaces.

Zhao et al. [68] developed a system called CADEXCATS to integrate a CAD system and a generative system for cutting tool and condition selection for turned components. The system was based on an IGES neutral format and feature recognition approach. In this system, two separate drawings are generated for finishing and roughing geometries. The roughing drawing contains the roughing profile, finishing profile, and blank space and size. The finishing drawing represents the exact profile of the final component. Both projections and cross-sections were accepted for description of a turned component in 2D. IGES data of a turned part CAD model was extracted and processed as geometry-based data that represent an input to CAPP system in the next stage. There are limitations in this system, for example (1) only two dimensions of data are used, (2) only symmetrical cylinder components are used, and (3) the prismatic parts are not included and also their interactions.

Zhang et al. [69] developed an intelligent approach for coordinate measuring machine (CMM). First step in this approach was feature extraction. The extracted features had been done through IGES or Step files. The components were represented in 2D by three views (front, top, and side). After the features had been extracted, a 3D shape was created. The recognized features were very limited. The approach needs improvement to achieve its objectives.

Xinhui et al. [65] developed a method to integrate CAD and CMM. This method was developed for labeling the tolerance items of 3D workpieces. First, CAD data of 2D workpiece are extracted through an IGES file, then 3D of the recognized entities was constructed with the information required to make the inspection process. This method dealt with simple prismatic parts features and it did not support rotational features. Moreover, the intersection between different types of features has not been developed.

No matter which approach is adopted, the geometric information always needs to be constructed early in order to advance the feature recognition. The literature review addressed in this chapter is categorized according to feature recognition method, representation type, standard type, dimension type, and feature type. From the system implementation point of view, the best system should be independent of the format, i.e., any kind of data format can be used for the input information and the internal geometric feature representation can be constructed based on the input data. B-rep is the most independent representation because other data formats have several limitations and disadvantages. Therefore, this methodology adopts the solid modeling representation. On the other hand, the CSG technique is used as a tool to construct the designed parts, so the hybrid CSG/B-rep is used in this book.

9.3 Intelligent Feature Recognition Methodology

In the following subsection, a methodology for feature analysis and extraction of prismatic parts for CAM applications is developed and presented. This approach aims to achieve integration between CAD and CAM. Different CAD or geometric modeling packages store the information related to the design in their own databases. The structures of these databases are different from each other. As a result, no common or standard structure has so far been developed that can be used by all CAD packages. For that reason, this technique develops an intelligent feature recognition methodology (IFRM) which has the ability to communicate with the different CAD/CAM systems.

The part design is introduced through CAD software and is represented as a solid model by using CSG technique as a design tool. The solid model of the part design consists of small and different solid primitives combined together to form the required part design. The CAD software generates and provides the geometrical information of the part design in the form of an ASCII file (IGES format) that is used as standard format which provides the proposed methodology the ability to communicate with the different CAD/CAM systems. The boundary (B-rep) geometrical information of the part design is analyzed by a feature recognition program that is created specifically to extract the features from the geometrical information based on the geometric reasoning and object-oriented approaches. The feature recognition program can recognize these features: slots (through, blind, and round corners), pockets (through, blind, and round corners), inclined surfaces, holes (blind and through), and steps (through, blind, and round corners), etc. These features are called manufacturing information that is mapped to process planning as an application for CAM. Figure 9.9 shows the structure of the proposed methodology.

The IFRM presented in this chapter consists of three main phases: (1) a data file converter, (2) an object form feature classifier, and (3) a manufacturing features classifier (production rules). The first phase converts a CAD data in IGES/B-rep format into the proposed OODS. The second phase classifies different part geometric features obtained from the data file converter (phase 1) into different feature groups. The third phase maps the extracted features to process planning's point of view. Figure 9.10 shows a basic flowchart of the proposed system. The sections that follow describe the steps of feature extraction in details.

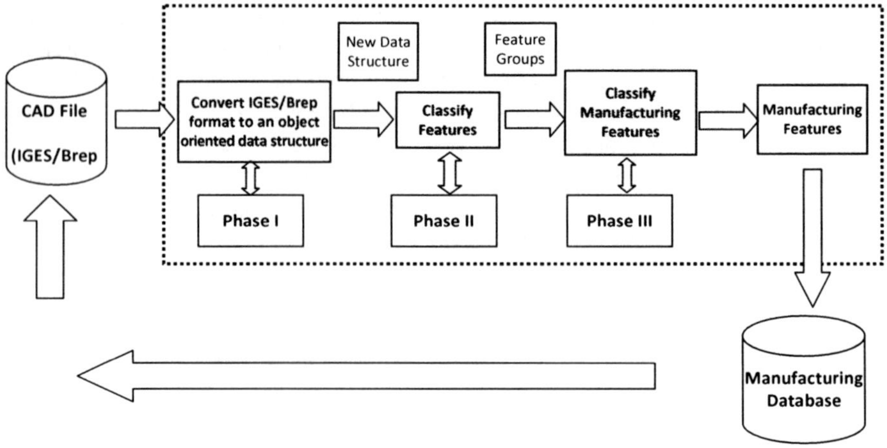

Fig. 9.10 Flowchart of extraction and classification of features

9.4 Conversion of CAD Data Files to OODS

The IGES is a standard file format for the data defining the object drawing in 3D CAD systems in the B-rep structure. The entry fields in IGES format consist of an object's geometric and topological information. The geometric information includes the definition of lines, planes, circles, and other geometric entities for a given object. The topological information defines the relationships between the object's geometric components, for example, in terms of loops (external loop and internal loop).

An external loop gives the location of major geometric shapes and an internal loop represents a protrusion (boss) or a depression (pocket or hole) on an external loop. The fundamental IGES entities, which are related to representing a solid object in B-rep structure, are discussed in the following subsection to understand how these entities are defined in IGES 5.3 [63].

9.4.1 Basic IGES Entities

9.4.1.1 Line (Entity 110)

A line in IGES file is defined by its end points. The coordinates of start point and terminate point are included in parameter data section of this entity.

9.4.1.2 Circular Arc (Entity 100)

To represent a circular arc in modeling space, IGES provides the information including a new plane (X_T, Y_T) in which the circular lies the coordinates of center point, start point, and terminate point. A new coordinate system (X_T, Y_T, Z_T) is defined by transferring the original coordinate system (X_O, Y_O, Z_O) via a transformation matrix and all coordinates of points (center point, start point, and terminate point) related to this new coordinate system. The order of end points is counterclockwise about Z_T axis.

9.4.1.3 Transformation Matrix (Entity 124)

This entity can give the relative location information between two coordinate systems, X_O, Y_O, Z_O coordinate system and X_T, Y_T, Z_T coordinate system

$$\begin{pmatrix} R_{11} & R_{12} & R_{13} \\ R_{21} & R_{22} & R_{23} \\ R_{31} & R_{32} & R_{33} \end{pmatrix} \mathrm{X} \begin{pmatrix} X_O \\ Y_O \\ Z_O \end{pmatrix} + \begin{pmatrix} T_1 \\ T_2 \\ T_3 \end{pmatrix} = \begin{pmatrix} X_T \\ Y_T \\ Z_T \end{pmatrix}, \tag{9.1}$$

where

$$\begin{vmatrix} R_{11} & R_{12} & R_{13} \\ R_{21} & R_{22} & R_{23} \\ R_{31} & R_{32} & R_{33} \end{vmatrix} = 1.$$

9.4.1.4 Surface of Revolution (Entity 120)

A surface is created by rotating the generatrix about the axis of rotation from the start position to the terminal position. The axis of rotation is a line entity. The generatrix may be a conic arc, line, circular arc, or composite curve. The angles of rotation are counterclockwise about the positive direction of rotation axis.

9.4.1.5 Point (Entity 116)

A point is defined by its coordinates (X, Y, Z).

9.4.1.6 Direction (Entity 123)

A direction entity is a nonzero vector in 3D that is defined by its three components with respect to the coordinate axes. The normal vector of surface can be determined by this entity.

9.4.1.7 Plane Surface (Entity 190)

The plane surface is defined by a point on the plane and the normal direction to the surface.

9.4.1.8 Vertex List (Entity 502)

This entity is used to determine the vertex list that contains all the vertexes of the object.

9.4.1.9 Edge List (Entity 504)

This entity is used to determine the edge list that contains all the edges of the object.

9.4.1.10 Loop (Entity 508)

This entity is used to determine the loops that are involved in all faces of the object.

9.4.1.11 Face (Entity 510)

This entity is used to determine the faces that comprise the object.

9.4.1.12 Shell (Entity 514)

The shell is represented as a set of edge connected and oriented faces. The normal of the shell is in the same direction as the normal of the face.

9.4.1.13 Right Circular Cylindrical Surface (Entity 192)

The right circular cylindrical surface is defined by a point on the axis of the cylinder, the direction of the axis of the cylinder, and a radius.

9.4.1.14 Composite Curve (Entity 102)

A composite curve is a continuous curve that results from the grouping of certain individual constituent entities into a logical unit. A composite curve can be defined as an ordered list of entities consisting of point, connect point, and parameterized curve entities. The list of entities appears in the parameter data entry.

9.4.1.15 Conic Arc (Entity 104)

A conic arc is a bounded connected portion of a conic curve that has distinct start and end points. The parent conic curve can be an ellipse, a parabola, or a hyperbola. The definition space coordinate system is always chosen so that the conic arc lies in a plane either coincident with or parallel to the X_T, Y_T plane. Within such a plane, a conic is defined by the six coefficients in the following equation, where X_T and Y_T are the coordinates of a point in the X_T, Y_T plane

$$AX_T^2 + BX_TY_T + CY_T^2 + DX_T + EY_T + F = 0. \tag{9.2}$$

9.4.1.16 Parametric Spline Curve (Entity 112)

The parametric spline curve is a sequence of parametric polynomial segments. The type of the curve is presented in the parameter data entry section.

9.4.1.17 Parametric Spline Surface Curve (Entity 114)

The parametric spline surface is a grid of parametric polynomial patches. The type of the surface is presented in the parameter data entry section.

9.4.1.18 Ruled Surface (Entity 118)

A ruled surface is formed by moving a line connecting points of equal relative arc length or equal relative parametric value on two parametric curves from a start point to a end point on the curves.

9.4.1.19 Tabulated Cylinder (Entity 122)

A tabulated cylinder is a surface formed by moving a line segment called the generatrix parallel to itself along a curve called the directrix. The curve may be a line, circular arc, etc.

9.4.1.20 Rational B-Spine Curve (Entity 126)

The rational B-spline curve may represent analytic curves.

9.4.1.21 Rational B-Spline Surface (Entity 128)

The rational B-spline surface represents various analytic surfaces.

9.4.1.22 Offset Curve (Entity 128)

The offset curve entity defines the data required to determine the curve offset from a given base curve. This entity points to the base curve to be offset and contains the offset distance.

9.4.1.23 Offset Surface (Entity 140)

The offset surface is a surface defined in terms of an existing surface.

9.4.1.24 Right Circular Conical Surface (Entity 194)

The right circular conical surface is defined by a point on the axis of the cone, the direction of the axis of the cone, the radius of the cone at the axis point, and the cone semiangle.

9.4.1.25 Spherical Surface (Entity 196)

The spherical surface is defined by the center point and the radius.

9.4.1.26 Toroidal Surface (Entity 196)

The toroidal surface is defined by the center point, the axis direction, and the major and minor radii.

9.5 The Overall OODS of the Proposed Methodology

In order to have a good generic representation of the designed object for CAM applications, especially for process planning, the overall designed object description and its features need to be represented in a suitable structured database. An object-oriented representation will be used in this methodology. The first step toward automatic feature extraction will be achieved by extracting the geometric and topological information from the (IGES/Brep) CAD file and redefining it as a new OODS as demonstrated in Fig. 9.11.

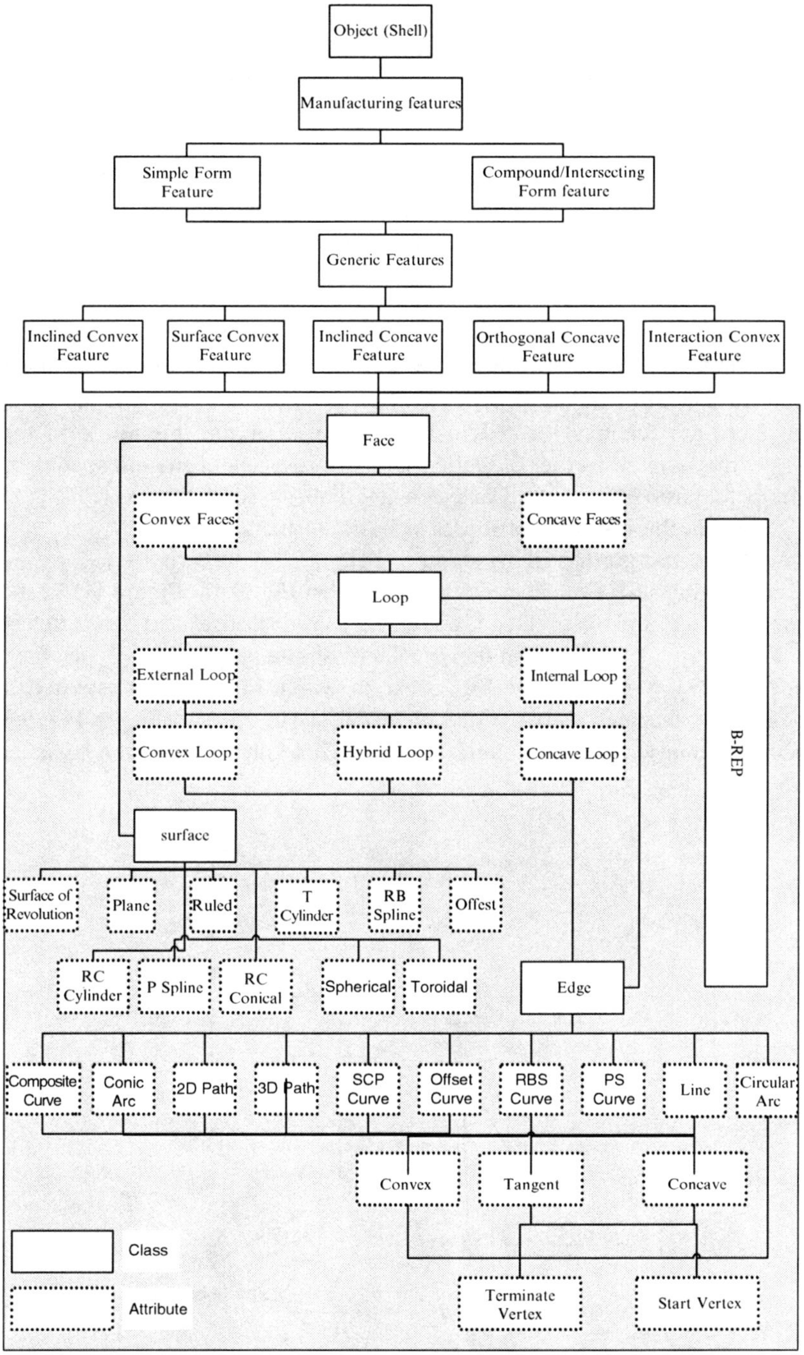

Fig. 9.11 Hierarchy of classes and attributes of the designed object

In this hierarchy, the highest level data class is the designed object (shell). An object consists of manufacturing features that can be classified into form features composed of either simple or compound/intersecting features. A simple feature is the result of two intersecting general geometric surfaces, while a compound/intersecting feature is one that results from the interaction of two or more simple features (slot and pocket) as shown in Fig. 9.12. Features are further classified into concave or convex as attributes in the generic feature class. Concave features consist of two or more concave faces, and convex features are composed of either one or more convex faces or the interaction between other features in the object as shown in Fig. 9.13.

Because of the attributes of the geometric entities of form features (FF), they will be classified into interior form feature ($FF_{interior}$), which is located inside the basic surface, and exterior form feature ($FF_{exterior}$), which is formed by the entire basic surface with its adjacent surfaces. The basic surface refers to the surface in which there are features located in that surface. For the interior form features ($FF_{interior}$), they can be further classified into two low-level categories, convex interior form feature ($FF_{interior\_convex}$) and concave interior form feature ($FF_{interior\_concave}$). $FF_{interior\_convex}$ is the convex portion in a basic surface, while $FF_{interior\_concave}$ is the concave geometric portion in the surface [14].

Examples of these FF categories are shown in Fig. 9.14. Figure 9.14a shows a convex portion of the top surface investigated (basic surface) and hence this constitutes a $FF_{interior\_convex}$ (boss), while Fig. 9.14b shows a $FF_{interior\_concave}$ in the basic surface. A $FF_{exterior}$ is shown in Fig. 9.14c in which $FF_{exterior}$ is constituted by the entire basic surface and its two adjacent surfaces. The $FF_{interior\_concave}$ in Fig. 9.14b is a through cylindrical hole. A blind cylindrical hole or a pocket in a basic surface is also a $FF_{interior\_concave}$.

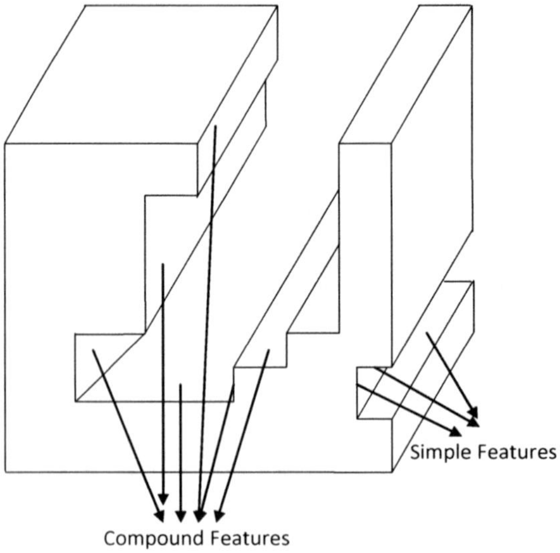

Fig. 9.12 Simple and compound features

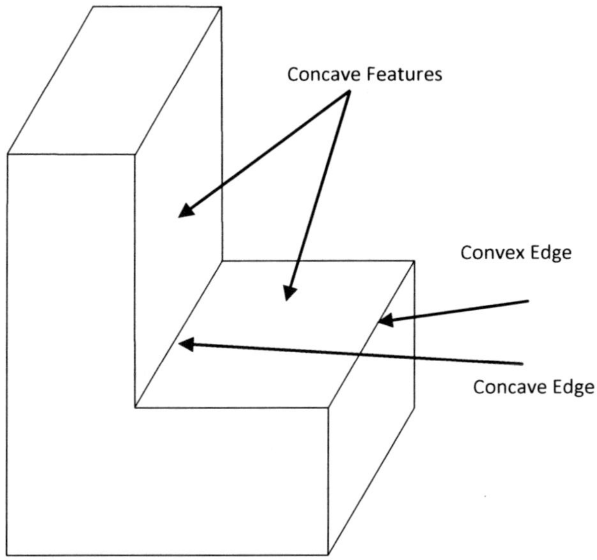

Fig. 9.13 Convex and concave features and edges

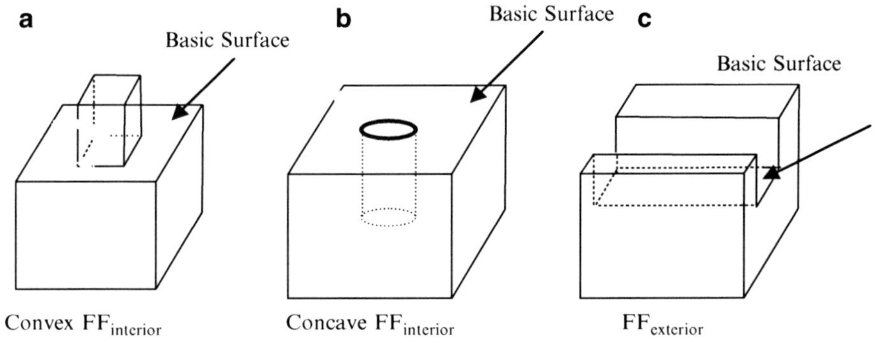

Fig. 9.14 Classification of interior and exterior form features

In this chapter, multilevel form feature taxonomy is proposed based on the feature geometry and topological characteristics. The lowest level features in the feature hierarchy are the features defined in CAM-I as shown in Fig. 9.15. Defining concave features is basically equivalent to identifying concave faces which are simply defined by a concave edge that joins two adjacent faces. A concave edge is determined by the concavity test, which is explained later. In general, the edge is defined by a pair of vertices described in the part drawing properties in terms of coordinates (*X*, *Y*, *Z*). On the other hand, convex features can be defined and classified as either inclined, interaction, or surface as shown in Fig. 9.16.

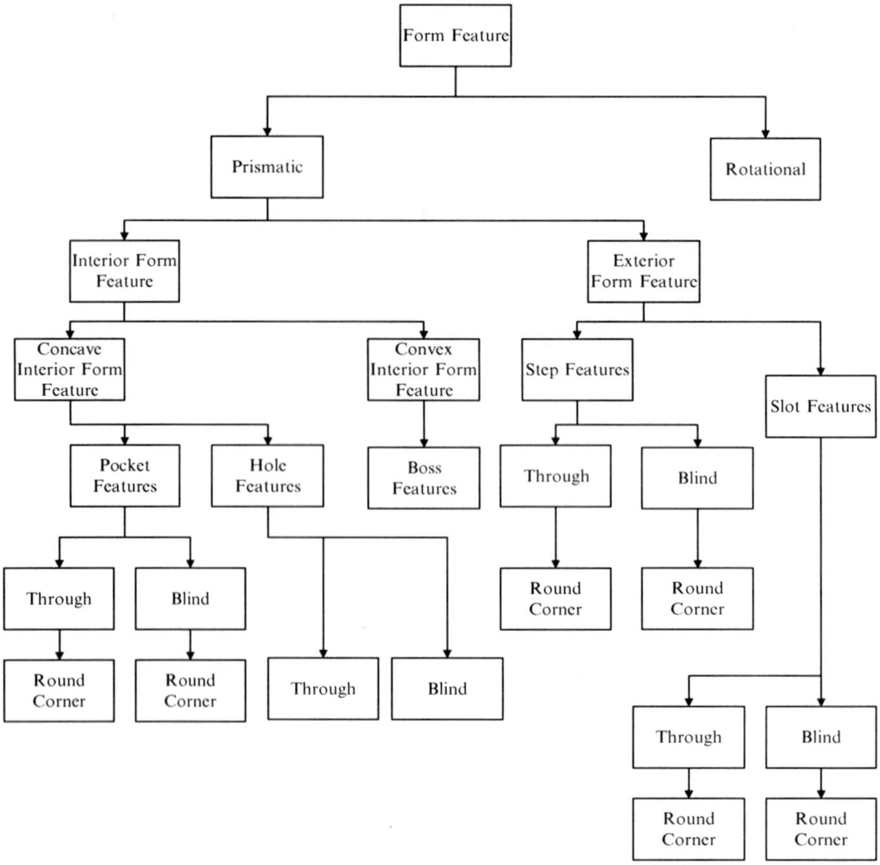

Fig. 9.15 Hierarchy of form features

Inclined convex features are defined by a set of convex faces that are not parallel or perpendicular to minimal enclosing box. The second type of convex feature (namely interaction) results from the interaction of two or more features. Surface convex features are features that lie on the minimal enclosing box as seen in Fig. 9.16.

9.6 Geometry and Topology of B-rep

The basic geometric entities of a 3D CAD model based on B-Rep description are vertex, edge, and face. The compound entities, which consist of basic geometric entities, are shell and loops. Shells and loops are the topological entities because

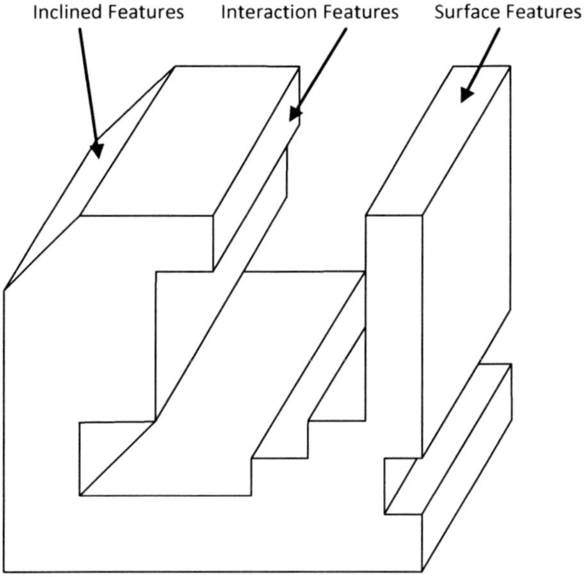

Fig. 9.16 Classifications of convex features

only topological and not geometric information is assigned to them. A solid machining object (O) model can be expressed as follows:

$$O = (V \rightarrow v \in \text{Vertex}, E \rightarrow e \in \text{Edge}, F \rightarrow f \in \text{Faces}), \tag{9.3}$$

where V, E, and F are the sets of object vertices, edges, and faces, respectively, and v, e, f are their respective elements. In this expression, each edge has two vertices and is shared by two adjacent surfaces. Each face has certain edges enclosing a specific 2D or 3D shape. The enclosed chain of edges in a surface can form one or more loops. The external loop bounds the face and the others (internal) are located inside the face.

9.6.1 Classification of Edges

Edges constitute the wireframe of a 3D solid model and they are the intersection boundaries of two adjacent faces. The following proposed edge categories facilitate the representation of the proposed methodology as shown in Fig. 9.11. To represent edge categories, the normal vectors of the two faces connected by that edge and the edge direction are determined as shown in Fig. 9.17. Then by applying the connectivity test that will be discussed later, the edges are classified into convex, concave, and tangent edges as shown in Fig. 9.18. Also, the angle between the two faces (Φ) is determined.

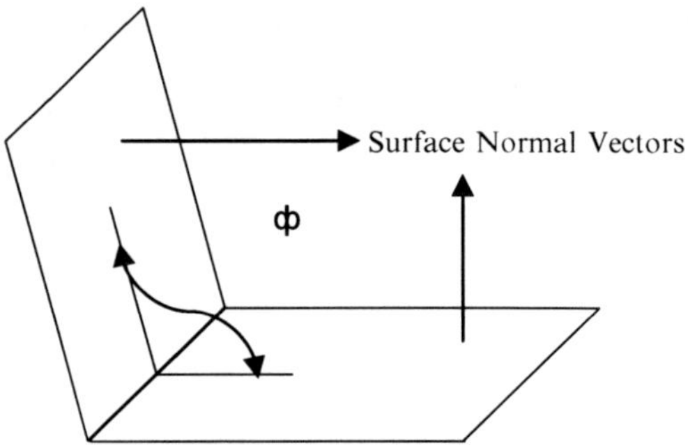

Fig. 9.17 The surface normal vectors

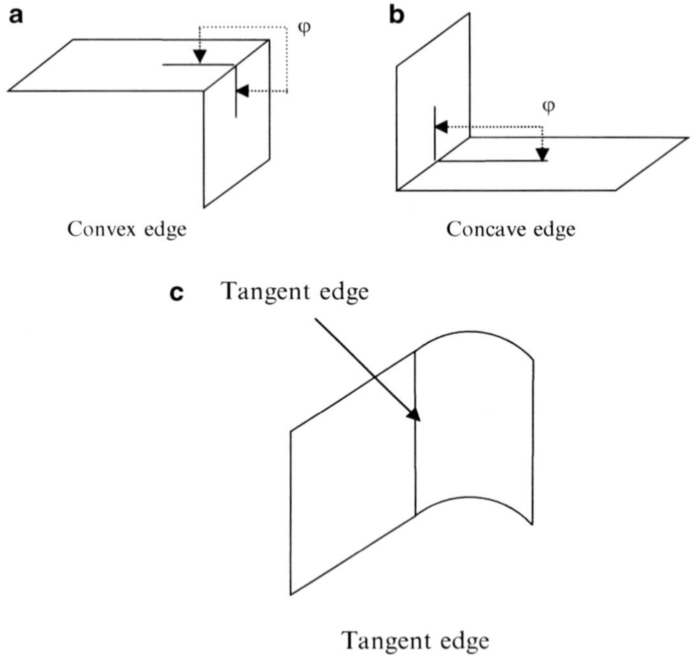

Fig. 9.18 Classification of edges

9.6.2 Classification of Loops

A loop is the border of a surface. It is also the intersection boundary of the surface with its adjacent surfaces. In this book, a loop is used as a fundamental reference to identify the interior and exterior form features. The loop can be classified as proposed in this book into the external loop and internal loop. The external loop is the outside boundary of a basic surface of which the loop is investigated, while internal loops are located inside the basic surface. In a basic surface, an external loop is the maximum boundary of the basic surface and the internal loop is the

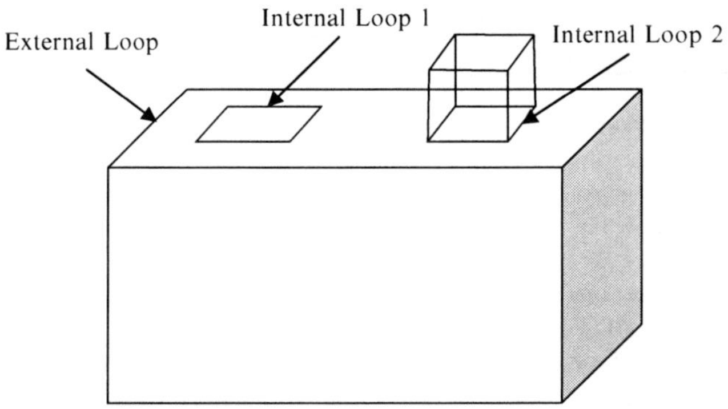

Fig. 9.19 Classification of loops

Table 9.1 Classification of loops

Type	Definition	Case
Convex loop (loop class type = convex)	All the edges are convex	If a convex loop is an internal loop, it is called an internal convex loop (Sub_type = Internal) (loop 2 in Fig. 9.19). The external convex loop is represented as (Sub_type = External)
Concave loop (loop class type = concave)	All the edges are concave	If a concave loop is an internal loop, it is called an internal concave loop (Sub_type = Internal) (loop 1 in Fig. 9.19). The external concave loop is represented as (Sub_type = External)
Hybrid loop (loop class type = hybrid)	The edges have convex and concave types	If a hybrid loop is internal loop, it is called a hybrid internal loop (Sub_type = Internal); otherwise, it is hybrid external loop is represented as (Sub_type = External)

internal interaction boundary of the basic surface with its internal features. In Fig. 9.19, there are three loops in the basic surface, one is an external loop and the other two are internal loops, which are the intersection boundaries of the basic surface with its internal convex and concave features [14]. The loops can be further classified into other categories as shown in Table 9.1.

9.7 Data Fields for Proposed Data Structure

Generally, faces are the basic entities that constitute the features, which are further defined by edges that are represented in terms of vertices, which are defined in terms of coordinates in CAD file. Therefore, the hierarchy of the designed object that was described in the previous section represents a multilevel of different classes. All classes, except for the super class that represents an object as a whole, are objects of classes that are higher up in the data structure. For example, each edge object is represented in terms of vertex objects [14]. Table 9.2 displays the data attributes required for each class in the OODS defined earlier.

Table 9.2 Definitions of classes and attributes

Class name	Attribute	Type
Point	X_Coordinates	(Real)
	Y_Coordinates	(Real)
	Z_Coordinates	(Real)
Vertex	Inherits point	(Point)
	Vertex_ID	(Integer)
Vertex_List	Vertex_Count	(Integer)
	Vertex_List	(Vector of vertex pointers)
Edge	Edge_ID	(Integer)
	Edge_Type	(Enumerated constants)
	Start_Vertex	(Vertex pointer)
	Terminate_Vertex	(Vertex pointer)
	Concavity	(Enumerated constants)
	Face_Pointers [2]	(Array of face pointers)
	Loop_Pointers [2]	(Array of loop pointers)
	Dimension	(Real)
Edge_List	Edge_Count	(Integer)
	Edge_List	(Vector of edge pointer)
Loop	Loop_ID	(Integer)
	Loop_Concavity	(Enumerated constants)
	Loop_Type (External or Internal)	(Enumerated constants)
	Edge_List	(An edge list of loop edges)
	Face_Pointers	(Pointer to face class)
Surface	Surface_Type	(Enumerated constants)
Face	Face_ID	Number
	Surface_Pointer	(Pointer to the surface)
	External_Loop	(Loop pointer)
	Internal_Loop_Count	(Number)
	Internal_Loop_List	(Vector of loop pointers)
Shell	Vertex_List	(Object of Vertex_List class)
	Edge_List	(Object of Edge_List class)
	Loop_List	(Vector of loop pointers)
	Surface_List	(Vector of surface pointers)
	Face_List	(Vector of face pointers)
	Name	(String)
	IGES_File	(Object of IGES_File Class)
Feature	Feature_ID	(Number)
	Feature_Type	(Enumerated constants)
	Feature_Origin	(Vertex pointer)
	Length	(Real)
	Width	(Real)
	Height	(Real)
Feature	Radius	(Real)
	Edge_List	(Vector of edge pointers)
	Face_List	(Vector of face pointers)
	Is_Subfeature	(Boolean)
	Direction	(Vector of up to three direction objects)

(continued)

Table 9.2 (continued)

Class name	Attribute	Type
Complex feature	Feature_ID	(Number)
	Feature_Type	(Enumerated constants)
	Feature_List	(Vector of Feature Pointers)
	Feature_Origin	(Vertex Pointer)
	Length	(Real)
	Width	(Real)
	Height	(Real)
	Radius	(Real)

9.8 Algorithms for Extracting Geometric Entities from CAD File

The IGES file is sequentially read (on a line basis) and parsed into appropriate entry classes known as **DEntry** and **PEntry**. The most important and useful sections of the IGES are the directory entry section and the parameter data section. **DEntry** represents an entry in the directory entry section while **PEntry** represents an entry in the parameter data section. The collection of directory entry classes is contained in a container class called **DSection.** Similarly, the parameter entry classes are contained in the **PSection** class. A **Parser class** object is created using these classes to parse the information present in the entries and classify the information into different classes that are used to represent different entities of the diagram described by the IGES file. Two algorithms for extraction of data from the IGES file into a proper set of data structures are defined in the following subsections.

9.8.1 Algorithm for Extracting Entries from Directory and Parameter Sections

// Algorithm to extract the directory entries
// and the parameter section entries from the IGES file.
// This process takes place during the construction of an object of IGESFile class.
// Each such object represents one IGES file.

1. Create a file descriptor IGES file.
2. Create an empty dSection1 class (container to store dEntry objects).
3. Create an empty pSection1 class (container to store pEntry objects).
4. Open the IGES file for reading using IGES File file descriptor.

// Read the file to scan and extract the directory and parameter sections.

5. While ReadLine line1 from the IGES file.

5.1 If line1 belongs to Directory section
  5.1.1 If line1 is the first line of Dsection
    5.1.1.1 Set dIndex to 1.
  5.1.2 ReadLine line2 from the IGES file.
  5.1.3 Create an object dEntry1 of class DEntry.
  5.1.4 Set dEntry1 index using dIndex.
  5.1.5 Initialize dEntry1 using string Line1 + Line2.
  5.1.6 Add dEntry1 to dSection1 class.
  5.1.7 Set dIndex = dIndex + 1.
5.2 If line1 belongs to Parameter Section
  5.2.1 If line1 is the first line of PSection
    5.2.1.1 Set pIndex to 1.
  5.2.2 Create an empty string Line2.
  5.2.3 while pEntry data incomplete
    5.2.3.1 Read Line Line3 from the IGES file.
    5.2.3.2 Append Line3 to Line2.
  5.2.4 Create an object pEntry1 of class PEntry.
  5.2.5 Set pEntry1 index equal to pIndex.
  5.2.6 Initialize pEntry1 using string Line1 + Line2.
  5.2.7 Add pEntry1 to pSection1 class.
  5.2.8 Set pIndex = pIndex + 1.
5.3 If line1 belongs to Terminate Section.
  5.3.1 Exit while loop.

6. End of while loop.

9.8.2 *Algorithm for Extracting the Basic Entities of the Designed Part*

// Part-extraction module, contained in the Shell class.
// This module extracts entities and groups them into lists.
// For example, it creates a list of all faces in the object represented by the IGES file.

Procedure to extract entities.

1. Create an object vertexList1 of vertexList class.
2. Create an object edgeList1 of edgeList class.
3. Create a vector of pointers loopList to Loop class.
4. Create a vector of pointers faceList to Face class.
5. Begin parsing entities from the IGES File object.
   5.1 Initialize counter i = 1.
   5.2 For i = 1 to size of dSection1 object from IGESFile object
   5.3 dEntry1 = DEntry object of index i from dSection1

5.4 if dEntry1 is a vertex list.
  5.4.1 pEntry1 = PEntry object pointed to by dEntry1, obtained from pSection1 of corresponding IGESFile object
  5.4.2 Initialize counter j = 1.
  5.4.3 For each vertex present in the pEntry1 object do.
  5.4.4 Instantiate a new vertex1 object of class Vertex.
  5.4.5 Assign id as j to vertex1.
  5.4.6 Initialize the object with vertex data from pEntry1.
  5.4.7 Add to vertexList1 a pointer to vertex1.
  5.4.8 Increment j by 1.
  5.4.9 End of For loop.
5.5 End For
5.6 Initialize counter i = 1.
5.7 For i = 1 to size of dSection1 object from IGESFile object
5.8 dEntry1 = DEntry object of index i from dSection1
5.9 If dEnry1 is an edge list.
  5.9.1 pEntry1 = PEntry object pointed to by dEntry1, obtained from pSection1 of corresponding IGESFile object
  5.9.2 Initialize counter j = 1.
  5.9.3 For each edge present in the pEntry1 class do.
  5.9.4 Instantiate a new object edge1 of class Edge.
  5.9.5 Assign id as j to edge1.
  5.9.6 Retrieve dEntry2 object that contains edge specific data.
  5.9.7 Retrieve pEntry2 object corresponding to dEntry2
  5.9.8 Instantiate a new object edgeS that is specific to the edge type.
  5.9.9 Initialize the edgeS object with edge1 and data from pEntry2.
  5.9.10 Assign start and terminate vertex to the edge using pointers from the vertexList1 object.
  5.9.11 Add to edgeList object a pointer to edges.
  5.9.12 Increment j by 1.
  5.9.13 End For.
5.10 End For.
5.11 Initialize counter i = 1.
5.12 For i = 1 to size of dSection1 object from IGESFile object.
5.13 dEntry1 = DEntry object of index i from dSection1.
5.14 If dEntry1 is a loop
  5.14.1 pEntry1 = PEntry object pointed to by dEntry1, obtained from pSection1 of corresponding IGESFile object.
  5.14.2 Create an instance loop1 of Loop class.
  5.14.3 Assign an id to loop1 using (size of loopList + 1).
  5.14.4 For each edge in the pEntry1
  5.14.5 If next edge is a Vertex
    5.14.5.1 Add to the loop1 a pointer to the vertex.
  5.14.6 Else Add to the loop1 a pointer to the edge.
  5.14.7 End For

5.14.8 Add to loopList the pointer to loop1.
5.15 End For
5.16 Initialize counter i = 1.
5.17 For i = 1 to size of dSection1 object from IGESFile object
5.18 dEntry1 = DEntry object of index i from dSection1.
5.19 If dEntry1 is a Face
5.19.1 pEntry1 = PEntry object pointed to by dEntry1, obtained from pSection1 of corresponding IGESFile object.
5.19.2 Create an instance face1 of Face class
5.19.3 Assign an id to face1 using (size of faceList + 1).
5.19.4 Obtain dEntry2 containing the DEntry of the Surface type of the Face from dSection1.
5.19.5 Instantiate a surface1 object specific to the type of the surface of the face.
5.19.6 Add to face1 a pointer to that surface object
5.19.7 For each Loop on the Surface
5.19.7.1 Add to face1 a pointer to loop object from loopList that represents this loop.
5.19.7.2 For each Edge in the loop add a pointer to face1.
5.19.8 End For
5.19.9 Add to faceList the pointer to face1
5.20 End For
6. End of extract entities procedure.

Figures 9.20–9.25 represent flowcharts of different modules of these two algorithms.

9.9 Extracting Form Features from CAD Files

The edge direction and the face direction are the basic entities of information that is used to extract both simple and compound form features from the object data structure. The edge directions in object models can be defined such that, when one walks along an edge, its face is always on the left-hand side. When an edge is in the external loop of a face, its direction will be in a counterclockwise direction relative to the surrounding face [19, 28]. On the other hand, when an edge is in the internal loop of a face, its direction will be clockwise as shown in Fig. 9.26.

The concave edge test used in this book is based on a cross product of the normal vectors of the two faces joined by a given edge. This is done by applying vector geometry to the face and edge direction vectors. Figure 9.27 shows the symbols used in this test where the *i*th face is designated as F_i, its corresponding normal direction vector is defined as $\mathbf{N}_i$ in the upward direction with respect to the given face, and the *k*th edge is designated as E_k.

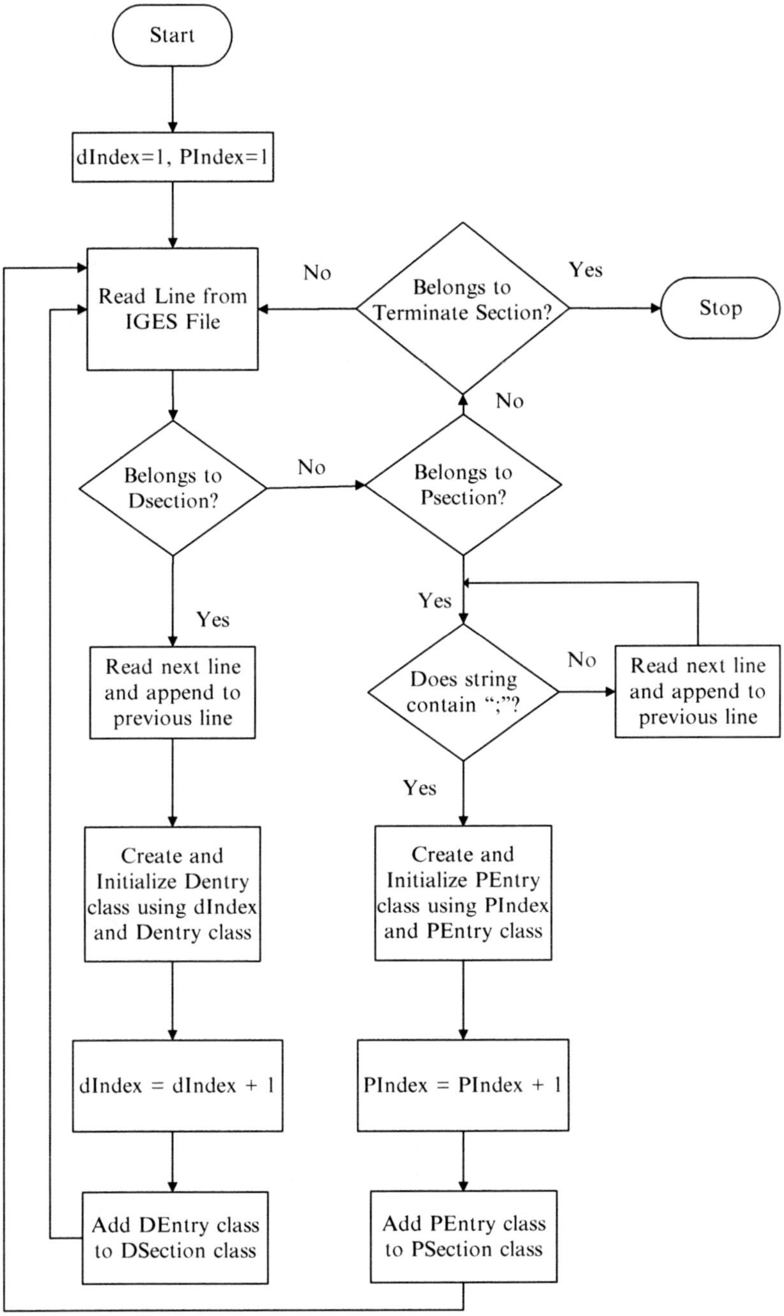

Fig. 9.20 Flowchart for creation of an IGES file object

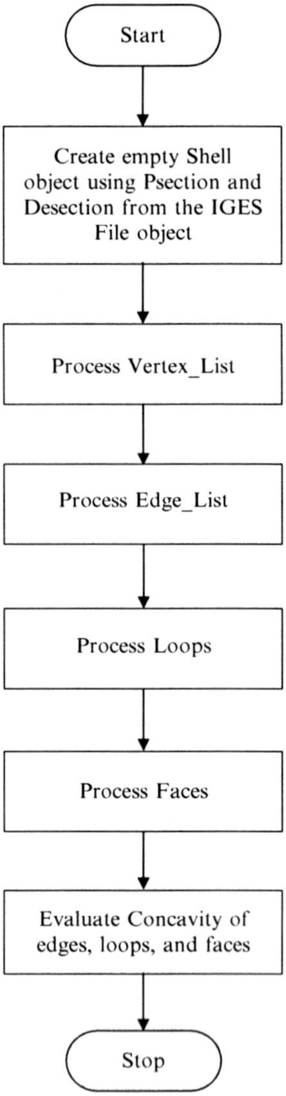

Fig. 9.21 Flowchart for shell object creation

The edge (E_K) shared by two faces (F_i and F_j) where the order is right to left from the left side of the edge view perspective. The direction vectors of the faces are as described above ($\mathbf{N}_i$ and $\mathbf{N}_j$). Finally, the edge's directional vector is given with respect to face F_i using loop L_i that contains the edge (E_k). The following is the methodology for concavity test [28]:

1. The cross (vector) product (**V**) of the directional vectors of the faces is determined as follows:

$$\mathbf{V} = \mathbf{N}i \times \mathbf{N}j \tag{9.4}$$

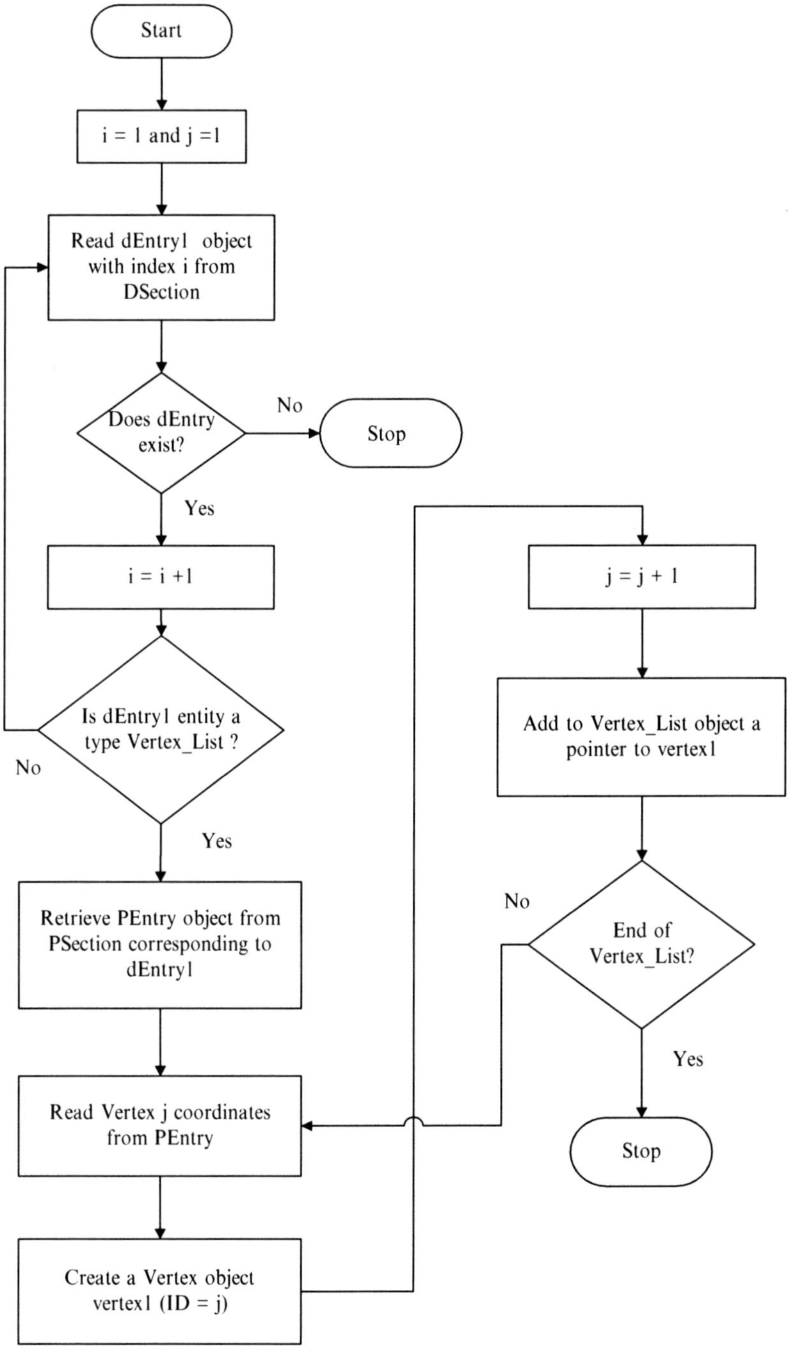

Fig. 9.22 Flowchart for vertex extraction

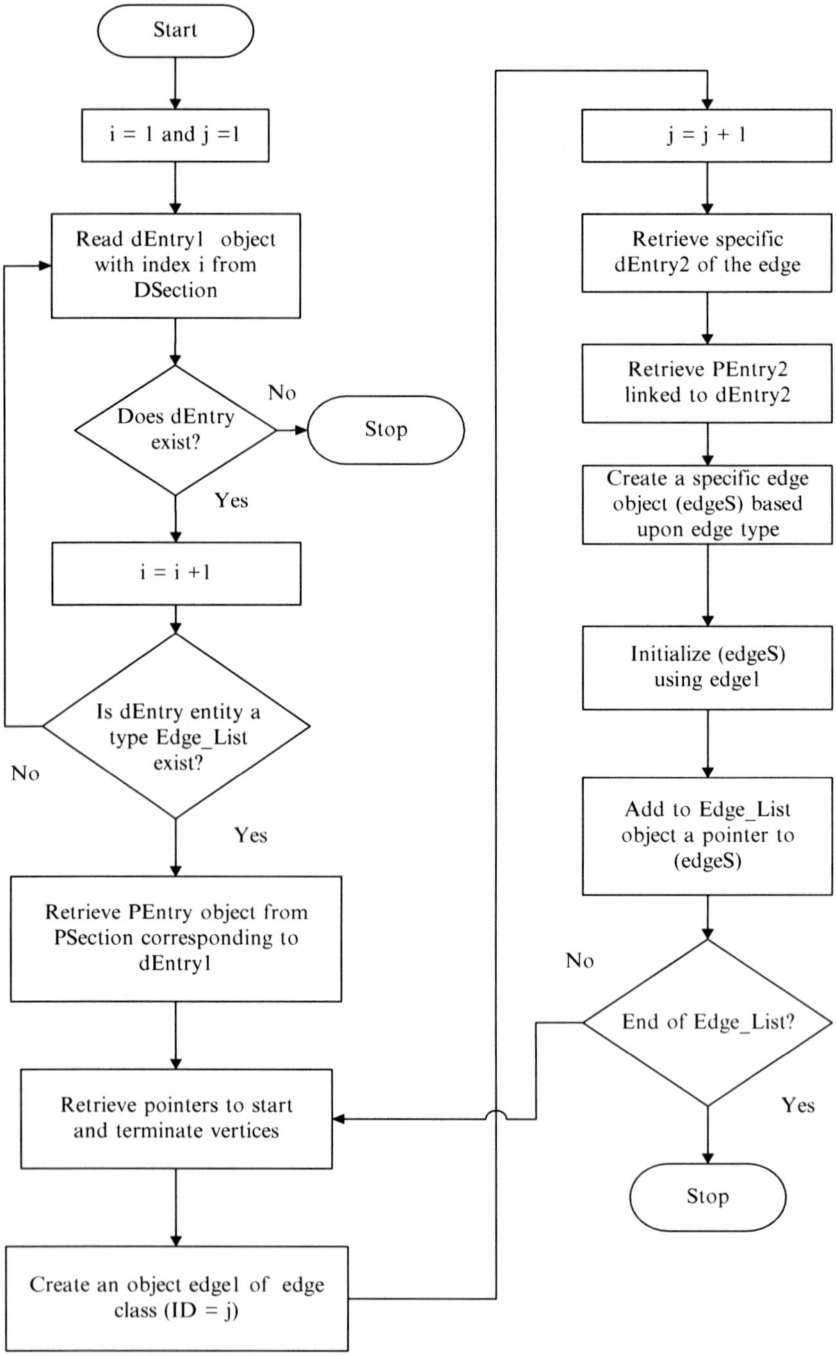

Fig. 9.23 Flowchart for edge extraction

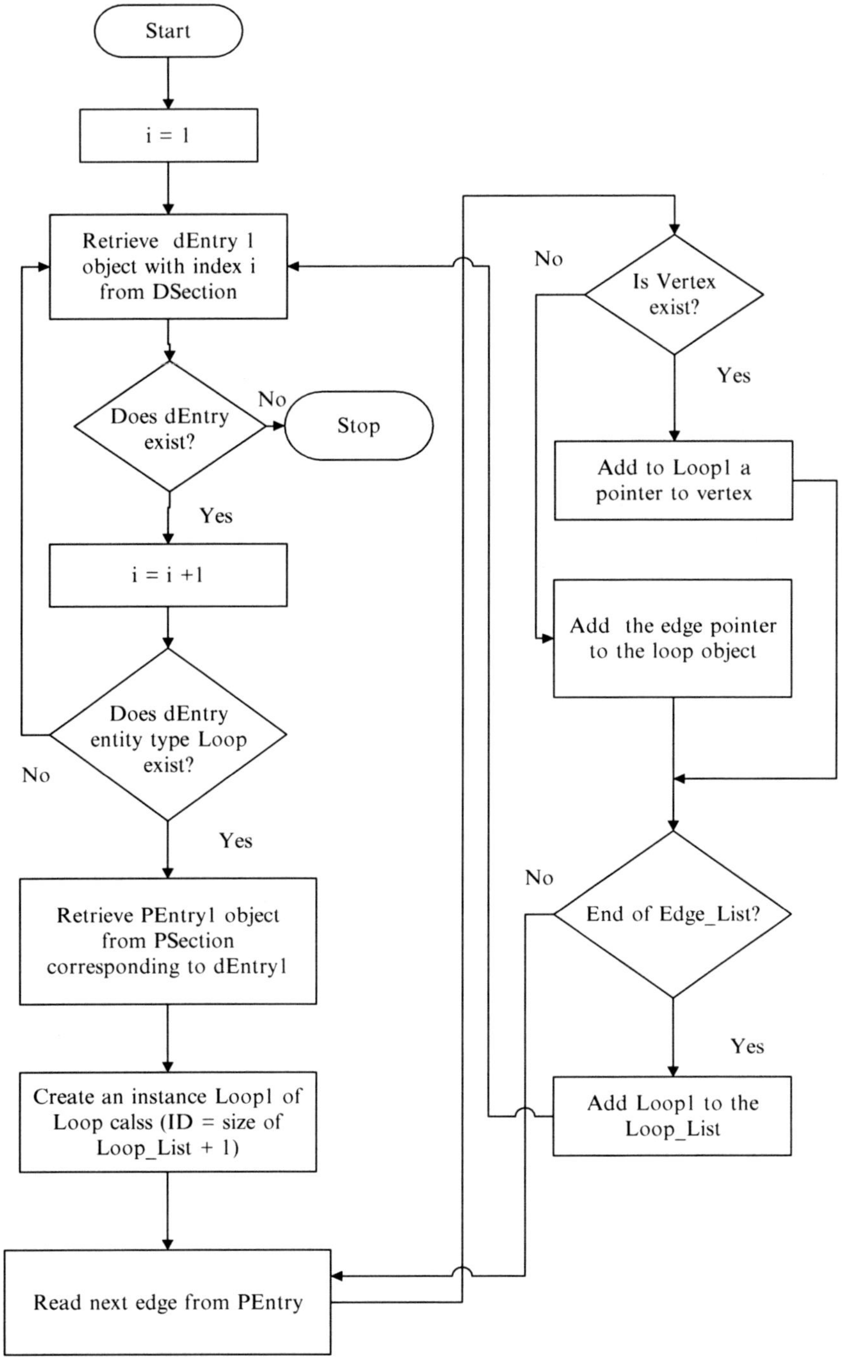

Fig. 9.24 Flowchart for loop extraction

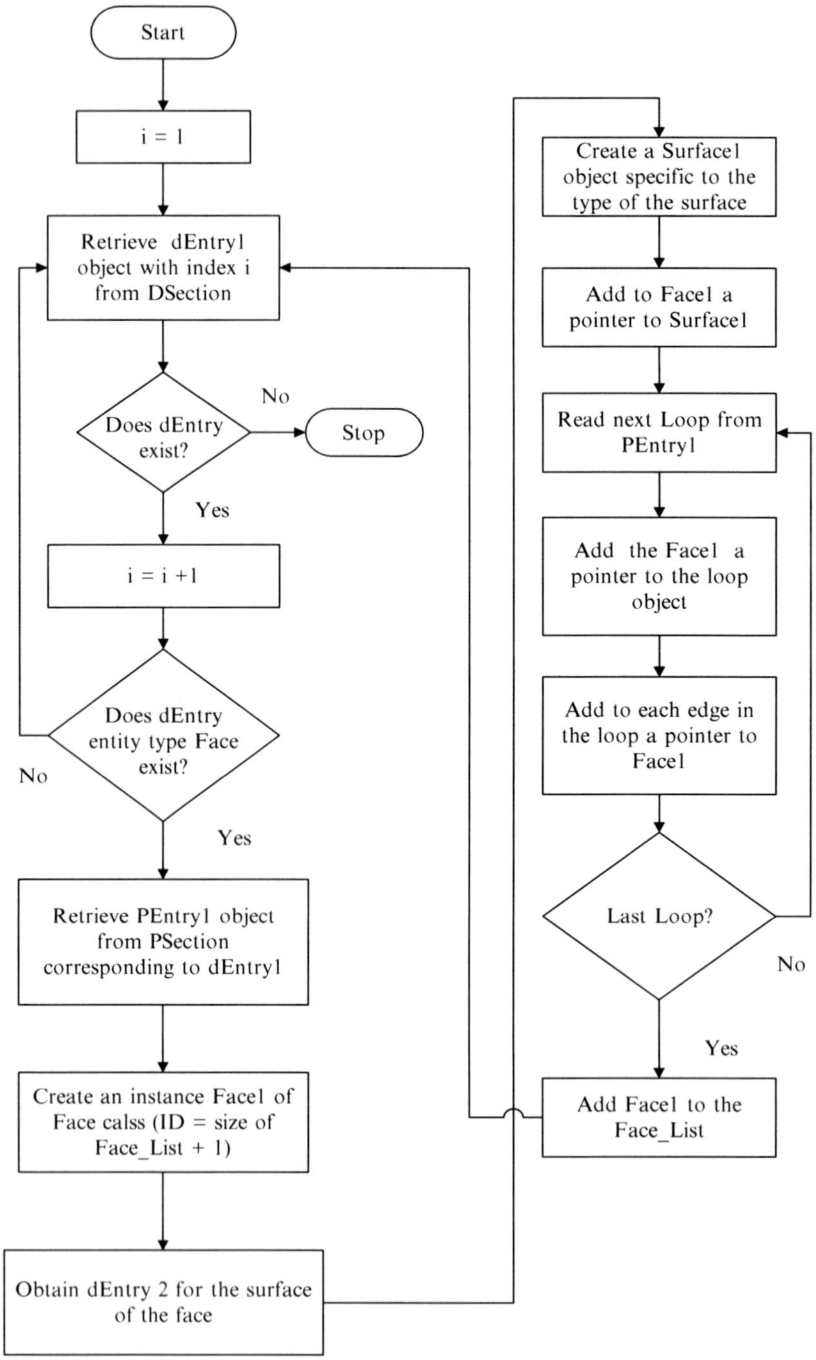

Fig. 9.25 Flowchart for face extraction

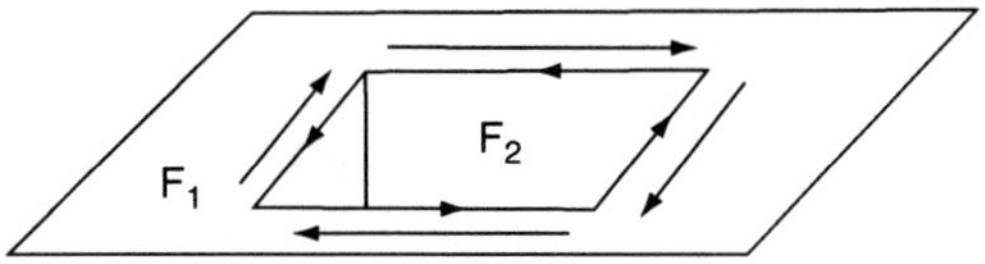

Fig. 9.26 The direction of edge

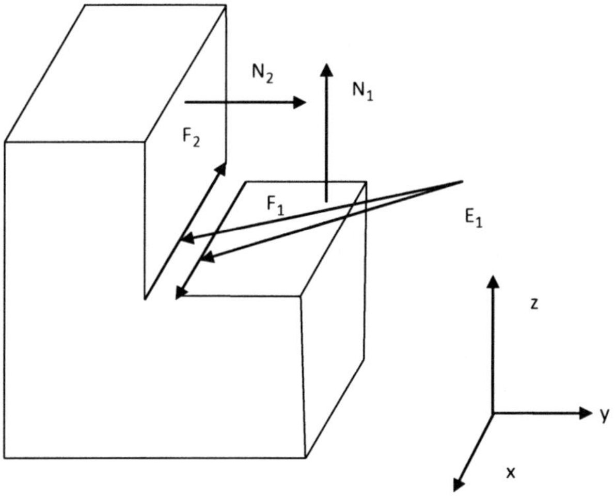

Fig. 9.27 A concave edge example

2. The direction of the edge E_k with respect to the face F_i is determined. The normal vector $\mathbf{N}_i$ of face F_i must be the first component in the cross product of step 1.
3. If the direction vector of edge E_k from step 2 is in the same direction of cross product **V**, then the edge E_k is a convex edge that concludes F_i and F_j are convex faces, otherwise, it will be a concave edge and F_i and F_j are concave faces. Also, if the cross product vector **V** is a zero vector that means the edge is a tangent category.

9.9.1 An Example for Finding the Concave Edge/Faces

The following is an example for finding the concave edge/faces by using the example in Fig. 9.27:

1. Find the face normal vectors $\mathbf{N}_1 = [0\ 0\ 1]$ and $\mathbf{N}_2 = [0\ 1\ 0]$.
2. Find the edge direction vector for $\mathbf{E}_1$ with respect to $F_1 = [1\ 0\ 0]$
3. Find the cross product $(\mathbf{V}) = [0\ 0\ 1]\ \mathrm{X}\ [0\ 1\ 0] = [-1\ 0\ 0]$

4. The edge direction of $\mathbf{E}_1$ and $\mathbf{V}$ have the opposite direction, so the edge $\mathbf{E}_1$ is a concave edge and F_1 and F_2 are concave faces.This procedure is applied to all the edges of the object to find the concave or convex faces. Moreover, concave features are identified by the premise that concave faces include at least one concave edge with adjacent concave faces forming a concave face group. Each concave face group defines a concave feature. Similarly, adjacent convex faces form a convex face group.

9.9.2 Algorithm for Determination of the Concavity of the Edge

The following algorithm is used to find the concavity of an edge entity. This algorithm uses the edge class to represent the edge entity.
Define:

startX = Edge.start-vertex.X-coordinate.
startY = Edge.start-vertex.Y-coordinate.
startZ = Edge.start-vertex.Z-coordinate.
termX = Edge.term-vertex.X-coordinate
termY = Edge.term-vertex.Y-coordinate.
termZ = Edge.term-vertex.Z-coordinate.
(Note: variables in italics are not attributes of classes)

1. Length of line = $\sqrt{[(\text{startX-termX})^2 + (\text{startY} - \text{termY})^2 + (\text{startZ} - \text{termZ})^2]}$
2. Edge.Concavity = UNKNOWN
3. If Edge.Face-Pointer [1] surface type == PLANE and Edge.Face-Pointer [2] surface type == PLANE
   3.1.1 Assign *crossDir* = cross product of Edge.Face-Pointer [1] normal vector and Edge.Face-Pointer [2] normal vector

3.2 If *crossDir* == 0
   3.2.1 Edge.Concavity = TANGENT

3.3 Else
   3.3.1 Calculate the direction vector *edgeDir* for the line with respect to the loop.
      3.3.1.1 If *crossDir* is in the same direction as *edgeDir*
         3.3.1.1.1 Edge.Concavity = CONVEX
      3.3.1.2 Else
         3.3.1.2.1 Edge.Concavity = CONCAVE

4. If Edge.Face-Pointer [1] surface type == RCCSURFACE and Edge.Face-Pointer [2] surface type == PLANE
   4.1 Find *dir1* (direction) that is orthogonal to the plane containing the edge and the axis of Edge.Face-Pointer [1].
   4.2 If *dir1* and normal of Edge.Face-Pointer [2] are orthogonal to each other

4.2.1 Edge.Concavity = TANGENT

The algorithm starts with assigning "unknown" to Edge.Concavity. Then the surface types of the first face and second face are checked. If both faces are planes, a new variable is assigned named *crossDir* which indicates the cross product of the normal vectors of the two faces. If the cross product equals zero, the concavity of the common edge of the two faces is tangent. Otherwise, the direction vector of the common edge (*edgeDir*) w.r.t the first face is calculated. If the cross product (*crossDir*) has the same value as the edge direction (*edgeDir*), the concavity of that edge is convex; otherwise, the concavity of the edge is concave. If the type of the surface of the first face is right cylinder surface (RCCSURFACE) and the surface type of the second face is plane, the direction (*Dir1*) that is perpendicular to the plane containing the edge and the axis of the first face is calculated. If the direction (*Dir 1*) is perpendicular to the normal vector of the first face, the concavity is tangent. Otherwise, the algorithm stops and all the previous procedure is repeated until all the edges in edge list class are checked. After classifying the edges into different categories as described in the previous algorithm, the classification of the loops is presented in the next subsection. Figure 9.28 represents a flowchart for determination of the concavity of the edge.

9.9.3 Algorithm for Determination of the Concavity of the Loop

The following algorithm is used to find the concavity of a loop entity (Loop.Loop-Concavity):

1. Set Loop.Loop-concavity = UNKNOWN.
2. Set *concave_count* = 0.
3. Set *convex_count* = 0.
4. For each edge in the loop (Loop.Edge-List [2])
   4.1. If Edge.Concavity is concave
      4.1.1. increment *concave_Count* by 1.
   4.2. If Edge.Concavity is convex
      4.2.1. increment *convex_Count* by 1.
5. End For
6. If *concave_count* == *edge_Count*
   6.1. Set Loop.Loop-concavity = CONCAVE.
7. Else If concave_count > 0
   7.1. Set Loop.Loop-concavity = HYBRID.
8. Else If convex_count == edge_Count
   8.1. Set Loop.Loop-concavity = CONVEX.

The algorithm starts with assigning "unknown" to Loop.Loop-Concavity. Also, two counters are set and initialized which are counters for the number of concave edge and convex edge (*concave_count*) and (*convex_count*), respectively. A third

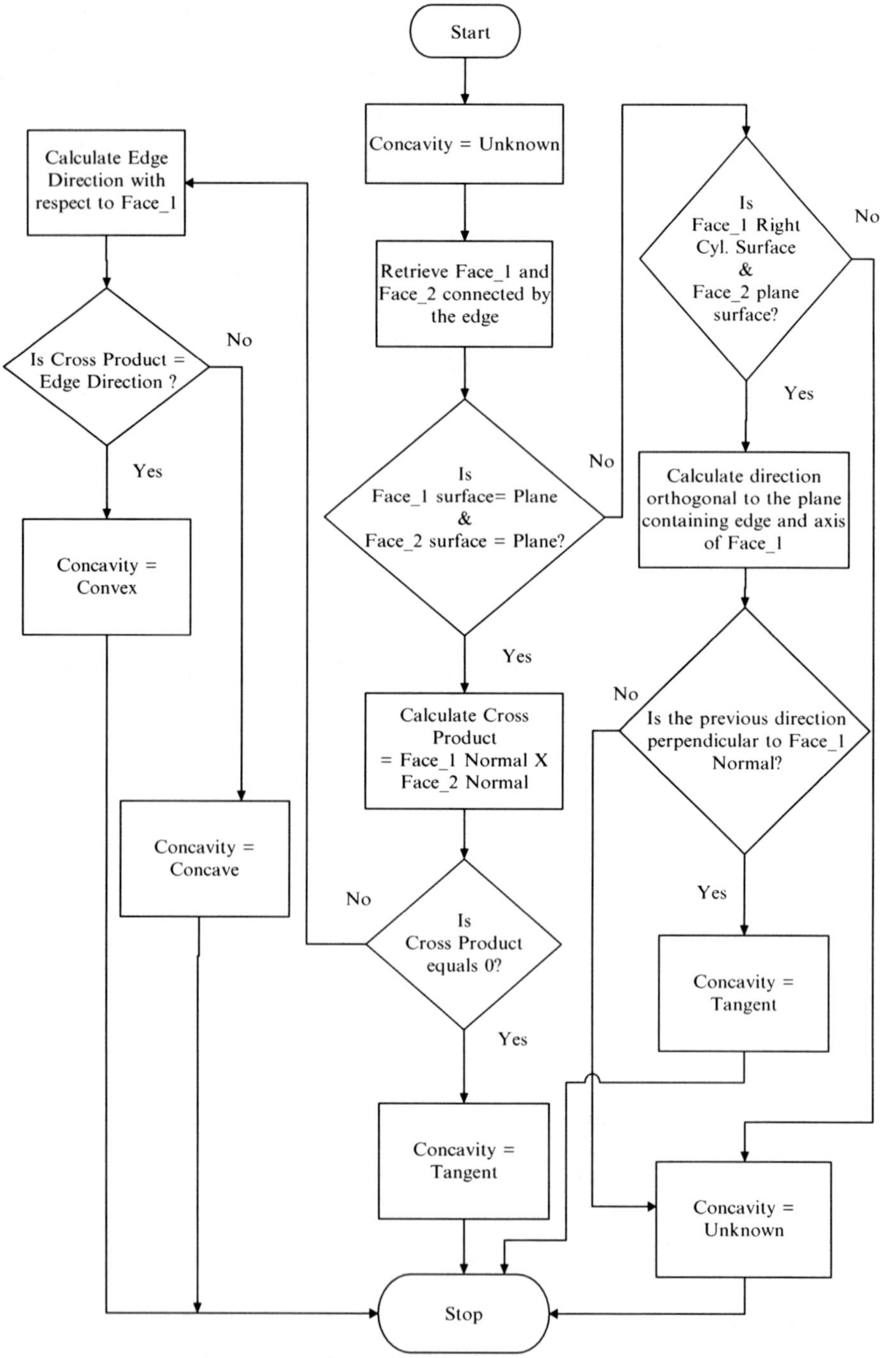

Fig. 9.28 Flowchart for determining the concavity of the edge

counter is assigned called *Edge_Count* which has the number of the edges in the loop that constitutes the specific face. If the edge in the loop is concave, the *concave_count counter* is incremented by 1. Similarly, if the edge in the loop is convex, the *convex_count counter* is incremented by 1. Then, the *concave_count counter* is checked. If the *concave_count counter* is equal to the *edge_count counter*, it implies that the concavity of the loop is concave. Similarly, if the *convex_count counter* is equal to the *edge_count counter*, it implies that the concavity of the loop is set to be convex. Finally, if the *concave_count counter* is greater than zero, the concavity of the loop is set to be hybrid. Figure 9.29 represents a flowchart for concavity analysis.

9.9.4 Algorithms for Feature Extraction (Production Rules)

The proposed methodology is able to extract many manufacturing features. Each feature has its own algorithm (production rule). The production rules are as follows:
Rule 1: Feature: **STEP THROUGH** (Fig. 9.30)

1. For every concave edge (e_1) of type Line in the edge list.
2. If the two common faces (face1 and face2) of the edge (e_1) are **plane** and **orthogonal** to each other
   2.1. If outer loop concave edge count equals 1 in both the faces
      2.1.1. STEP THROUGH found.
      2.1.2. Create a new StepT object and add to feature list.
3. End For

Rule 2: Feature **STEP BLIND** (Fig. 9.31)

1. For every three concave connecting edges (e_1, e_2, e_3) of type Line in the edge list.
2. If the three edges have a common vertex.
   2.1. If the every pair (e_1e_2, e_1e_3, e_2e_3) in the three edges have a common face of type plane surface (face1, face2, face3).
      2.1.1. If all the common faces are orthogonal to each other.
         2.1.1.1. If concave edge count of the outer loop of each such face equals 2
            2.1.1.1.1. STEP BLIND found
            2.1.1.1.2. Create a new StepB object and add to feature list
3. End For

Rule 3: Feature **SLOT THROUGH** (Fig. 9.32)

1. For every two concave edges of type Line in the edge list.
2. If the two edges have a common face (face1) connected to them and concave edge count of the outer loop of the face equals 2.

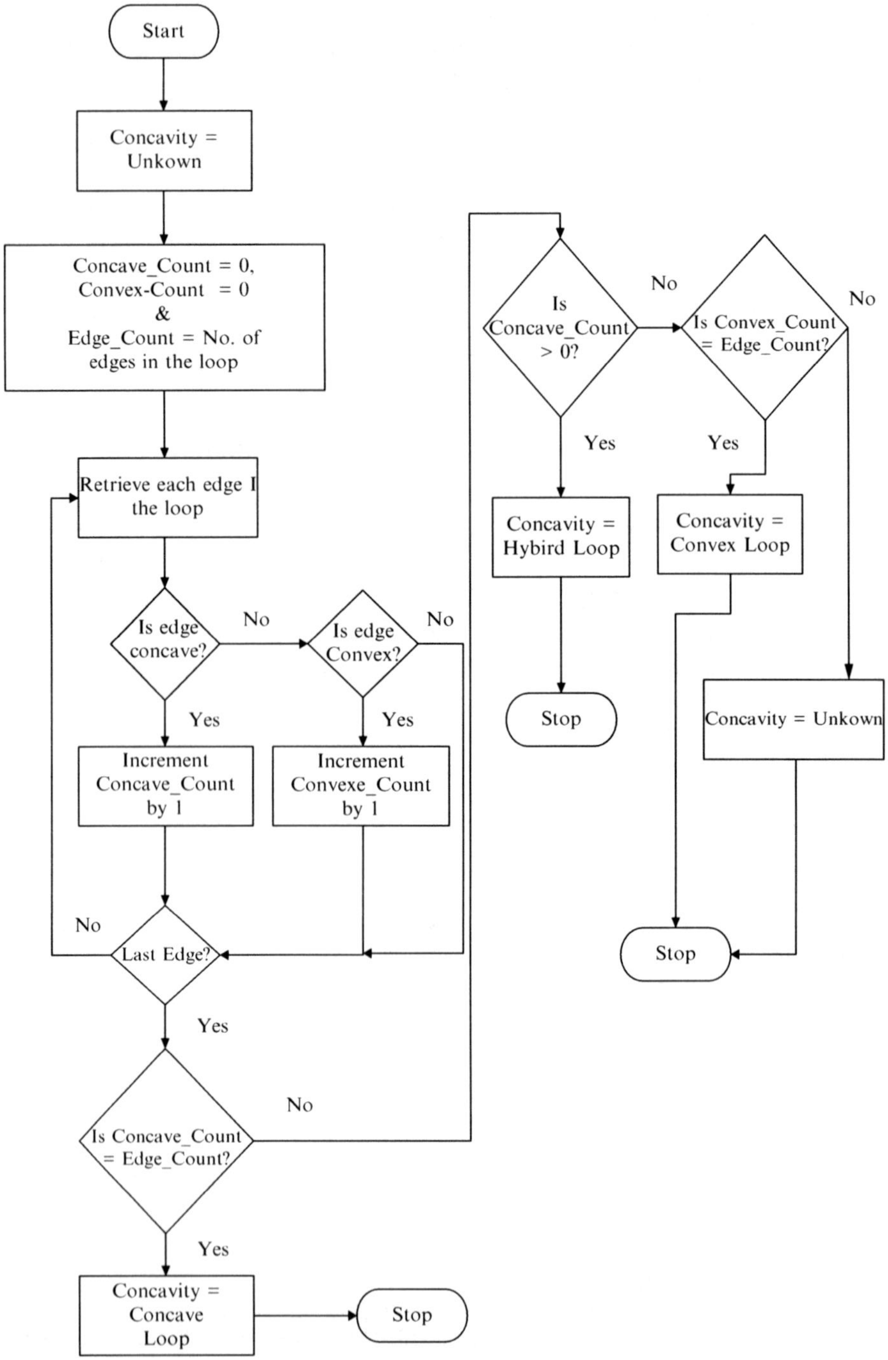

Fig. 9.29 Flowchart for determination of loop type

Fig. 9.30 Step through

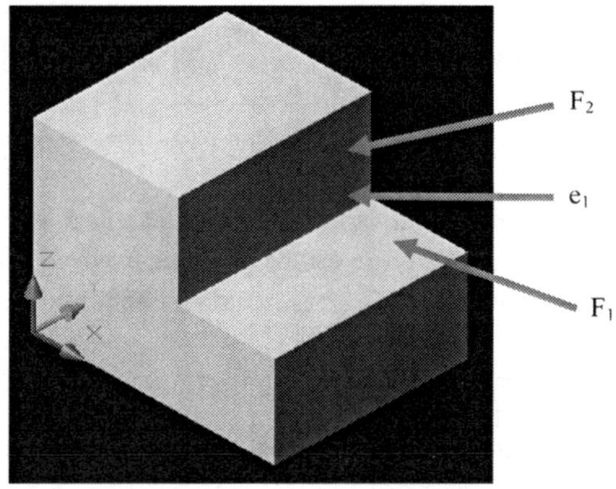

Fig. 9.31 Step blind

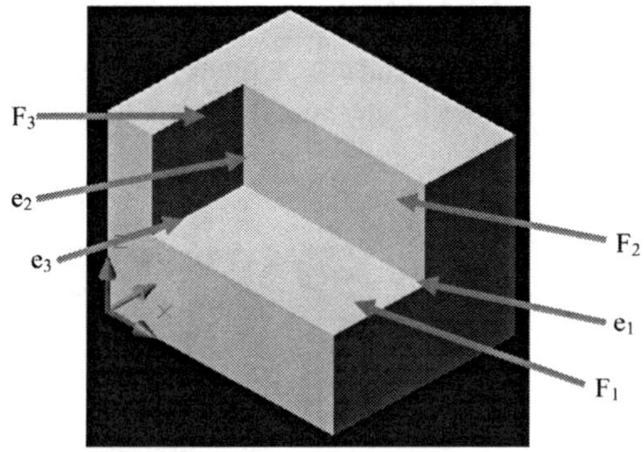

Fig. 9.32 Slot through

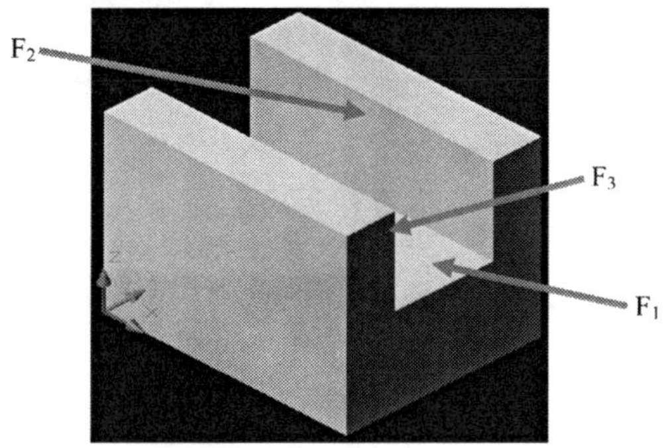

2.1. If the other faces (face2, face3) connected to edges are perpendicular to the common face and parallel to each other with concave edge count of their outer loops equals 1.

2.1.1. SLOT THROUGH found.
2.1.2. Create a new SlotT object and add to feature list.
3. End For

Rule 4: Feature **SLOT BLIND** (Fig. 9.33)

1. For every four faces of surface type plane (named as face1 to face4).
2. If face1, face2 concave edge count of outer loop equals 3 each and face3, face4 concave edge count of outer loop equals 2 each, and face1 connected to face2 by one edge.
   2.1. If face1 perpendicular to face2, face3, and face4 also face 3 and face4 parallel to each other.
      2.1.1. SLOT BLIND found
      2.1.2. Create a new SlotB object and add to feature list.
3. End For

Rule 5: Feature **POCKET THROUGH** (Fig. 9.34)

1. For every four faces of surface type plane (named as face1 to face4).
2. If concave edge count of outer loop equals 2 for each face.

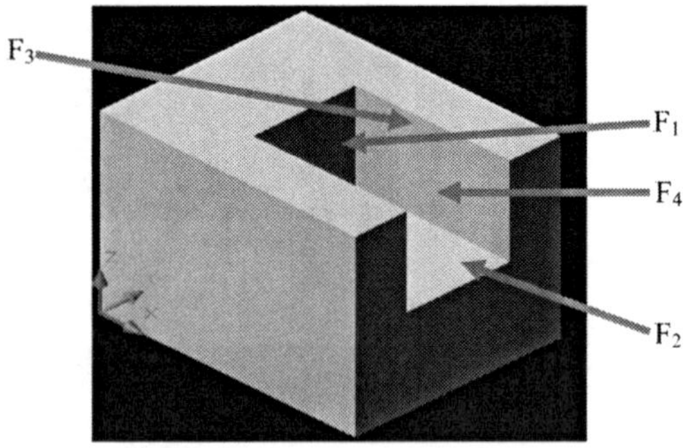

Fig. 9.33 Slot blind

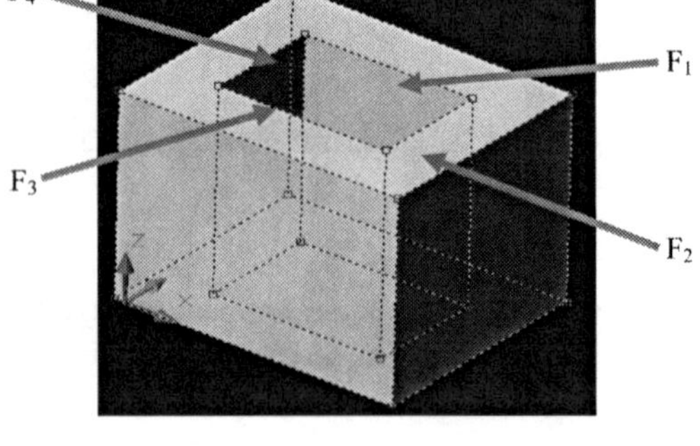

Fig. 9.34 Pocket through

2.1. If each consecutive pair of faces (F_1F_2, F_2F_3, F_3F_4, F_4F_1) has one common concave edge of type Line and the pair of faces is perpendicular to each other.
    2.1.1. POCKET THROUGH found.
    2.1.2. Create a new PocketT object and add to feature list.
3. End For

Rule 6: Feature **POCKET BLIND** (Fig. 9.35)

1. For every five faces of surface type plane (named as face1 to face5).
2. If face1 connects to the rest of faces (F_2, F_3, F_4, F_5) and is perpendicular to them
    2.1. If adjacent faces are perpendicular and nonadjacent faces parallel.
        2.1.1. Face1 concave edge count of outer loop equals 4 and concave edge count of other faces' outer loop equals 3 each.
            2.1.1.1. POCKET BLIND found.
            2.1.1.2. Create a new PocketB object and add to feature list.
3. End For

Rule 7: Feature **HOLE THROUGH** (Fig. 9.36)

1. For every two edges of type Line
2. If the common faces of two edges are the same (half cylindrical surfaces)
    2.1. If the two faces form cylinder.
        2.1.1. If the two loops formed by the circular edges of the outer loops of the two faces are internal loops.
            2.1.1.1. HOLE THROUGH found
            2.1.1.2. Create a new HoleT object and add to feature list.
3. End For

Rule 8: Feature **HOLE BLIND** (Fig. 9.37)

1. For every two edges of type Line
2. If the common faces of two edges are the same (cylindrical surfaces)

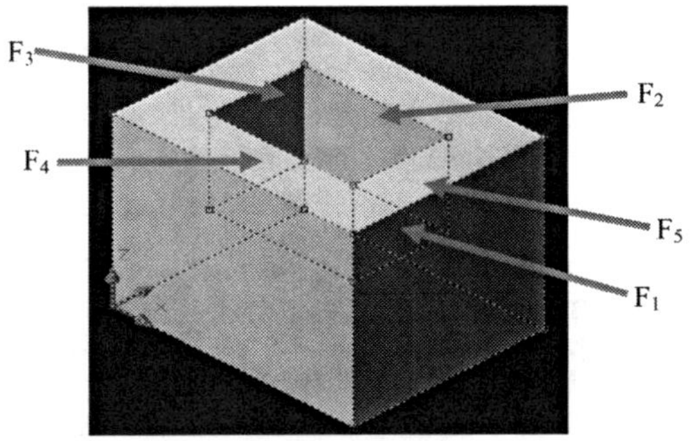

Fig. 9.35 Pocket blind

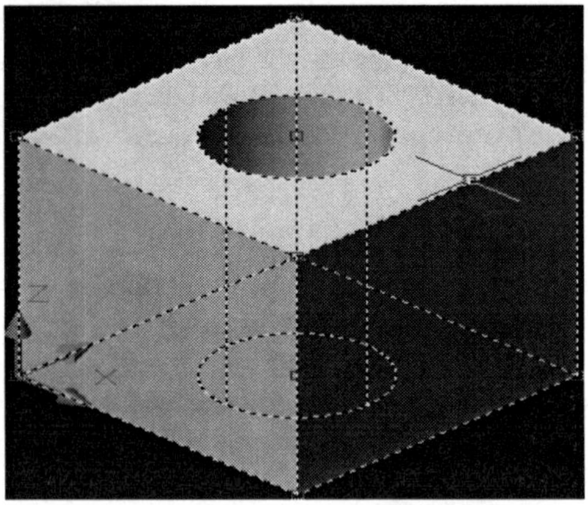

Fig. 9.36 Hole through

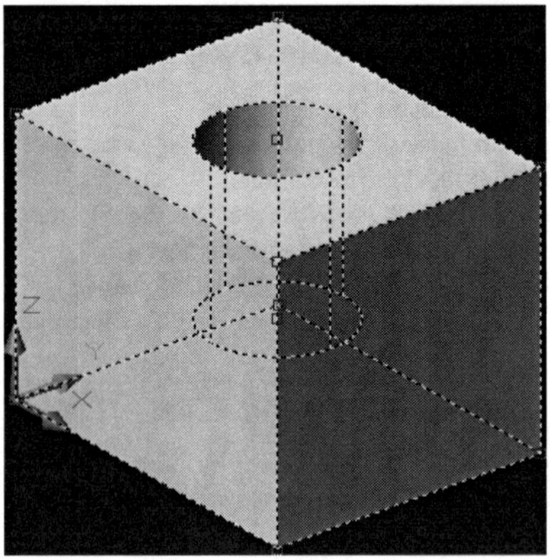

Fig. 9.37 Hole blind

2.1. If the two faces form a cylinder
   2.1.1. If only one of the loops formed by the circular edges of the outer loops of the two faces is an external loop.
      2.1.1.1. HOLE BLIND found.
      2.1.1.2. Create a new HoleB object and add to feature list.
3. End For

Rule 9: Feature **STEP THROUGH ROUND CORNER** (Fig. 9.38)

1. For every three tangent edges of type Line in the edge list (e_1, e_2, e_3)
2. If the common face of the two edges is quarter cylindrical surface (F_2, F_3).

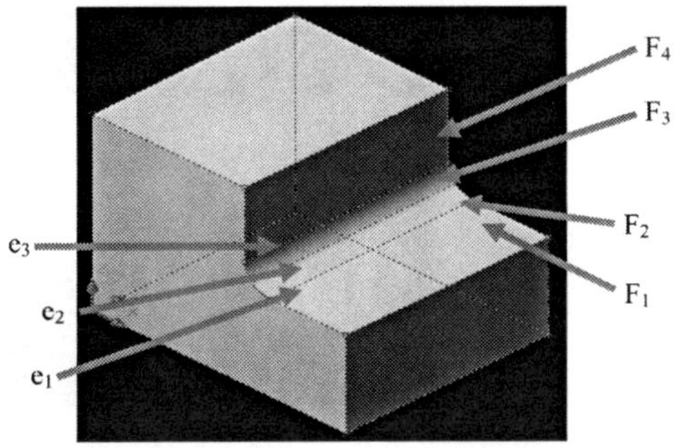

Fig. 9.38 Step through round corner

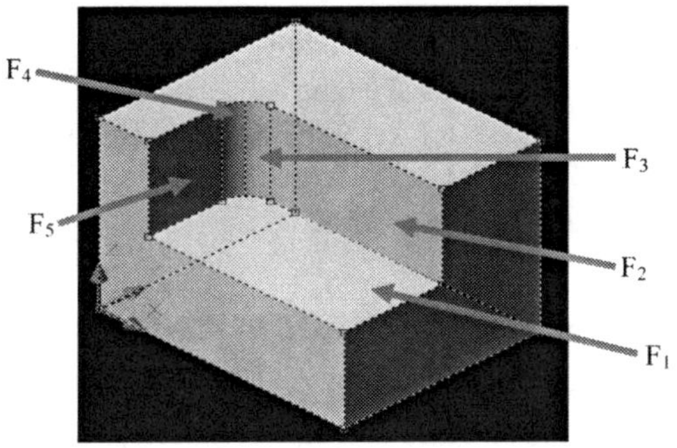

Fig. 9.39 Step blind round corner

2.1. The other two faces (F_1, F_4) connected to edges are perpendicular to each other.
 2.1.1. If concave edge count of the outer loops of the four faces equals 0 each.
  2.1.1.1. STEP THROUGH ROUND CORNER found.
  2.1.1.2. Create a new StepT_RC object and add to feature list.

3. End For

Rule 10: Feature **STEP BLIND ROUND CORNER** (Fig. 9.39)

1. For every five faces (named as face1 to face5).
2. If face3 and face4 are ¼ of a cylinder and the other three (face1, face2, face5) are plane.
 2.1. If face3 connects to face2 and face4 by tangent lines and to face1 by a circular arcs and is perpendicular to it.
 2.2. If face4 connects to face3 and face5 by tangent lines and to face1 by a circular arcs and is perpendicular to it.
  2.2.1. If face1, face2, and face5 are perpendicular to each other.

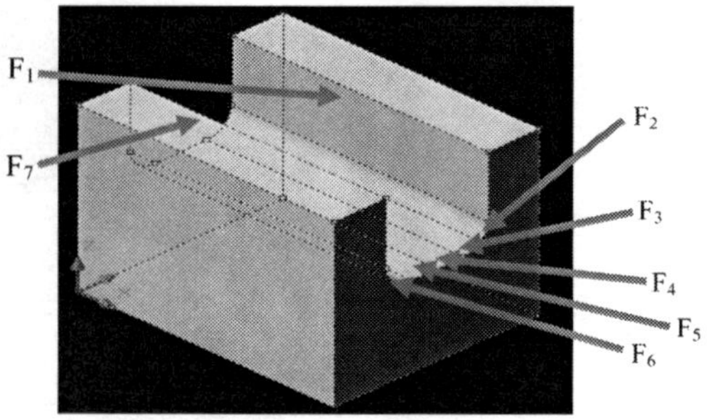

Fig. 9.40 Slot through round corner

2.2.1.1. If concave edge count of face2, face5 equals 1 and face1 equals 2 each.
  2.2.1.1.1. STEP BLIND ROUND CORNER found
  2.2.1.1.2. Create a new StepT_BC object and add to feature list.

3. End For

Rule 11: Feature **SLOT THROUGH ROUND CORNER** (Fig. 9.40)

1. For every seven faces (named as face1 to face7).
2. If face2, face3, face5, and face6 are ¼ of a cylinder.
   2.1. If all faces (face1 to face7) connected by tangent lines
     2.1.1. If face1, face7 perpendicular to face4 and parallel to each other
       2.1.1.1. If concave edge count of the outer loop of all faces equals 0 each.
         2.1.1.1.1. SLOT THROUGH ROUND CORNER found
         2.1.1.1.2. Create a new SlotT_RC object and add to feature list.
3. End For

Rule 12: Feature **SLOT BLIND ROUND CORNER** (Fig. 9.41)

1. For every eight faces (named as face1 to face8).
2. If face3, face 4, face6, and face7 have ¼ of a cylinder.
   2.1. If all faces and edges connected to each other.
     2.1.1. If face2 and face8 are perpendicular to face5 and parallel to each other
       2.1.1.1. If face1 is perpendicular to all other faces and concave edge count of outer loop of face1 equals 3.
       2.1.1.2. If (face3, face4) and (face6, face7) connected to each other and face2, face5, and face8 are connected to them by tangent edges.
       2.1.1.3. If concave edge count of outer loop of face2, face5, and face8 equals 1 each and of face3, face4, face6, and face-7equals 0 each.

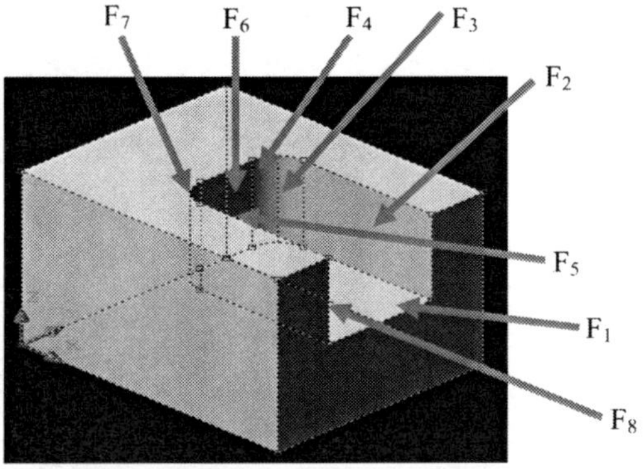

Fig. 9.41 Slot blind round corner

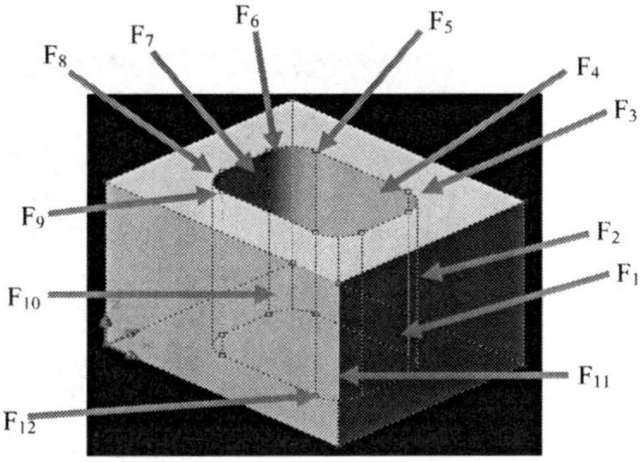

Fig. 9.42 Pocket through round corner

2.1.1.3.1. SLOT BLIND ROUND CORNER found.
2.1.1.3.2. Create a new SlotB_RC object and add to feature list.

3. End For

Rule 13: Feature **POCKET THROUGH ROUND CORNER** (Fig. 9.42)

1. For every 12 faces (named as face1 to face12).
2. If faces (F_2, F_3), (F_5, F_6), (F_8, F_9), (F_{11}, F_{12}) ¼ of a cylinder and all other faces are plane surface.
   2.1. If all the adjacent faces are connected by tangent lines, and concave edge count of the outer loop of all the faces equals 0 each.
      2.1.1. POCKET THROUGH ROUND CORNER found
      2.1.2. Create a new PocketT_RC object and add to feature list.
3. End For

Rule 14: Feature **POCKET BLIND ROUND CORNER** (Fig. 9.43)

1. For every 13 faces (named as face1 to face13).

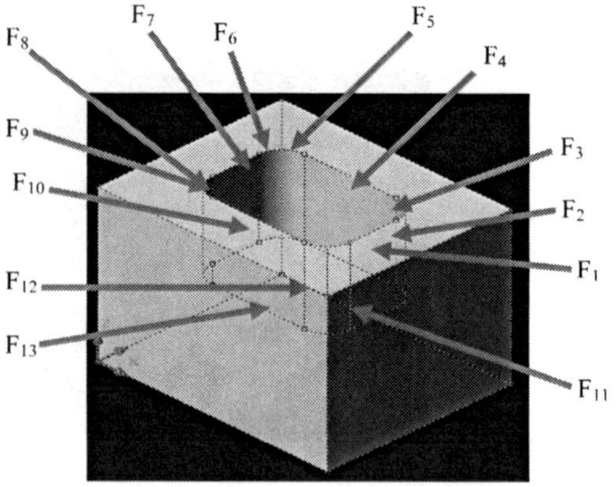

Fig. 9.43 Pocket blind round corner

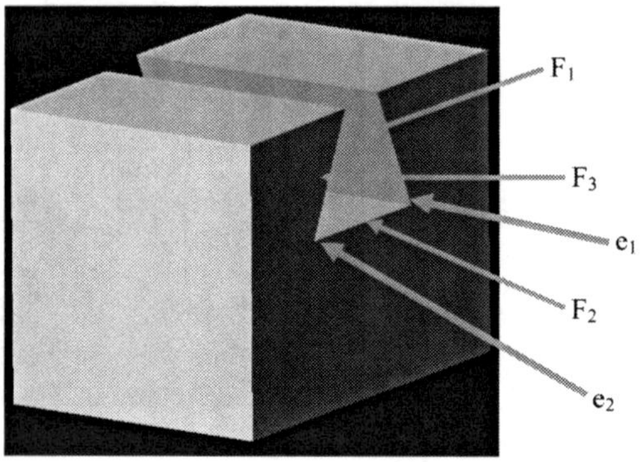

Fig. 9.44 Dovetail slot

1.1. If faces (F_2, F_3), (F_5, F_6), (F_8, F_9), (F_{11}, F_{12}) ¼ of a cylinder and all other faces are plane surface.
1.2. If all faces from face1 to face12 are connected by tangent lines and all faces are perpendicular to face13.
 1.2.1. If all faces except face13 have concave edge count of outer loop equal to 1 and for face13 equals 4.
  1.2.1.1. POCKET BLIND ROUND CORNER found.
  1.2.1.2. Create a new PocketB_RC object and add to feature list.
2. End For

Rule 15: Feature **DOVETAIL SLOT** (Fig. 9.44)

1. For every three faces (named as face1, face2, face3).
2. If face1 is connected to face2 and face2 is connected to face3.
 2.1. If the connecting edges (e_1 and e_2) between the faces are concave.
  2.1.1. If the angle at the edges is less than 90°.

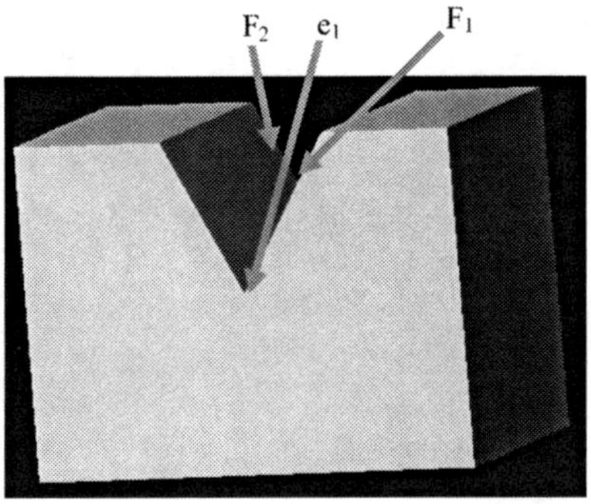

Fig. 9.45 V Slot through

2.1.1.1. DOVTAIL SLOT THROUGH found.
2.1.1.2. Create a new DOV SLOT THROUGH feature and add to the feature list.

3. End For

Rule 16: Feature **V SLOT THROUGH** (Fig. 9.45)

1. For every two faces (named as face1, face2).
2. If face1, face2 are connected to each other.
   2.1. If the common edge (e_1) is concave.
      2.1.1. If the angle between face1 and adjacent faces where the common edge (e_1) is not directly connected to (e_1) is greater than 90°.
      2.1.2. If the angle between face2 and adjacent faces where the common edge (e_1) is not directly connected to (e_1) is grater than 90°.
         2.1.2.1. V SLOT THROUGH found.
         2.1.2.2. Create a new V SLOT THROUGH feature and add to the feature list.
3. End For

Rule 17: Feature **W SLOT THROUGH** (Fig. 9.46)

1. For every three faces (named as face1, face2, face3).
2. If face1 is connected to face2 and face2 is connected to face3.
   2.1. If the connecting edges (e_1 and e_2) between the faces are concave.
      2.1.1. If the angle at the edges is greater than 90°.
      2.1.2. W SLOT THROUGH found.
      2.1.3. Create a new W SLOT THROUGH feature and add to the feature list.
3. End For

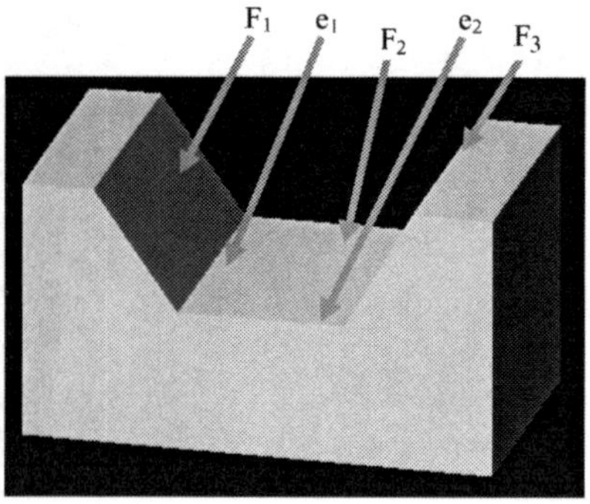

Fig. 9.46 W Slot through

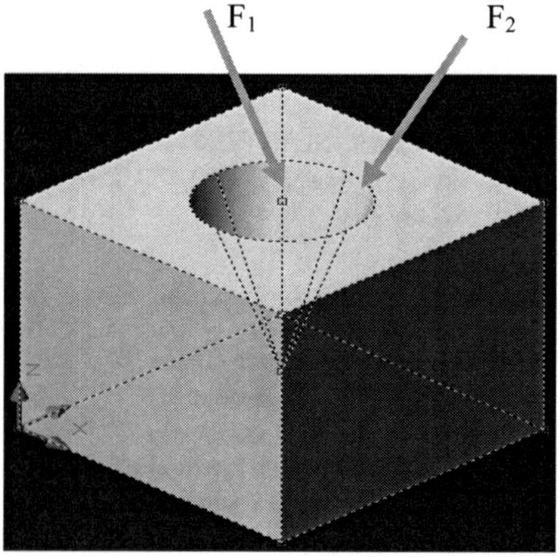

Fig. 9.47 Cone

Rule 18: Feature **CONE** (Fig. 9.47)

1. For every two faces (named as face1, face2).
2. If face1 and face2 have surface of type surface of revolution and face1 is connected to face2 at exactly two common edges.
   2.1. If the common edges are of type line (straight) and all the noncommon edges are circular arcs.
   2.2. If both the common edges on each face intersect each other.
      2.2.1. If the angle between generatrix and axis is less than 90° for both faces, and corresponding circular arcs have the same radius
         2.2.1.1. CONE found.
         2.2.1.2. Create a new CONE feature and add to feature list
3. End For

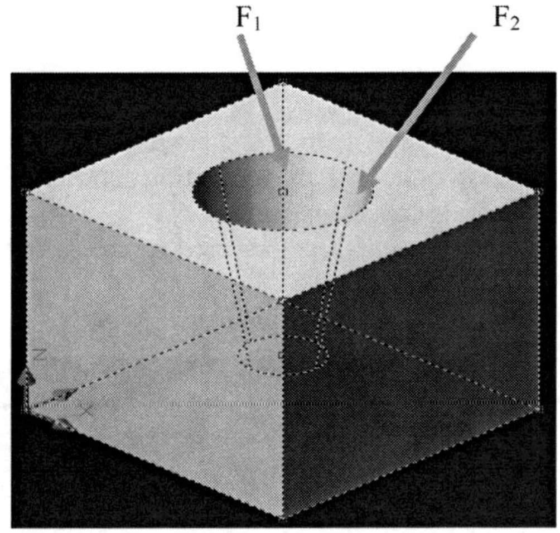

Fig. 9.48 Sink

Rule 19: Feature **SINK** (Fig. 9.48)

1. For every two faces (named as face1, face2).
2. If face1 and face2 have surface of type surface of revolution and face1 is connected to face2 at exactly two common edges.
   2.1. If the common edges are of type line (straight) and all the noncommon edges are circular arcs.
   2.2. If both the connecting edges on each face are separated by at least one other edge.
      2.2.1. If the angle between generatrix and axis is less than 90° for both faces, and corresponding circular arcs have the same radius.
         2.2.1.1. SINK found.
         2.2.1.2. Create a new SINK feature and add to feature list.
3. End For

9.10 Summary

In this chapter, the literature review shows the major contributions of the methods related to feature-based design. The representation of an object can be modeled in various forms. Many techniques have been developed to acquire information about the object in order to build the model and to recognize the object. Some of the literature mentions the object recognition; some uses feature recognition or part database to gain more knowledge about the object and then reconstruct the object or build a higher level model of the part.

9.11 Review Questions

9.1 What is the current technology of feature representations?
9.2 List all the feature recognition techniques.
9.3 Explain the syntactic pattern recognition approach.
9.4 Explain the logic-based approach.
9.5 Develop a rule by using the logic-based approach for the following component:

9.6 Explain the graph-based approach.
9.7 Describe briefly the AAG.
9.8 Construct the AAG graph for the following component:
9.9 Define the ES and explain in detail its components.
9.10 What are volume decomposition and composition approaches?
9.11 Explain hint generation process for the viewer-centered approach.
9.12 Draw the proposed diagram for IFRM presented in this chapter.
9.13 What are the three main phases for the IFRM presented in this chapter?
9.14 What are the basic IGES entities used for the IFRM? Explain.
9.15 Develop an alternative hierarchy of classes and attributes of the designed object.
9.16 What is the difference between interior form feature ($FF_{interior}$) and exterior form feature ($FF_{exterior}$)? Support your answer with a sketch.
9.17 Distinguish between convex features and concave features.
9.18 Discuss the geometry and topology of B-rep.
9.19 Define the loop and discuss its classification.
9.20 Develop an algorithm for extracting entries from directory and parameter sections of IGES file.
9.21 Develop an algorithm for extracting the basic entities of the designed part.
9.22 Explain how the direction of the edge can be defined.
9.23 What is concavity test?
9.24 Perform the concavity test for the following component:

9.25 Develop an alternative algorithm for determination of the concavity of the edge.
9.26 Develop an alternative algorithm for determination of the concavity of the loop.
9.27 Develop alternative production rules for the following features:

References

1. Ahmad N, Haque AFM (2001) Manufacturing feature recognition of parts using DXF files. Fourth international conference on mechanical engineering, Dhaka, Bangladesh, 1(1):111–115.
2. Ahmad N, Haque AFM, Hasin AA (2001) Current trend in computer aided process planning. Proceedings of the 7th annual paper meeting and 2nd international conference of the Institution on Engineering, Dhaka, Bangladesh, 1(10):81–92.
3. Alan C, Shou L (1998) A volume decomposition approach to process planning for prismatic parts with depression and protrusion design features. Int J Comput Integr Manuf 11(6):548–563.
4. Allada V, Anand S (1996) Machine understanding of manufacturing features. Int J Prod Res 34(7):1791–1820.
5. Aslan E, Seker U, Alpdemir N (1999) Data extraction from CAD model for rotational parts to be machined at turning centers. Turk J Eng Environ Sci 23(5):339–347.
6. Bartholomew O, Nnaji BO, Liu HC (1990) Feature reasoning for automatic robotic assembly and machining in polyhedral representation. Int J Prod Res 2(3):517–540.
7. Chang T-C, Wysk RA (1984) Integrated CAD and CAM through automated process planning. Int J Prod Res 1(1):158–172.
8. Choi BK, Barash MM, Anderson DC (1984) Automatic recognition of machined surfaces from 3D solid models. Comput-Aided Des 16(2):81–86.
9. Chuang S-H, Henderson M (1994) Using subgraph isomorphism to recognize and compose boundary representation features. J Mech Des 116(1):793–800.
10. Devireddy CR, Ghosh K (1999) Feature-based modeling and neural networks-based CAPP for integrated manufacturing. Int J Comput Integr Manuf 12(1):61–74.
11. Dong J, Parsaei HR (1994) An automated process planning system. Intell Automat Soft Comput: Trend Res 1(1):57–62.
12. Ferreira JCE, Hinduja S (1992) Convex hull-based feature-recognition method for 2.5D components. Comput-Aided Des 22(1):41–49.
13. Floriani LD (1987) A graph-based approach to object-oriented feature recognition, Proc. 3rd ACM Symp. Computational Geometry, 100–109.
14. Fu MW Ong SK Lu WF Lee IBH, Nee AYC (2003) An approach to identify design and manufacturing features from a data exchanged part model. Comput-Aided Des 35(1): 979–993.
15. Gavankar P, Henderson MR (1990) Graph based extraction of protrusion and depression from boundary representations. Comput-Aided Des 22(7):42–450.
16. Gossard DC, Zuffante RP, Sakurai H (1988) Representing dimensions, tolerances and features in MCAE system. IEEE Comput Graph Appl 51–59.
17. Henderson MR, Anderson DC (1984) Computer recognition and extraction of form feature: a CAD/CAM link. Comput Ind 5(4):329–339.
18. Henderson MR, Chang GJ (1989) FRAPP: automated feature recognition and process planning from solid model data. ASME Comput Eng Proc 529–536.
19. Hwang J (1988) Rule-based feature recognition: concepts, primitives and implementation. Thesis, Arizona State University.
20. Joshi S, Chang TC (1988) Graph-based heuristics for recognition of machined features from a 3D solid model. Comput-Aided Des 20(2):58–66.
21. Jung H, Inho H (1999) Manufacturable feature recognition and its integration with process planning. Fifth symposium on solid modeling, Ann Arbor, Michigan, 108–118.

22. Kang TS, Nnaji BO (1993) Feature representation and classification for automatic process planning. J Manuf Syst 12(2):133–145.
23. Kao CY, Kumara SRT (1993) 3-D manufacturing feature recognition using super relation graph method. Proceedings of 2nd industrial engineering research conference, 614–618.
24. Kao C-Y, Kumara SRT, Kasturi R (1995) Extraction of 3D object features from CAD boundary representation using super relation graph method. IEEE Trans Patt Anal Mach Intell 17(12):1228–1233.
25. Kayacan MC, Celik SA (2003) Process planning system for prismatic parts. Integr Manuf Syst 14(2):75–86.
26. Li RK, Bedworth DD (1988) A framework for integration of computer-aided design and computer aided process planning. Comput Ind Eng 14(4):395–413.
27. Liu C-H, Perng D-B, Chen Z (1994) Automatic form feature recognition and 3D part recognition from 2D CAD data. Comput Ind Eng 14(4):689–707.
28. Liu S, Gonzalez M, Chen J (1996) Development of an automatic part feature extraction and classification system taking CAD data as input. Comput Ind 29(1):137–150.
29. Luger GF, Stubblefield WA (1989) Artificial intelligence and the design of expert systems. Benjamin/Cummings, Menlo Park.
30. Madurai SS, Lin L (1992) Rule-based automatic part feature extraction and recognition from CAD data. Comput Ind Eng 22(1):49–62.
31. Mansour S (2002) Automatic generation of part programs for milling sculptured surfaces. J Mater Process Technol 127(1):31–39.
32. Meeran S, Pratt MJ (1993) Automatic feature recognition from 2D drawings. Comput-Aided Des 25(1):7–17.
33. Munns A, Li Y, Wang XC (1995) A rule-based feature extraction from CSG representations and an application in construction. Proc SPIE – Int Soc Opt Eng 2620(1):269–276.
34. Nagaraj HS, Gurumoorthy B (2002) Machinable volume extraction for automatic process planning. IIE Trans 34(1):393–410.
35. Nasser B, Hamid P, Leep HR (1996) A prototype solid modeling based automated process planning system. Comput Eng 31(1/2):169–172.
36. Perng DB, Chen Z, Li RK (1990) Automatic 3D machining feature extraction from 3D CSG solid input. Comput-Aided Des 22(1):285–295.
37. Ping C, Chang C (2000) An integrated artificial intelligent computer aided process planning system. Int J Comput Integr Manuf 13(6):483–497.
38. Requicha AAG, Chan SC (1986) Representation of geometric features, tolerances, and attributes in solid modelers based on constructive geometry. IEEE J Robot Automat RA-2(3):56–166.
39. Rozenfeld H, Kerry HT (1999) Automated process planning for parametric parts. Int J Prod Res 37(17):3981–3993.
40. Roy U, Liu CR (1988) Feature-based representational scheme of a solid modeler for providing dimension and tolerancing information. Robot Comput-Integr Manuf 4(3/4):335–354.
41. Salomons OW et al (1993) Review of research in feature based design. J Manuf Syst 12(2):113–132.
42. Sakurai H (1995) Volume decomposition and feature recognition: part I-polyhedral objects. Comput-Aided Des 27(11):833–843.
43. Sakurai H, Dave P (1996) Volume decomposition and feature recognition: part II-curved objects. Comput-Aided Des 28(6/7):519–537.
44. Sashikumar V, Milind S, Vinay K (2001) A graph-based framework for feature recognition. Appearing in ACM symposium on solid modeling and applications, Ann Arbor, Michigan, 194–205.
45. Schulte RM, Padmanabhan S, Devugun MS (1992) Feature-driven, process-based approach to the integration of CAD/CAM in wireframe models. Int J Prod Res 30(5):1005–1028.
46. Shah JJ, Roger MT (1988) Expert form feature modeling shell. Comput-Aided Des 20(9):515–524.

47. Shah JJ, Shen Y, Shirur (1994) Determination of machining volumes from extensible sets of design features. In: Shah JJ, Mantyla M, Nau D (eds) Advances in feature based manufacturing. Elsevier, Amsteram, pp 281–297.
48. Sharma R, Gao JX (2002) Implementation of STEP application protocol 224 in an automated manufacturing planning system. Proc Inst Mech Eng, Part B, J Eng Manuf 216(1):1 277–1452.
49. Sheu JJ (1998) A computer integrated manufacturing system for rotational parts. Int J Comput Integr Manuf 11(6):538–547.
50. Somashekar S, Michael W (1995) An overview of automatic feature recognition techniques for computer-aided process planning. Comput Ind 26(1):1–21.
51. Sommerville MGL, Cleark DER, Corney JR (1995) Viewer-centered feature recognition. Proceedings: symposium on solid modeling and applications, 125–129.
52. Srikantappa AB, Crawford RH (1994) Automatic part coding based on the interfeature relationships. Advances in feature based manufacturing. Elsevier, Amsterdam, pp 215–237.
53. Staley SM, Henderson MR, Anderson DC (1983) Using syntactic pattern recognition to extract feature information from a solid geometric database. Comput Mech Eng 2(2):61–66.
54. Tseng YJ, Joshi SB (1994) Recognizing multiple interpretations of interacting machining features. Comput-Aided Des 26(9):667–688.
55. Tzong SK, Bartholomew ON (2001) Feature representation and classification for automatic process planning systems. J Manuf Syst 12(2):133–145.
56. Qiang J, Michael M (1997) Machine interpretation of CAD data of manufacturing applications. ACM Comput Surv 24(3):264–311.
57. Vandenbrande JH, Requicha AAG (1993) Spatial reasoning for the automatic recognition of machinable feature in solid models. IEEE Trans Patt Anal Mach Intell 15(12):1269–1285.
58. Vandenbrande JH, Requicha AAG (1994) Geometric computation for the recognition of spatially interacting machinable features. In: Shah JJ, Mantyla M, Nau D (eds) Advances in feature based manufacturing. Elsevier, Amsterdam, pp 83–106.
59. Wang HP (1986) Intelligent reasoning for process planning. Ph.D. Dissertation, Pennsylvania State University.
60. Wang, M-T (1990) A geometric reasoning methodology for manufacturing feature extraction from a 3D CAD Model. Ph.D. Dissertation, Purdue University.
61. Wang N, Ozsoy TM (1991) A scheme to represent features, dimensions, and tolerances in geometric modeling. J Manuf Syst 10(3):233–240.
62. Wang HP, Wysk RA (1986) An expert system of machining data selection. Comput Ind Eng J 10(2):99–107.
63. William B (1996) Initial graphics exchange specification IGES 5.3, ANS US PRO/IPO-100.
64. Woo TC (1976) Computer aided recognition of volumetric design. Advances in computer aided manufacturing, PROLAMAT 76. North-Holland, Amsterdam, 195–204.
65. Xinhui M, Guoxiong Z, Shugi L, Xu W (2003) Measuring information integration model for CAD/CMM. Chin J Mech Eng 16(1):59–61.
66. Yong K, Wang E, Hyung MR (2001) Geometry-based machining precedence reasoning for feature-based process planning. Int J Prod Res 39(10):2077–2103.
67. Zhao Z, Ghosh SK, Link D (1990) Recognition of machined surfaces for manufacturing based on wireframe models. J Mater Process Technol 24(1):137–145.
68. Zhao Y, Ridgway K, Al-Ahmari AMA (2002) Integration of CAD and a cutting tool selection system. Comput Ind Eng 42(1):17–34.
69. Zhang GX, Liu SG, Ma XH, Wang YQ (2002) Toward the intelligent CMM. CIRP Annu Manuf Technol 51(1):437–442.

Part III
Rapid Design and Manufacturing

Chapter 10
Engineering Materials: An Overview

Abstract This chapter provides an overview of material used for manufacturing. One of the most important aspects of material science involves the study of material's structure. Introducing a new product or changing an existing one involves many decisions. Although these decisions may seem to be independent, they will have a significant impact on the overall product life cycle, which will in turn influence cost, performance, and service.

Key words Materials • Structure • Product life cycle • Performance

10.1 Introduction

Manufacturing operations require a vast amount of knowledge from many different disciplines. The required disciplines include design, material, processes, marketing, purchasing, production, and quality control necessary to develop the product required by the consumer. This chapter presents an overview of materials used for manufacturing. One of the most important aspects of materials science involves the study of material's structure. Introducing a new product or changing an existing one involves many decisions. Even though these decisions may seem to be independent, they will have a significant impact on the overall product life cycle which will influence cost, performance, and service.

10.1.1 Materials in Engineering Applications

The most common method used for material classification is based on metallic and nonmetallic [1]. Metallic materials are those which are metals. Examples include iron, steel, brass, bronze, copper, aluminum, magnesium, lead, tin, and zinc. Nonmetallic materials include wood, brick, rubber, and plastic. The above classification is further categorized as following [2].

A.K. Kamrani and E.A. Nasr, *Engineering Design and Rapid Prototyping*, DOI 10.1007/978-0-387-95863-7_10,

- **Metals and alloys:** This category has relatively high strength, high stiffness, good ductility or formability, and shock resistance. They are particularly used for structural or load-bearing applications. Examples include steel, aluminum, magnesium, zinc, cast iron, titanium, copper, and nickel. Metals possess several properties including yield strength or maximum stress that can be applied without causing plastic, or permanent, deformation, ultimate tensile strength or maximum stress that can be applied without causing necking, stiffness, or resistance to deformation or deflection, electrical and thermal conductivities, toughness or ability to absorb energy without fracture, and hardness or resistance to deformation by indentation, among others.
- **Ceramics, glass, and glass–ceramics:** This category consists of the inorganic crystalline materials. Ceramics do not conduct heat well and are heated to very high temperatures before melting. They are strong and hard, but also very brittle. Ceramics have exceptional strength under compression. Glass is an amorphous material derived from molten liquid. They can be thermally treated to make them stronger. Examples of this category include barium titanate, silica, SiO_2, and Al_2O_3.
- **Polymers:** Polymers are organic materials. They are produced using polymerization process. They have very good electrical resistivity and provide good thermal insulation. Polymers have lower strength, although they provide a very good strength-to-weight ratio. Polymers are further classified into three groups:

  1. Thermoplastics have a linear structure and can be re-softened by heating and then reshaped.
  2. Thermosets have cross-linked structures that develop upon curing. They cannot be reshaped after they are cured.
  3. Elastomers have a cross-linked structure and possess elastic behavior similar to rubber.

- **Semiconductors:** This category provides a level of conductivity that can be controlled. They are used to develop electronic devices. Examples include silicon and GaAs.
- **Composites:** Composites are the combination of two or more materials with improved properties that are not found in a single material. Examples are graphite–epoxy, tungsten carbide–cobalt, and titanium–clad steel.

10.1.2 Atomic Bonding and Crystalline Structure

Materials are selected according to the desired properties of the finished part. Materials in each of the above groups possess different structures and properties. The structure is based on the arrangements of the atoms, the properties, and the specific characteristics based on the crystalline structure [1, 2]. Strong bonds between adjacent atoms resulting from the transfer or sharing of outer orbital electrons are known as *primary bonds*. Primary bonds allow for the formation of molecules.

In *ionic bonding*, the valence electrons are given up by atoms that have only a few valence atoms, usually one or two, and taken by the atoms that have a nearly complete outer shell and need only one or two electrons to complete the outer shell (NaCl). In *covalent bonding*, the valence electrons are shared between two particular atoms. In this case, an atom may share electrons with more than one atom (CO_2 and SiO_2). *Metallic bonding* occurs when the valence electrons are not associated with a particular atom or ion, but exist as an electron cloud ion centered around the core. *Metallic bonds* are the electrostatic attraction between the valence electrons and the positively charged ion cores. It allows for easy movement of electrons, and therefore materials with metallic bonding have both good thermal and electrical conductivity (Fe). *Secondary bonds* are weak bonds and typically join molecules to one another. These are attraction forces between molecules. *Van der Waals bonding* is the resulting force where atoms or molecules have either induced or permanent dipole moment for attracting each other. These forces have their origin in the presence of a dipole in polar molecules. *Hydrogen bonds* are Keeson interaction between molecules in which hydrogen atoms are involved. Hydrogen atoms represent one polarized region from the permanent polarized interactions (H_2O). *Dipole forces* arise in a molecule where two atoms have similar and opposite electrons (HCl). *London forces* are category of *Van der Waals* forces that occur between molecules that do not have a permanent dipole moment (carbon tetrachloride).

Materials are also classified as either amorphous (glass and polymer) or crystalline. Amorphous materials have no long-range order and the atoms are less closely packed. Crystalline materials have both long- and short-range order and if the exact atom arrangement in one location is known, the atom arrangement in another position can be predicted. The crystalline structure of a material consists of a three-dimensional arrangement of points in space where each lattice point has identical surroundings. A lattice is a regular array of lattice points in three dimensions. Unit cell is a subdivision of a lattice that consists of the overall characteristics of the entire lattice. The material's structure and properties can change due to temperatures. Crystalline lattice structures are as follows [1, 2]:

- **Body centered cubic (BCC):** Unit cells have one atom located at each lattice point and one atom located at the center of the cubic structure. Each corner atom is shared by eight unit cells. Metals have poor ductility but low carbon steels have good ductility. BCC has the most slip direction. Examples include iron, chromium, tungsten, manganese, and vanadium. Figure 10.1 illustrates the BCC structure.
- **Face centered cubic (FCC):** FCC cells have one atom located at each lattice point, and one located at the center of each face of the cubic structure. A face center atom is shared between the two unit cells. Metals that have this structure are generally more ductile due to the number of slip direction. Examples include aluminum, copper, gold, silver, nickel, and lead. Figure 10.2 illustrates the BCC structure.
- **Hexagonal close packet (HCP):** Metals that present HCP crystalline structure have a hexagonal unit cell. They have poor ductility and have difficulties to form and deform at room temperature. Examples include magnesium, titanium, and zinc. Figure 10.3 illustrates the HCP structure.

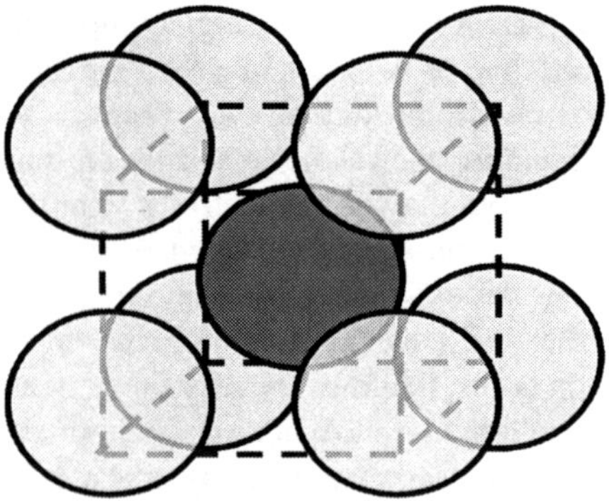

Fig. 10.1 BCC lattice structure

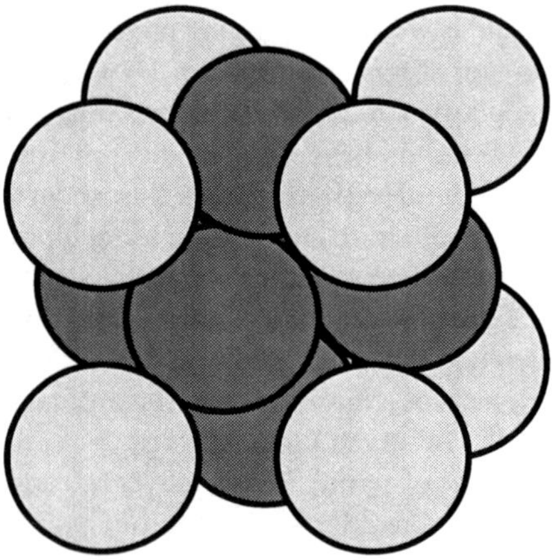

Fig. 10.2 FCC lattice structure

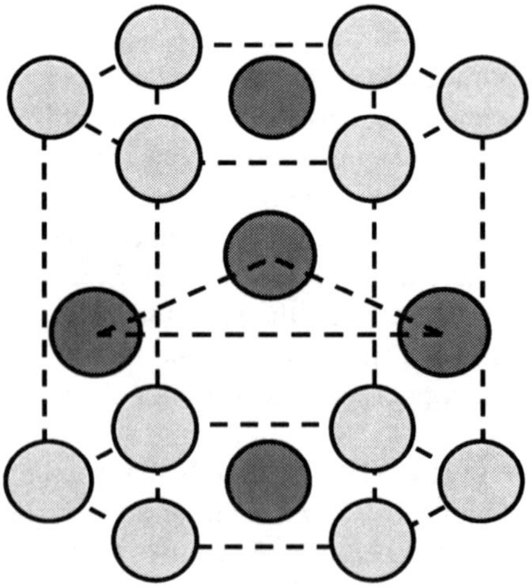

Fig. 10.3 HCP lattice structure

The developing of metallic grains begins when metals start to solidify. This will result in the formation of the first unit lattice. After the initial formation, other atoms tend to attach themselves to the existing lattice and to bring about the growth in a repeated pattern and configuration. The resulting solid is the product of nucleation of initial unit lattice and the subsequent growth of each unit cell. The number and size of the grains are based on the rate of nucleation and grain growth. The greater the rate of nucleation, the larger the grain size and likewise, the smaller the rate of grain growth, the smaller the size of the resulting grains.

The behavior within single crystals is based on the lattice type, the inter-atomic forces, the spacing between the planes of atoms, and the density of atoms on various planes. There are both elastic and plastic deformation behaviors in materials. In elastic deformation, no permanent displacement of the atoms with respect to each other occurs. The displacement that occurs is slight warping of the lattice unit. The strain is nearly proportional to stress. On other hand, in plastic deformation a permanent displacement takes place by the slipping of one plane of atoms over an adjacent plane. Normal stresses, resulting from tension and compression, produce shear stresses and thus produce plastic deformation. Vacancy is a missing atom or ion from its regular crystallographic site lattice structure and dislocations are line imperfections in an otherwise perfect crystal. These are introduced into the crystal during solidification of the material or when the material is deformed permanently. Dislocations are very useful in expanding deformation and strengthening metallic materials. Types of dislocations are screw, edge, and mixed dislocations.

10.2 Mechanical Properties

When a material is designed for a given application, a number of factors must be considered. The material must have the desired mechanical properties, the capability of being shaped and manufactured into a desired shape (manufacturability), and economically feasible [3]. Mechanical properties of a material determine its behavior when subjected to mechanical stresses. The properties include elastic modulus, ductility, hardness, and various measures of strength. Elastic strain is defined as a fully recoverable strain resulting from an applied stress. The strain is elastic if (1) it develops instantaneously, (2) remains as long as the stress is applied, and (3) is removed as soon as the force is withdrawn. It is nonpermanent. Materials subjected to an elastic strain do not show any permanent deformation. In many materials, the relationship between elastic stress and strain is linear. The slope of a tensile stress–strain curve in the linear regime is defined by the Young's modulus or modulus of elasticity (E) of a material. Materials can be subjected to mechanical loadings. Several tests have been developed to study and measure the impact of mechanical loads on material and how it can be used for both functionality and manufacturability analysis. An overview of some of these tests is discussed below [1–3].

10.2.1 Stress–Strain Relationship

10.2.1.1 Tensile Properties

Tensile test is the most common procedure for studying the stress–strain relationship. In the test, a force is applied (pulling) tending to elongate. This will reduce the diameter of the test specimen. Figure 10.4 illustrates the load and displacement curve for metals. This diagram consists of two major regions. The linear elastic region starts at the beginning of deformation and is followed by a plastic region (Fig. 10.5a).

- **Elastic region:** As stress increases, some point in the linear relationship is finally reached at which the material begins to yield. This yield point (Y) of the material can be identified by changing the slope at the end of the linear region. The yield point is a strength characteristic of the material, and is therefore often referred to as the yield strength.

Fig. 10.4 Stress–strain curve

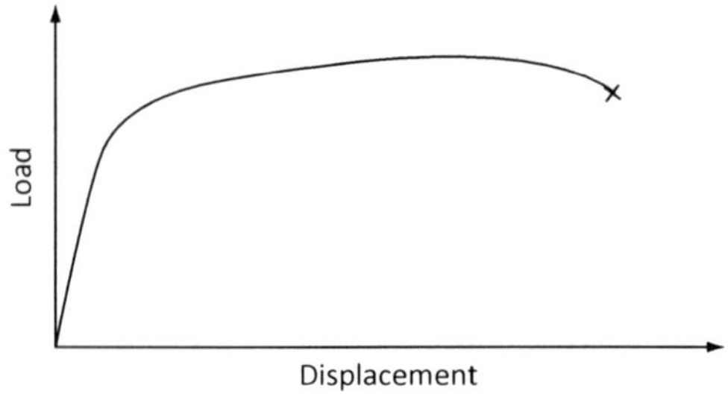

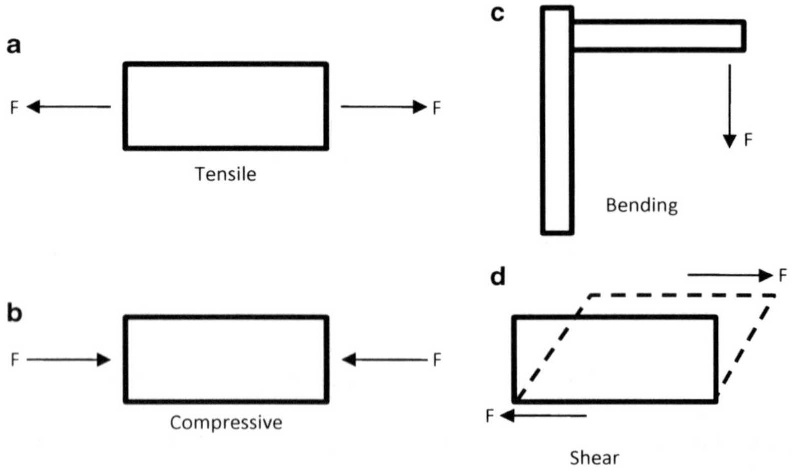

Fig. 10.5 Tests of measuring the mechanical properties

- **Plastic region:** The yield point marks the transition to the plastic region and the start of plastic deformation of the material. Elongation is accompanied by a uniform reduction in cross-sectional area, consistent with maintaining constant volume.

10.2.2 Other Properties

- A compression test applies a load that squeezes a cylindrical specimen between the two plates (Fig. 10.5b).
- The bending test is used to test the strength of materials. Brittle materials do not flex. They deform elastically until immediately before fracture. The strength value derived from this test is called the transverse rupture strength (Fig. 10.5c).
- Shear involves application of stresses in opposite directions on either side of the test element to deflect it (Fig. 10.5d).

10.2.3 Hardness

The harness of a material is defined as its resistance to permanent indentation. For most metals, hardness is closely related to strength. Because the method of testing for hardness is usually based on resistance to indentation, which is a form of compression, there is a good correlation between hardness and strength properties determined in a compression test. Polymers have the lowest hardness among the three types of engineering materials. Tables 10.1 and 10.2 list the equations used for measuring sample properties [1–3].

10.3 Nanomaterial

Over the past decade, nanomaterials have been the subject of enormous interest. These materials have the extremely small feature size and have the potential for wide ranging industrial, biomedical, and electronic applications. Nanomaterials can be metals, ceramics, polymeric materials, or composite materials. Their defining characteristic is a very small feature size in the range of 1–100 nm (nm = billionths of a meter). One nano meter spans three to five atoms lined up in a row. By comparison, the diameter of a human hair is about five orders of magnitude larger than a nanoscale particle. The nanoworld lays midway between the scale of atomic and quantum phenomena, and the scale of bulk materials [4–6].

Materials that can be manufactured at the scale of 1–100 nm can be classified as nanomaterials. Some examples of nanomaterials include gold nanoparticles, gold nanoshells, and silver plasmon resonant particles, carbon nanotubes (CNT), and single-wall carbon nanotubes (SWNT). The classification of the nanomaterials include (1) carbon based nanomaterials (CBN), (2) nanocomposites, (3) metals and

Table 10.1 Summary of tensile properties

Mechanical property	Parameter	Equations and definitions	
Tensile	Engineering stress (δ)	$\frac{P}{A_0}$	P = load A_0 = initial cross-section area
	Engineering strain (ε)	$\frac{l-l_0}{l_0}$	l = final length l_0 = initial length
	Hooks law	$\sigma_{eng} = E\varepsilon_{eng}$ (elastic) $\sigma_{eng} = K\varepsilon^{n}_{eng}$ (plastic)	E = modulus of elasticity K = coefficient n = strain exponent
	Tensile strength	$\frac{P_{max}}{A_0}$	P_{max} = maximum force
	Ductility (%)	$\frac{l_f - l_0}{l_0}$	Ability of a material to plastically strain without fracture
	True stress	$\frac{P}{A} = \sigma_{eng}(1+\varepsilon_{eng})$	
	True strain	$\ln\frac{l}{l_0} = \frac{\ln A}{A_0} = \ln(1+\varepsilon_{eng})$	
	Bending (TRS)	$3PL/2bt_2$	TRS = transverse rupture strength P = load L = length of the specimen b and t = cross-section dimension
	Shear stress (τ)	$\frac{P}{A}$	P = load A = cross-section area
	Shear strain (γ)	$\frac{\delta}{b}$	δ = deflection b = deflection distance

alloys, (4) biological nanomaterials, (5) nano-polymers, (6) nano-glasses, and (7) nano-ceramics [6]. Only CBNs are discussed in this chapter.

10.3.1 Carbon-Based Nanomaterials and Applications

Carbon based nanomaterials are defined as materials in which the "nanocomponent" is pure carbon. The CBN is further classified as carbon nanoparticles (fullerene), graphite, and single-wall carbon nanotubes (SWCNT) shown in Fig. 10.6 and multiwall carbon nanotubes (MWCNT) shown in Fig. 10.7.

The classification also includes carbon films and carbon nanocrystals. The CNTs are sheets of graphite rolled up to make a tube. The dimensions are variable (down to

Table 10.2 Summary of hardness property

Mechanical property	Parameter	Equations and definitions	
Hardness	Brinell test (BHN)		P = indentation load (500, 1,500, and 3,000 kg) D_0 = diameter of the ball D_i = diameter of the indentation on surface
	Vikers test (HV)	$\frac{1.85P}{D^2}$	P = indentation load (1–120 kg) D = diagonal indentation on surface
	Knoop test (HK)	$\frac{14.2P}{D^2}$	P = indentation load (5 g–5 kg) D = long diagonal indentation on surface
	Rockwell test (HR)		
	HRA (60 kg)	100–500d	t = indentation on the surface (diamond cone)
	HRC (150 kg) HRD (100 kg) HRB (100 kg)	130–500d	$\frac{1}{16}$ steel ball
	HRF (60 kg) HRG (150 kg) HRE (100 kg)	130–500d	$\frac{1}{8}$ steel ball

0.4 nm in diameter) and it also exists within nanotubes, leading to a distinction between MWCNT and SWCNT. Apart from good mechanical properties such as tensile strength, they also exhibit varying electrical properties depending on the way the graphite structure spirals around the tube. This capability allows nanotube to be insulating, semiconducting, or conducting [4–6]. The CNTs are 100 times stronger than steel at one-sixth of the weight which posses 200 GPa of tensile strength. It is also as stiff as diamond with the Young's modulus over 1 TPa. The CNTs are used as media for electrical energy storage and has shown to have good properties in electron emission. Other physical properties include heat transmission capacity, heat stability, and low density. Some of these properties are listed in Table 10.3 [4, 6, 8].

Fullerenes (C60) are a class of cage-like carbon compounds composed of fused, pentagonal, and/or hexagonal sp2 carbon rings (see Fig. 10.8).

Fullerenes are, however, like all pure carbon structures insoluble in water. The applications of fullerenes as drug deliverers either by the attachment of functional ligands to the carbon cage or by trapping molecules inside have been predicted [4, 5].

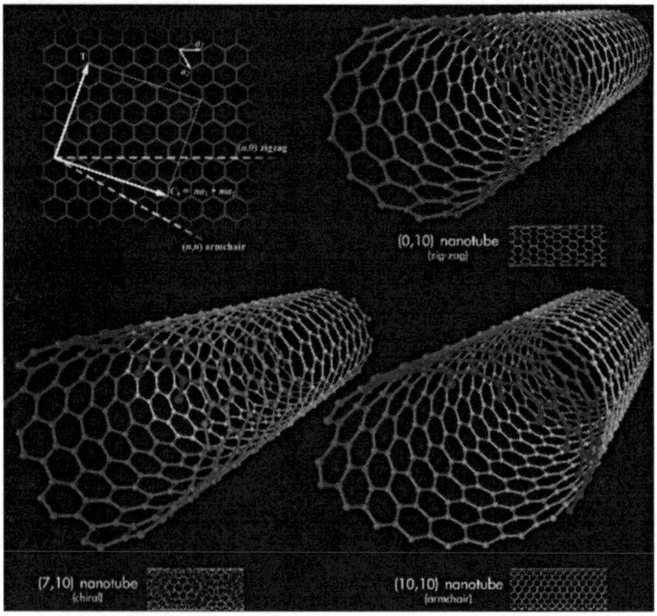

Fig. 10.6 Single-wall nanotube [7]

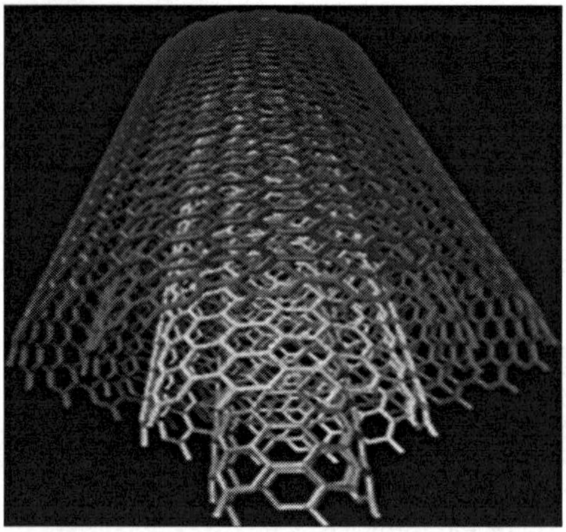

Fig. 10.7 Multiwall nanotube [7]

Table 10.3 Sample nanotube mechanical properties

Material	Young's modulus (GPa)	Tensile strength (GPa)
Single-wall nanotube	1,054	150
Multiwall nanotube	1,200	150
Steel	208	0.4
Epoxy	3.5	0.005
Wood	16	0.008

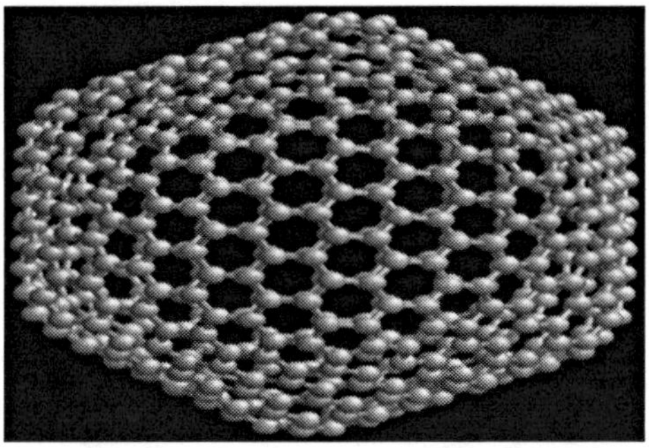

Fig. 10.8 C60 nanotube [7]

Solar cells and fuel cells are further potential areas of application of CBNs in the field of energy related applications.

10.3.2 Carbon Nanotube Manufacturing

Carbon nanotubes are generally produced by three main techniques, although more economical approach is being studied and developed. The following are current methods being used.

10.3.2.1 Arc Discharge Method

This method creates nanotubes through arc vaporization of two carbon rods placed end to end, separated by approximately 1 mm, in an enclosure that is usually filled with inert gas (helium, argon) at low pressure (50–700 mbar). A direct current of 50–100 A driven by approximately 20 V creates a high temperature discharge between the two electrodes. The discharge vaporizes one of the carbon rods and forms a small rod shaped deposit on the other rod (Fig. 10.9).

Producing nanotubes in high yield depends on the uniformity of the plasma arc and the temperature of the deposit form on the carbon electrode [3]. If SWCNTs are preferable, the anode has to be doped with metal catalyst, such as Fe, Co, Ni, Y, or Mo. The quantity and quality of the nanotubes obtained depend on various parameters, such as metal concentration, inert gas pressure, kind of gas, current, and system geometry [9]. Usually, the diameter is in the range of 1.2–1.4 nm.

Plasma Rotating Arc Discharge Method

A second possibly economical route to mass production of MWCNTs is synthesis by plasma rotating arc discharge technique. The centrifugal force caused by the rotation generates turbulence and accelerates the carbon vapor perpendicular to the anode. In

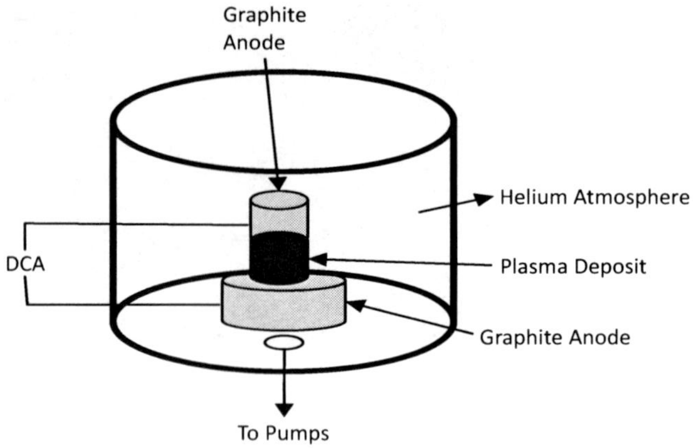

Fig. 10.9 Arch discharge apparatus

addition, the rotation distributes the micro discharges uniformly and generates stable plasma. Consequently, it increases the plasma volume and raises the plasma temperature. At a rotation speed of 5,000 rpm, a yield of 60% was found at temperature 1,025°C without the use of a catalyst. The yield increases up to 90% after purification if the rotation speed is increased and the temperature is enlarged to 1,150°C.

10.3.2.2 Chemical Vapor Deposition Method

The chemical vapor deposition method (CVD) has been widely used for the deposition of a large variety of materials. It has been widely used for semiconductor manufacturing.

Plasma Enhanced Chemical Vapor Deposition Method

The plasma enhanced chemical vapor deposition (PECVD) method generates a glow discharge in a chamber or a reaction furnace by a high frequency voltage applied to both electrodes. Figure 10.10 illustrates PECVD with a parallel plate electrode structure.

A substrate is placed on the grounded electrode. In order to form a uniform film, the reaction gas is supplied from the opposite plate. Catalytic metal, such as Fe, Ni, and Co, are used on, for example, a Si, SiO_2, or glass substrate using thermal CVD or sputtering. After nanoscopic fine metal particles are formed, CNTs will be grown on the metal particles on the substrate by glow discharge generated from high frequency power.

A carbon containing reaction gas, such as C_2H_2, CH_2, C_2H_2, C_2H_2, CO, is supplied to the chamber during the discharge [11]. The catalyst has a strong effect on the nanotube diameter, growth rate, wall thickness, morphology, and microstructure. Ni seems to be the most suitable pure-metal catalyst for the growth of aligned

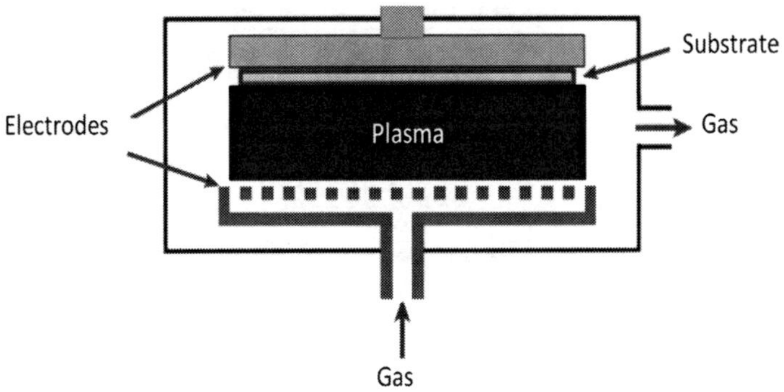

Fig. 10.10 Plasma CVD [10]

MWCNTs [2]. The diameter of the MWNTs is approximately 15 nm. The highest yield of CNTs achieved was about 50% and was obtained at relatively low temperatures (below 330°C).

Thermal Chemical Vapor Deposition

In this method, Fe, Ni, Co, or an alloy of the three catalytic metals is initially deposited on a substrate [11, 12]. After the substrate is etched in a diluted HF solution with distilled water, the specimen is placed in a quartz boat. The boat is positioned in a CVD reaction furnace, and nanometre-sized catalytic metal particles are formed after an additional etching of the catalytic metal film using NH_3 gas at a temperature of 750–1,050°C. As CNTs are grown on these fine catalytic metal particles in CVD synthesis, forming these fine catalytic metal particles is the most important process. Figure 10.11 illustrates the thermal CVD.

The diameter range of the CNTs depends on the thickness of the catalytic film during growing CNTs on a Fe catalytic film using thermal CVD. By using a thickness of 13 nm, the diameter distribution lies between 30 and 40 nm. When a thickness of 27 nm is used, the diameter range is between 100 and 200 nm. The CNTs formed are multiwalled [5].

10.3.2.3 Laser Ablation Method

In 1995, Smalley's group at Rice University [4, 14] reported the synthesis of CNTs by laser vaporization. The laser vaporization apparatus used by Smalley's group is shown in Fig. 10.12.

A pulsed or continuous laser is used to vaporize a graphite target in an oven at 1,200°C. The oven is filled with helium or argon gas in order to keep the pressure at 500 Torr. A very hot vapor plume forms, then expands and cools rapidly. As the vaporized species cool, small carbon molecules and atoms quickly condense to form

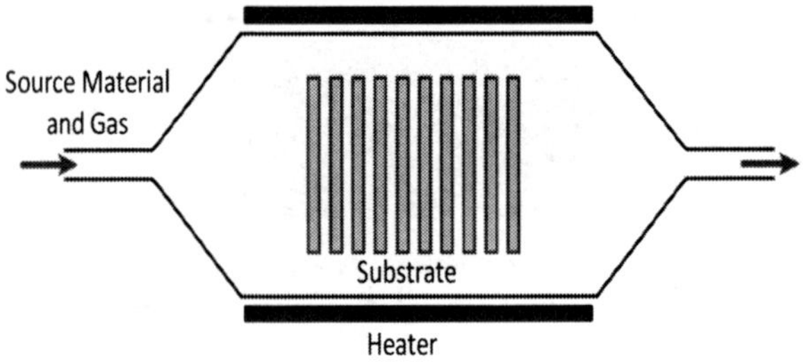

Fig. 10.11 Thermal CVD [13]

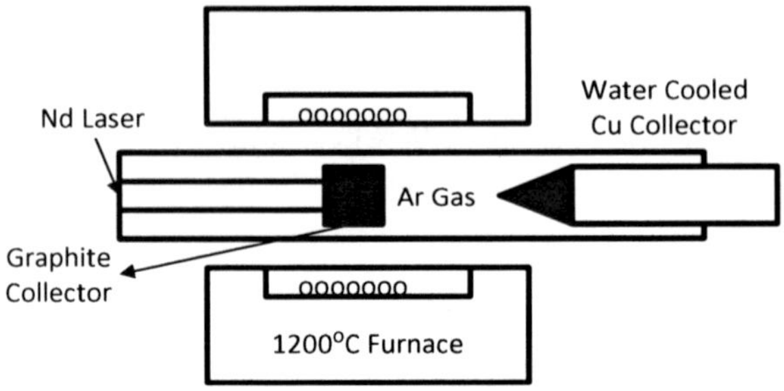

Fig. 10.12 Laser vaporization apparatus

larger clusters, possibly including fullerenes. The catalysts begins to condense, but more slowly at first, and attach to carbon clusters and prevent their closing into cage structures. From these initial clusters, tubular molecules grow into SWNT until the catalyst particles become too large, or until conditions have cooled sufficiently that carbon no longer can diffuse through or over the surface of the catalyst particles.

10.4 Purification of Thin Films

After the growth of thin films, the CNT contains many impurities. Hence the purification or cleaning process is important. The main impurities are graphite sheet, amorphous carbon, metal catalyst, and the smaller fullerene. These impurities will interfere with most of the desired properties of the CNTs. To purify the product, a two-stage gas-phase oxidation procedure must be taken [5, 9]. First, the sample is heated in a 15% air/Ar mixture at 400°C for 1 h. A wash in concentrated HCl acid (37%) dissolves the metal oxide that was formed during the oxidation process (Fig. 10.13).

In the second stage, the oxidation process was carried out at 450°C for 1 h and then the sample is washed again in HCl acid (acid treatment). The final purified product consists of mainly (>90%) CNT in bundle form (Fig. 10.14).

Final step is annealing at high temperatures (873–1,873 K) where the nanotubes will be rearranged and defects will be consumed (Fig. 10.15).

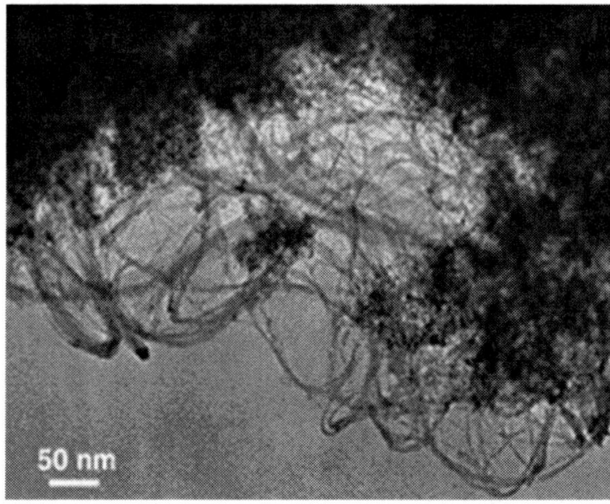

Fig. 10.13 Purification process

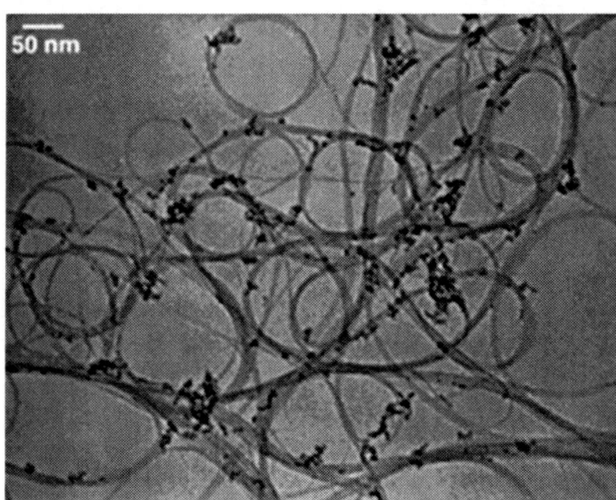

Fig. 10.14 Purification process – continued

10.5 Challenges and Limitations

The preparation process of nanotubes is based on a well-established technology (CVD, PECVD, and arc discharge). Therefore, there is considerable expertise at hand for the development of new materials. Most of these methods are at laboratory levels and industrial scale production requires new techniques.

There is need for the investigation of the composition and structure of the films produced as well as the physical properties of the film. Laser ablation produces the

Fig. 10.15 Purification process – completed

cleanest material, but the costs are still rather high. Arc discharge can produce low purity nanotubes. The CVD technique is still under development for large-scale CVD. The main purification techniques are oxidation, acidic treatment, annealing, and filtration techniques. These techniques have shown to be economically feasible and are ready to be applied at larger scale.

Nanotechnology is a revolution, not evolution, in science and technology. The nanorevolution will impact every aspect of human's life activities, not just in science and technology. Due to the great potential of CNTs, it is clear that novel technologies will emerge in the near future [2, 9, 14]. However, larger quantities of nanotubes are needed if composite materials are to be fabricated. In addition, electronic and optical applications need to be investigated in detail. It is highly likely that low-cost gas sensors, three-dimensional composite materials, nanotube fabrics, novel catalytic supports, heat exchangers, biological microfilters, and virus inhibitors will be produced using CNTs. Therefore, the field of nanotube should be exploited further.

10.6 Summary

In this chapter, an overview of material used for manufacturing is presented. One of the most important aspects of materials science involves the study of material's structure. Introducing a new product or changing an existing one involves many decisions. Even though these decisions may seem to be independent, they will have a significant impact on the overall product life cycle which will influence cost, performance, and service.

10.7 Review Questions

10.1 What are the classifications of materials used in engineering applications?
10.2 Discuss the following terms: atomic bonding and crystalline structure.
10.3 What are the structures of crystalline lattice?
10.4 Define the following terms: tensile stress and hardness.
10.5 What is nanomaterial?
10.6 Discuss briefly CBN and their applications.
10.7 Explain arc discharge method used in CNT manufacturing.
10.8 Explain plasma rotating arc discharge method used in CNT manufacturing.
10.9 Discuss thermal CVD method used in CNT manufacturing.

References

1. Askeleand D (2006) The science and engineering of materials. Thomson Publishing, Toronto.
2. DeGarmo P (1962) Materials and processes in manufacturing. Macmillan, New York.
3. Farag M (1997) Materials selection for engineering design. Prentice–Hall, New York.
4. Teo BK, Sun XH (2007) Chem Rev 107:1454–1532.
5. Terrones M (2003) Annu Rev Mater Res 33:419–501.
6. Ying JY (2001) Nanostructured materials. Academic Press, New York. 1st edition (December 28).
7. http://en.wikipedia.org/wiki/Carbon_nanotube
8. Zheng C, Cui Z (2006) Micro-Nanofabrication: Technologies and applications, Springer Berlin Heidelberg.
9. Ebbesen TW, Ajayan PM (1992) Nature 358:220–222.
10. http://en.wikipedia.org/wiki/File:PlasmaCVD.PNG
11. Zheng B, Li Y, Liu J (2002) Appl Phys A 74:345–348.
12. Park JB et al (2002) J Cryst Growth 244(2):211–217.
13. http://en.wikipedia.org/wiki/File:ThermalCVD.PNG
14. Ajayan PM, Lambert JM (1993) Chem Phys Lett 215:509–517.
15. Takikawa H, Kusano O, Sakakibara T (1999) Graphite cathode spot produces carbon nanotubes in arc discharge, J Phys D: Appl Phys 32:2433.

Chapter 11
Geometric Dimensioning and Tolerancing

Abstract In this chapter, an introductory discussion is provided on the topics of geometric dimension and tolerancing, and its impact on selecting the right manufacturing processes. Also, improvements in measurements provide a need to better understand that variation is unavoidable in manufacturing. However, acceptable levels of variation will result in a good assembly. It is important to understand the limits of this variation. After the development of GD&T, drawings became the main tool for communication among design, manufacturing, testing, etc.

Keywords Geometric dimensioning • Tolerancing • Variation • Manufacturing

11.1 Introduction

Most manufactured products are usually an assembly of several other products or components. The designer develops an ideal assembly of the product where these components/parts will fit together exactly and perfectly. Even if these parts are manufactured in different locations, they must still fit together based on defined fit and clearances criteria. Since every manufacturing process has inherent variations, these variations will impact how accurate parts are manufactured. It is nearly impossible to perfectly duplicate identical components. When parts are inspected, they must have sufficient level of accuracy in order to be considered as "good," otherwise they will fail the inspection and will be rejected. During design, the designer decides for each feature of the part the allowed amount of size, orientation, and location variations. This information is then communicated to the manufacturer and the best process sequences within the allowed levels of variation are selected for part manufacturing. This variation is defined using geometric dimensioning and tolerancing rules. Without Geometric Dimensioning and Tolerancing (GD&T) no spare part policy could be achieved. GD&T is a standard language used for communicating engineering design details and specifications across all relevant departments such as manufacturing, testing, packaging, etc. GD&T achieves this objective by a set of standard definitions, rules, symbols, and mathematical formulae.

A.K. Kamrani and E.A. Nasr, *Engineering Design and Rapid Prototyping*,
DOI 10.1007/978-0-387-95863-7_11,

It conveys both the perfect geometry and the tolerances for a part. Since GD&T is largely symbolic and numerical it is applied worldwide. GD&T incorporates the design based on the required functionality. It cannot specify manufacturing processes for making the parts, but processes or sequence of process is selected based on the required accuracy. GD&T bridges the gap between design requirements and measurable specifications. This chapter is intended to provide an introductory discussion to the topics of geometric dimension and tolerancing and its impact on selecting the right manufacturing processes [5–7].

Engineering drawings have existed for many centuries. The earliest known technical drawing was dated back to 4000 B.C. After the Industrial Revolution of 1800s, drawings were used to better support the manufacturing operations. However, drawings were mostly the pictorial representations of parts and not necessarily including details such as dimensions and feature relations. This was sufficient since most of the manufacturing was done manually by a skilled operator that performed the entire process of part production and fabrication. In this case, if an assembly problem occurred, the operator would remanufacture new parts until they have the proper assembly condition. With the advancement of assembly lines, the trial and error approach was not feasible. Also, improvements in measurements helped engineers to better understand that variation is unavoidable in manufacturing. However, acceptable levels of variation will result in a good assembly. It is important to understand the limits of this variation. After the development of GD&T, drawings became the main tool for communication among design, manufacturing, testing departments, etc.

11.2 GD&T Definition and Standard

Geometric Dimensioning and Tolerancing is a language for communicating engineering design specifications. GD&T includes all the symbols, definitions, mathematical formulae, and application rules necessary to embody a viable engineering language. At the same time, GD&T is a philosophy of defining a part based on how it functions. As its name implies, it conveys both the nominal dimensions (ideal geometry) and the tolerances for a part. Because GD&T is expressed using line drawings, symbols, and numerals, it is now the standard language used worldwide for design and manufacturing applications.

- GD&T reduces assumptions by maintaining uniformity in drawing specifications and interpretation.
- Since GD&T is a dimensioning philosophy based on function of the parts, it establishes tolerances based on functional requirements.
- GD&T can maintain extra tolerance for manufacturing provide cost savings in manufacturing and material usage.
- It provides a clear and concise technique for defining a reference coordinate system (datum's) on a component or assembly to be used throughout the manufacturing and inspection processes.

- Proper application of geometric dimensioning closely dovetails accepted and logical mechanical design process and design for manufacturing considerations.
- Geometric dimensioning dramatically reduces the need for drawing notes to describe complex geometry requirements on a component or assembly by the use of standard symbology that accurately and quickly defines design, manufacturing, and inspection requirements.
- GD&T concepts such as maximum material condition (MMC) when applied properly will facilitate and simplify the design of cost saving functional check gages, manufacturing fixtures and jigsThe American National Standards Institute (ANSI) administers the guidelines for standards creation in the United States [2]. The following standards are used to define GD&T rules and definitions:
- American Society of Mechanical Engineers (ASME) Y14.5M-1994, Dimensioning and Tolerancing [1].
- ASME Y14.5.1M-1994, Mathematical Definition of Dimensioning and Tolerancing Principles [2].

11.3 GD&T Terminologies

11.3.1 Functional Dimensioning

Functional dimensioning is a philosophy of dimensioning and tolerancing parts based on its function. Some benefits of functional dimensioning are as follows:

- Establishing the design objectives
- Better understanding of functions of components
- Identification of potential problems during the design stage
- Development of a method to evaluate and minimize change requests
- Maximize tolerances for ease of manufacturing
- Effective way of communication among design and manufacturing

The general term applied to a physical portion of a part, such as a surface or a hole is known as a *feature*. Usually, a part feature is a single surface having uniform shape. By definition, a feature does not exist unless a part is actually produced. A *datum* is a theoretically exact point, axis, or plane derived from the true geometric counterpart of a specified datum feature. A datum is the origin from which the location or geometric characteristics of features of a part are established. An actual feature of a part that is used to establish a datum is known as a datum feature. Datums are planes, points, lines, or axes derived from datum features.

A *feature of size* can be a shaft, a hole or two opposing planar features. Features of size can be measured for size. The geometric control for a feature is represented by a rectangular sign called as a *feature control frame*. The feature control frame is divided into section expressing feature information sequentially from left to right. The first section if feature control frame (FCF) contains a geometric characteristic symbol specifying the type of geometric control. This will be discussed later in this chapter. The second section contains the geometric tolerance value. The modifying symbols such as M (MMC), L (LMC), or R (RFS) appear in this section also adding special attributes to the geometric control. The maximum material condition (MMC) is the condition in which a feature of size contains the maximum amount of material within the stated limits of size. MMC can be thought as the condition where the most part material is present at the surface of a feature, or where the part weighs the most. This equates to the smallest allowable hole or the largest allowable pin, relative to the stated size limits.

Least material condition (LMC) is the condition in which a feature of size contains the least amount of material within the stated limits of size. LMC can be thought as the condition where the least part material is present at the surface of a feature, or where the part weighs the least. This equates to the largest allowable hole or the smallest allowable pin, relative to the stated size limits. Regardless of feature size (RFS) is the condition when a geometric tolerance or a datum applies independent of the feature size. The rest of FCF sections contains the primary, secondary, and tertiary datum references (3–2–1 rule), each of which may be followed by a material condition modifier symbol, if required. The feature control frame displays most of the information necessary to control a single geometric characteristic of the subject feature. Only basic dimensions are left out of the feature control frame.

11.3.2 Basic Dimensioning

A basic dimension is a numerical value used to describe the theoretically exact size, profile, orientation, or location of a feature or datum target [3]. It is the basis from which permissible variations are established by tolerances on other dimensions, in notes, or in feature control frames.

11.3.3 ASME Y14.5M-1994 Fundamental rules of GD&T

The ASME Y14.5M-1994 rules for Dimensioning and Tolerancing clearly define engineering intent and conform to the following:

- Each dimension shall have a tolerance, except for those dimensions specifically identified as reference, maximum, minimum, or stock. The tolerance may be applied directly to the dimension indicated by a general note or located in a supplementary block of the drawing format.
- Dimensioning and tolerancing shall be complete so there is full understanding of the characteristics of each feature. Neither scaling nor assumption of a distance or size is permitted.

- Each necessary dimension of an end product shall be shown. No more dimensions than those necessary for complete definition shall be given. The use of reference dimensions on a drawing should be minimized.
- Dimensions shall be selected and arranged to suit the function and mating relationship of a part and shall not be subject to more than one interpretation.
- The drawing should define a part without specifying manufacturing methods. Thus, only the diameter of a hole is given without indicating whether it is to be drilled, reamed, punched, or made by any other operation.
- It is permissible to identify certain processing dimensions that provide for finish allowance, shrink allowance, and other requirements provided the final dimensions are given on the drawing.
- Dimensions should be arranged to provide required information for optimum readability.
- A 90-degree basic angle applies where center lines and lines depicting features are shown on a drawing at right angles and no angle is specified.
- A 90-degree basic angle applies where center lines of features in a pattern or surfaces shown at right angles on the drawing are located or defined by basic dimensions and no angle is specified.
- All dimensions and tolerances apply in a free state condition. This principle does not apply to nonrigid parts.
- Unless otherwise specified, all geometric tolerances apply for full depth, length, and width of the feature.
- Dimensions and tolerances apply only at the drawing level where they are specified. A dimension specified for a given feature on one level of drawing is not mandatory for that feature at any other level.

11.4 Engineering Tolerance

Tolerance is the total amount of variation by which a specific dimension is allowed. Tolerances are expressed using bilaterally or unilaterally. An example of the bilateral is 2.500 ± 0.002. An example of the unilateral is 2.500–0.000 ± 0.004. The exact part dimension cannot be achieved due to the imperfection in manufacturing operations and the assigned tolerance allows for these variations. The higher the amount of tolerance, the more expensive the manufacturing operations used for its production. Selecting the best value is critical in achieving both required part functionality and minimizing costs. The allowance is the difference between the maximum limits of mating parts. It is an algebraic value and it is used to determine the type of the fit used for part assembly (functionality) of mating components. There are three major categories of assembly fit. These are clearance, transition, and interference fits. A clearance fit occurs when one part is always loose relative to the other and could slide freely. An interference fit is when one part is always forced tight into the other during assembly. The transition fit results in either close clearance fit or light interference fit. The allowance for clearance fit is the minimum clearance between the two components, whereas the allowance for the interference fit is the maximum interference between the two. To achieve interchangeable manufacturing fits, ANSI has developed tolerances and limits calculations. Five classes of fits are identified. These are listed in Tables 11.1 and 11.2.

Each symbol will provide details regarding the mating parts assembly and their functions. The tables are used for calculating the maximum and minimum required tolerances based on these standards. An example of using these tables is provided. RC1 (H5/g4) is selected for a design with the nominal dimension of 1.500″. RC1 is used for accurate locating of mating components. Using Tables 11.3–11.6, the range used is 1.19″–1.97″. The standard limits are retrieved from the table for the rest of calculation. The hole is calculated as 1.500 + 0.0004 = 1.5004″ for maximum and 1.500″ for minimum. The shaft calculation is 1.500 − 0.0004 = 1.4996″ for maximum and 1.500 − 0.0005 = 1.4993 for minimum. Figure 11.1 illustrates the results of the analysis.

Table 11.1 Five classes of fit

Grade	Definition
RC	Running or sliding
LC	Location clearance
LT	Location transition
LN	Location interference
FN	Force on shrink fit

Table 11.2 Fit classifications [8]

Symbol grade	Hole shaft		ANSI symbol grade	Hole shaft	
RC1	H5	g4	LT1	H7	j6
RC2	H6	g5	LT2	H8	j7
RC3	H7	f6	LT3	H7	k6
RC4	H8	f7	LT4	H8	k7
RC5	H8	e7	LT5	H7	n6
RC6	H9	e8	LT6	H7	n7
RC7	H9	d8	LN1	H6	n5
RC8	H10	c9	LN2	H7	p6
RC9	H11	c11	LN3	H7	r6
LC1	H6	h5	FN1	H6	n6
LC2	H7	h6	FN2	H7	s6
LC3	H8	h7	FN3	H7	t6
LC4	H10	h9	FN4	H7	u6
LC5	H7	g6	FN5	H8	x7
LC6	H9	f8			
LC7	H10	e9			
LC8	H10	d9			
LC9	H11	c10			
LC10	H12	c12			
LC11	H13	c13			

Table 11.3 Clearance running fits

	RC1			RC2			RC3		
		Standard limits			Standard limits		Limits	Standard limits	
Range	Limits of clearance	Hole H5	Shaft g4	Limits of clearance	Hole H6	Shaft g5	of clearance	Hole H7	Shaft f6
0–0.12	0.1	+0.2	−0.1	0.1	+0.25	−0.1	0.3	+0.4	−0.3
	0.45	0	−0.25	0.55	0	−0.3	0.95	0	−0.55
0.12–0.24	0.15	+0.2	−0.15	0.15	+0.3	−0.15	0.4	+0.5	−0.4
	0.5	0	−0.3	0.65	0	−0.35	1.12	0	−0.7
0.24–0.40	0.2	+0.25	−0.2	0.2	+0.4	−0.2	0.5	+0.6	−0.5
	0.6	0	−0.35	0.85	0	−0.45	1.5	0	−0.9
0.40–0.71	0.25	+0.3	−0.25	0.25	+0.4	−0.25	0.6	+0.7	−0.6
	0.75	0	−0.45	0.95	0	−0.55	1.7	0	−1.0
0.71–1.19	0.3	+0.4	−0.3	0.3	+0.5	−0.3	0.8	+0.8	−0.8
	0.95	0	−0.55	1.2	0	−0.7	2.1	0	−1.3
1.19–1.97	0.4	+0.4	−0.4	0.4	+0.6	−0.4	1.0	+1.0	−1.0
	1.1	0	−0.7	1.4	0	−0.8	2.6	0	−1.6
1.97–3.15	0.4	+0.5	−0.4	0.4	+0.7	−0.4	1.2	+1.2	−1.2
	1.2	0	−0.7	1.6	0	−0.9	3.1	0	−1.9
3.15–4.73	0.5	+0.6	−0.5	0.5	+0.90	−0.5	1.4	+1.4	−1.4
	1.5	0	−0.9	2.0	0	−1.1	3.7	0	−2.3
4.73–7.09	0.6	+0.7	−0.6	0.6	+1.0	−0.6	1.6	+1.6	−1.6
	1.8	0	−1.1	2.3	0	−1.3	4.2	0	−2.6
7.09–9.85	0.6	+0.8	−0.6	0.6	+1.2	−0.6	2.0	+1.8	−2.0
	2.0	0	−1.2	2.6	0	−1.4	5.0	0	−3.2
9.85–12.41	0.8	+0.9	−0.8	0.8	+1.2	−0.8	2.5	+2.0	−2.5
	2.3	0	−1.4	2.9	0	−1.7	5.7	0	−3.7

	RC4			RC5			RC6		
		Standard limits			Standard limits			Standard limits	
Range	Limits of clearance	Hole H8	Shaft f7	Limits of clearance	Hole H8	Shaft e7	Limits of clearance	Hole H9	Shaft e8
0–0.12	0.3	+0.6	−0.3	0.6	+0.6	−0.6	06	+1.0	−0.6
	1.3	0	−0.7	1.6	0	−1.0	2.2	0	−1.2
0.12–0.24	0.4	+0.7	−0.4	0.8	+0.7	−0.8	0.8	+1.2	−0.8
	1.6	0	−0.9	2.0	0	−1.3	2.7	0	−1.5
0.24–0.40	0.5	+0.9	−0.5	1.0	+0.9	−1.0	1.0	+1.4	−1.0
	2.0	0	−1.1	2.5	0	−1.6	3.3	0	−1.9
0.40–0.71	0.6	+1.0	−0.6	1.2	+1.0	−1.2	1.2	+1.6	−1.2
	2.3	0	−1.3	2.9	0	−1.9	3.8	0	−2.2
0.71–1.19	0.8	+1.2	−0.8	1.6	+1.2	−1.6	1.6	+2.0	−1.6
	2.8	0	−1.6	3.6	0	−2.4	4.8	0	−2.8
1.19–1.97	1.0	+1.6	−1.0	2.0	+1.6	−2.0	2.0	+2.5	−2.0
	3.6	0	−2.0	4.6	0	−3.0	6.1	0	−3.6
1.97–3.15	1.2	+1.8	−1.2	2.5	+1.8	−2.5	2.5	+3.0	−2.5
	4.2	0	−2.4	5.5	0	−3.7	7.3	0	−4.3
3.15–4.73	1.4	+2.2	−1.4	3.0	+2.2	−3.0	3.0	+3.5	−3.0
	5.0	0	−2.8	6.6	0	−4.4	8.7	0	−5.2
4.73–7.09	1.6	+2.5	−1.6	3.5	+2.5	−2.5	3.5	+4.0	−3.5
	5.7	0	−3.2	7.6	0	−5.1	10.0	0	−6.0
7.09–9.85	2.0	+2.8	−2.0	4.0	+2.8	−4.0	4.0	+4.5	−4.0
	6.6	0	−3.8	8.6	0	−5.8	11.3	0	−6.8
9.85–12.41	2.5	+3.0	−2.5	5.0	+3.0	−5.0	5.0	+5.0	−50
	7.5	0	−4.5	10.0	0	−7.0	13.0	0	−8.0

(continued)

Table 11.3 (continued)

Range	RC7			RC8			RC9		
		Standard limits			Standard limits			Standard limits	
	Limits of clearance	Hole H9	Shaft d8	Limits of clearance	Hole H10	Shaft c9	Limits of clearance	Hole H11	Shaft c11
0–0.12	1.0	+1.0	−1.0	2.5	+1.6	−2.5	4.0	+2.5	−4.0
	2.6	0	−1.6	5.1	0	−3.5	8.1	0	−5.6
0.12–0.24	1.2	+1.2	−1.2	2.8	+1.8	−2.8	4.5	+3.0	−4.5
	3.1	0	−1.9	5.8	0	−4.0	9.0	0	−6.0
0.24–0.40	1.6	+1.4	−1.6	3.0	+2.2	−3.0	5.0	+3.5	−5.0
	3.9	0	−2.5	6.6	0	−4.4	10.7	0	−7.2
0.40–0.71	2.0	+1.6	−2.0	3.5	+2.8	−3.5	6.0	+4.0	−6.0
	4.6	0	−3.0	7.9	0	−5.1	12.8	0	−8.8
0.71–1.19	2.5	+2.0	−2.5	4.5	+3.5	−4.5	7.0	+5.0	−7.0
	5.7	0	−3.7	10.0	0	−6.5	15.5	0	−10.5
1.19–1.97	3.0	+2.5	−3.0	5.0	+4.0	−5.0	8.0	+6.0	−8.0
	7.1	0	−4.6	11.5	0	−7.5	18.0	0	−12.0
1.97–3.15	4.0	+3.0	−4.0	6.0	+4.5	−6.0	9.0	+7.0	−9.0
	8.8	0	−5.8	13.5	0	−9.0	20.5	0	−13.5
3.15–4.73	5.0	+3.5	−5.0	7.0	+5.0	−7.0	10.0	+9.0	−10.0
	10.7	0	−7.2	15.5	0	−10.5	24.0	0	−15.0
4.73–7.09	6.0	+4.0	−6.0	8.0	+6.0	−8.0	12.0	+10.0	−12.0
	12.5	0	−8.5	18.0	0	−12.0	28.0	0	−18.0
7.09–9.85	7.0	+4.5	−7.0	10.0	+7.0	−10.0	15.0	+12.0	−15.0
	14.3	0	−9.8	21.5	0	−14.5	34.0	0	−22.0
9.85–12.41	8.0	+5.0	−8.0	12.0	+8.0	−12.0	18.0	+12.0	−18.0
	16.0	0	−11.0	25.0	0	−17.0	38.0	0	−26.0

Table 11.4 Clearance location fits

Range	LC1			LC2			LC3		
		Standard limits			Standard limits			Standard limits	
	Limits of clearance	Hole H6	Shaft h5	Limits of clearance	Hole H7	Shaft h6	Limits of clearance	Hole H8	Shaft h7
0–0.12	0	+0.25	0	0	+0.4	0	0	+0.6	0
	0.45	0	−0.2	0.65	0	−0.25	1	0	−0.4
0.12–0.24	0	+0.3	0	0	+0.5	0	0	+0.7	0
	0.5	0	−0.2	0.8	0	−0.3	1.2	0	−0.5
0.24–0.40	0	+0.4 0	0	0	+6	0	0	+0.9	0
	0.65		−0.25	1.0	0	−0.4	1.5	0	−0.6
0.40	0	+0.4	0	0	+0.7	0	0	+1.0	0
0.71	0.7	0	−0.3	1.1	0	−0.4	1.7	0	−0.7
0.71–1.19	0	+0.5	0	0	+0.8	0	0	+1.2	0
	0.9	0	−0.4	1.3	0	−0.5	2	0	−0.8
1.19–1.97	0	+0.6	0	0	+1.0	0	0	+1.6	0
	1.0	0	−0.4	1.6	0	−0.6	2.6	0	−1
1.97–3.15	0	+0.7	0	0	+1.2	0	0	+1.8	0
	1.2	0	−0.5	1.9	0	−0.7	3	0	−1.2
3.15–4.73	0	+0.9	0	0	+1.4	0	0	+2.2	0
	1.5	0	−0.6	2.3	0	−0.9	3.6	0	−1.4
4.73–7.09	0	+1.0	0	0	+1.6	0	0	+2.5	0
	1.7	0	−0.7	2.6	0	−1.0	4.1	0	−1.6
7.09–9.85	0	+1.2	0	0	+1.8	0	0	+2.8	0
	2.0	0	−0.8	3.0	0	−1.2	4.6	0	−1.8
9.85–12.41	0	+1.2	0	0	+2.0	0	0	+3.0	0
	2.1	0	−0.9	3.2	0	−1.2	5	0	−2.0

(continued)

Range	LC4 Limits of clearance	LC4 Standard limits Hole H10	LC4 Standard limits Shaft h9	LC5 Limits of clearance	LC5 Standard limits Hole H7	LC5 Standard limits Shaft g6	LC6 Limits of clearance	LC6 Standard limits Hole H9	LC6 Standard limits Shaft f8
0–0.12	0	+1.6	0	0.1	+0.4	−0.1	0.3	+1.0	−0.3
	2.6	0	−1.0	0.75	0	−0.35	1.9	0	-0.9
0.12–0.24	0	+1.8	0	0.15	+0.5	−0.15	0.4	+1.2	−0.4
	3.0	0	−1.2	0.95	0	−0.45	2.3	0	−1.1
0.24–0.40	0	+2.2	0	0.2	+0.6	−0.2	0.5	+1.4	−0.5
	3.6	0	−1.4	1.2	0	−0.6	2.8	0	−1.4
0.40–0.71	0	+2.8	0	0.25	+0.7	−0.25	0.6	+1.6	−0.6
	4.4	0	−1.6	1.35	0	−0.65	3.2	0	−1.6
0.71–1.19	0	+3.5	0	0.3	+0.8	−0.3	0.8	+2.0	−0.8
	5.5	0	−2.0	1.6	0	−0.8	4.0	0	−2.0
1.19–1.97	0	+4.0	0	0.4	+1.0	−0.4	1.0	+2.5	−1.0
	6.5	0	−2.5	2.0	0	−1.0	5.1	0	−2.6
1.97–3.15	0	+4.5	0	0.4	+1.2	−0.4	1.2	+3.0	−1.2
	7.5	−0	−3	2.3	0	−1.1	6.0	0	−3.0
3.15–4.73	0	+5.0	0	0.5	+1.4	−0.5	1.4	+3.5	−1.4
	8.5	0	−3.5	2.8	0	−1.4	7.1	0	−3.6
4.73–7.09	0	+6.0	0	0.6	+1.6	−0.6	1.6	+4.0	−1.6
	10	0	−4	3.2	0	−1.6	8.1	0	−4.1
7.09–9.85	0	+7.0	0	0.6	+1.8	−0.6	2.0	+4.5	−2.0
	11.5	0	−4.5	3.6	0	−1.8	9.3	0	−4.8
9.85–12.41	0	+8.0	0	0.7	+2.0	−0.7	2.2	+5.0	−2.2
	13	0	−5	3.9	0	−1.9	10.2	0	−5.2

Range	LC7 Limits of clearance	LC7 Standard limits Hole H10	LC7 Standard limits Shaft e9	LC8 Limits of clearance	LC8 Standard limits Hole H10	LC8 Standard limits Shaft d9	LC9 Limits of clearance	LC9 Standard limits Hole H11	LC9 Standard limits Shaft c10
0–0.12	0.6	+1.6	−0.6	1.0	+1.6	−1.0	2.5	+2.5	−2.5
	3.2	0	−1.6	3.6	0	−2.0	6.6	0	−4.1
0.12–0.24	0.8	+1.8	−0.8	1.2	+1.8	−1.2	2.8	+3.0	−2.8
	3.8	0	−2.0	4.2	0	−2.4	7.6	0	−4.6
0.24–0.40	1.0	+2.2	−1.0	1.6	+2.2	−1.6	3.0	+3.5	−3.0
	4.6	0	−2.4	5.2	0	−3.0	8.7	0	−5.2
0.40–0.71	1.2	+2.8	−1.2	2.0	+2.8	−2.0	3.5	+4.0	−3.5
	5.6	0	−2.8	6.4	0	−3.6	10.3	0	−6.3
0.71–1.19	1.6	+3.5	−1.6	2.5	+3.5	−2.5	4.5	+5.0	−4.5
	7.1	0	−3.6	8.0	0	−4.5	13.0	0	−8.0
1.19–1.97	2.0	+4.0	−2.0	3.0	+4.0	−3.0	5.0	+6.0	−5.0
	8.5	0	−4.5	9.5	0	−5.5	15.0	0	−9.0
1.97–3.15	2.5	+4.5	−2.5	4.0	+4.5	−4.0	6.0	+7.0	−6.0
	10.0	0	−5.5	11.5	0	−7.0	17.5	0	−10.5
3.15–4.73	3.0	+5.0	−3.0	5.0	+5.0	−5.0	7.0	+9.0	−7.0
	11.5	0	−6.5	13.5	0	−8.5	21.0	0	−12.0
4.73–7.09	3.5	+6.0	−3.5	6.0	+6.0	−6.0	8.0	+10.0	−8.0
	13.5	0	−7.5	16.0	0	−10.0	24.0	0	−14.0
7.09–9.85	4.0	+7.0	−4.0	7.0	+7.0	−7.0	10.0	+12.0	−10.0
	15.5	0	−8.5	18.5	0	−11.5	29.0	0	−17.0
9.85–12.41	4.5	+8.0	−4.5	7.0	+8.0	−7.0	12.0	+12.0	−12.0
	17.5	0	−9.5	20.0	0	−12.0	32.0	0	−20.0

(continued)

Table 11.4 (continued)

	LC10			LC11		
	Limits of	Standard limits		Limits of	Standard limits	
Range	clearance	Hole H12	Shaft c12	clearance	Hole H13	Shaft c13
0–0.12	4.0	+4.0	−4.0	5.0	+6.0	−5.0
	12	0	−8.0	17.0	0	−11.0
0.12–0.24	4.5	+5.0	−4.5	6.0	+7.0	−6.0
	14.5	0	−9.5	20.0	0	−13.0
0.24–0.40	5.0	+6.0	−5.0	7.0	+9.0	−7.0
	17	0	−11.0	25.0	0	−16.0
0.40–0.71	6.0	+7.0	−6.0	8.0	+10.0	−8.0
	20.0	0	−13.0	28.0	0	−18.0
0.71–1.19	7.0	+8.0	−7.0	10.0	+12.0	−10.0
	23.0	0	−15.0	34.0	0	−22.0
1.19–1.97	8.0	+10.0	−8.0	12.0	+16.0	−12.0
	28.0	0	−18.0	44.0	0	−28.0
1.97–3.15	10.0	+12.0	−10.0	14.0	+18.0	−14.0
	34.0	0	−22.0	50.0	0	−32.0
3.15–4.73	11.0	+14.0	−11.0	16.0	+22.0	−16.0
	39.0	0	−25.0	60.0	0	−38.0
4.73–7.09	12.0	+16.0	−12.0	18.0	+25.0	−18.0
	44.0	0	−28.0	68.0	0	−43.0
7.09–9.85	16.0	+18.0	−16.0	22.0	+28.0	−22.0
	52.0	0	−34.0	78.0	0	−50.0
9.85–12.41	20.0	+20.0	−20.0	28.0	+30.0	−28.0
	60.0	0	−40.0	88.0	0	−58.0

11.5 Geometric Characteristic Symbols

There are five categories of geometric characteristic. In addition to these geometric symbols, there are another set of characters which are known as the modifying symbols. These are listed in Tables 11.7 and 11.8.

11.5.1 Form Controls

The four form characteristics are straightness, flatness, circularity, and cylindricity. All controlled features are individual and are not related to datums. They are assessed by comparison to a perfect geometric counterpart of themselves (a feature formed perfectly flat, straight, circular, or cylindrical).

Table 11.5 Transition fits

	LT1			LT2			LT3		
		Standard limits			Standard limits			Standard limits	
Range	Limits of fit	Hole H7	Shaft js6	Limits of fit	Hole H8	Shaft js7	Limits of fit	Hole H7	Shaft k6
0–0.12	−0.10	+0.4	+0.10	−0.2	+0.6	+0.2			
				+0.50	0	−0.10	+0.8	0	−0.2
0.12–0.24	−0.15	+0.5	+0.15	−0.25	+0.7	+0.25			
				+0.65	0	−0.15	+0.95	0	−0.25
0.24–0:40	−0.2	+0.6	+0.2	−0.3	+0.9	+0.3	−0.5	+0.6	+0.5
	+0.8	0	−0.2	+1.2	0	−0.3	+0.5	0	+0.1
0.40–0.71	−0.2	+0.7	+0.2	−0.35	+1.0	+0.35	−0.5	+0.7	+0.5
	+0.9	0	−0.2	+ 1.35	0	−0.35	+0.6	0	+0.1
0.71–1.19	−0.25	+0.8	+0.25	−0.4	+1.2	+0.4	−0.6	+0.8	+0.6
	+1.05	0	−0.25	+1.6	0	−0.4	+0.7	0	+0.1
1.19–1.97	−0.3	+1.0	+0.3	−0.5	+1.6	+0.5	−0.7	+1.0	+0.7
	+1.3	0	−0.3	+2.1	0	−0.5	+0.9	0	+0.1
1.97–3.15	−0.3	+1.2	+0.3	−0.6	+1.8	+0.6	−0.8	+1.2	+0.8
	+1.5	0	−0.3	+2.4	0	−0.6	+1.1	0	+0.1
3.15–4.73	−0.4	+1.4	+0.4	−0.7	+2.2	+0.7	−1.0	+1.4	+1.0
	+1.8	0	−0.4	+2.9	0	−0.7	+1.3	0	+0.1
4.73–7.09	−0.5	+1.6	+0.5	−0.8	+2.5	+0.8	−1.1	+1.6	+1.1
	+2.1	0	−0.5	+3.3	0	−0.8	+1.5	0	+0.1
7.09–9.85	−0.6	+1.8	+0.6	−0.9	+2.8	+0.9	−1.4	+1.8	+1.4
	+2.4	0	−0.6	+3.7	0	−0.9	+1.6	0	+0.2
9.85–12.41	−0.6	+2.0	+0.6	−1.0	+3.0	+1.0	−1.4	+2.0	+1.4
	+2.6	0	−0.6	+4.0	0	−1.0	+ 1.8	0	+0.2
12.41–15.75	−0.7	+2.2	+0.7	−1.0	+3.5	+1.0	−1.6	+2.2	+1.6
	+2.9	0	−0.7	+4.5	0	−1.0	+2.0	0	+0.2
15.75–19.69	−0.8	+2.5	+0.8	−1.2	+4.0	+1.2	−1.8	+2.5	+1.8
	+3.3	0	−0.8	+5.2	0	−1.2	+2.3	0	+0.2

	LT4			LT5			LT6		
		Standard limits			Standard limits			Standard limits	
Range	Limits of fit	Hole H8	Shaft K7	Limits of fit	Hole H7	Shaft n6	Limits of fit	Hole H7	Shaft n7
0–0.12				−0.5	+0.4	+0.5	−0.65	+0.4	+0.65
				+0.15	0	+0.25	+0.15	0	+0.25
0.12–0.24				−0.6	+0.5	+0.6	−0.8	+0.5	+0.8
				+0.2	0	+0.3	+0.2	0	+0.3
0.24–0.40	−0.7	+0.9	+0.7	−0.8	+0.6	+0.8	−0.1	+0.6	+1.0
	+0.8	0	+0.1	+0.2	0	+0.4	+0.2	0	+0.4
0.40–0.71	−0.8	+1.0	+0.8	−0.9	+0.7	+0.9	−1.2	+0.7	+1.2
	+0.9	0	+0.1	+0.2	0	+0.5	+0.2	0	+0.5
0.71–1.19	−0.9	+1.2	+0.9	−1.1	+0.8	+1.1	−1.4	+0.8	+1.4
	+1.1	0	+0.1	+0.2	0	+0.6	+0.2	0	+0.6
1.19–1.97	−1.1	+1.6	+1.1	−1.3	+1.0	+1.3	−1.7	+1.0	+1.7
	+1.5	0	+0.1	+0.3	0	+0.7	+0.3	0	+0.7
1.97–3.15	−1.3	+1.8	+1.3	−1.5	+1.2	+1.5	−2.0	+1.2	+2.0
	+1.7	0	+0.1	+0.4	0	+0.8	+0.4	0	+0.8

(continued)

Table 11.5 (continued)

	LT4			LT5			LT6		
		Standard limits			Standard limits			Standard limits	
Range	Limits of fit	Hole H8	Shaft K7	Limits of fit	Hole H7	Shaft n6	Limits of fit	Hole H7	Shaft n7
3.15–4.73	−1.5	+2.2	+1.5	−1.9	+1.4	+1.9	−2.4	+1.4	+2.4
	+2.1	0	+0.1	+0.4	0	+1.0	+0.4	0	+1.0
4.73–7.09	−1.7	+2.5	+1.7	−2.2	+1.6	+2.2	−2.8	+1.6	+2.8
	+2.4	0	+0.1	+0.4	0	+1.2	+0.4	0	+1.2
7.09–9.85	−2.0	+2.8	+2.0	−2.6	+1.8	+2.6	−3.2	+1.8	+3.2
	+2.6	0	+0.2	+0.4	0	+1.4	+0.4	0	+1.4
9.85–12.41	−2.2	+3.0	+2.2	−2.6	+2.0	+2.6	−3.4	+2.0	+3.4
	+2.8	0	−0.2	+0.6	0	+1.4	+0.6	0	+1.4
12.41–15.75	−2.4	+3.5	+2.4	−3.0	+2.2	+3.0	−3.8	+2.2	+3.8
	+3.3	0	+0.2	+0.6	0	+1.6	+0.6	0	+1.6
15.75–19.69	−2.7	+4.0	+2.7	−3.4	+2.5	+3.4	−4.3	+2.5	+4.3
	+3.8	0	+0.2	+0.6	0	+1.8	+0.7	0	+1.8

	LN1			LN2			LN3		
		Standard limits			Standard limits			Standard limits	
Range	Limits of interference	Hole H6	Shaft n5	Limits of interference	Hole H7	Shaft p6	Limits of interference	Hole H7	Shaft r6
0–0.12	0	+0.25	+0.45	0	+0.4	+0.65	0.1	+0.4	+0.75
	0.45	0	+0.25	0.65	0	+0.4	0.75	0	+0.5
0.12–0.24	0	+0.3	+0.5	0	+0.5	+0.8	0.1	+0.5	+0.9
	0.5	0	+0.3	0.8	0	+0.5	0.9	0	+0.6
0.24–0.40	0	+0.4	+0.65	0	+0.6	+1.0	0.2	+0.6	+1.2
	0.65	0	+0.4	1.0	0	+0.6	1.2	0	+0.8
0.40–0.71	0	+0.4	+0.8	0	+0.7	+1.1	0.3	+0.7	+1.4
	0.8	0	+0.4	1.1	0	+0.7	1.4	0	+1.0
0.71–1.19	0	+0.5	+1.0	0	+0.8	+1.3	0.4	+0.8	+1.7
	1.0	0	+0.5	1.3	0	+0.8	1.7	0	+1.2
1.19–1.97	0	+0.6	+1.1	0	+1.0	+1.6	0.4	+1.0	+2.0
	1.1	0	+0.6	1.6	0	+1.0	2.0	0	+1.4
1.97–3.15	0.1	+0.7	+1.3	0.2	+1.2	+2.1	0.4	+1.2	+2.3
	1.3	0	+0.7	2.1	0	+1.4	2.3	0	+1.6
3.15–4.73	0.1	+0.9	+1.6	0.2	+1.4	+2.4	0.6	+1.4	+2.9
	1.6	0	+1.0	2.5	0	+1.6	2.9	0	+2.0
4.73–7.09	0.2	+1.0	+1.9	0.2	+1.6	+2.8	0.9	+1.6	+3.5
	1.9	0	+1.2	2.8	0	+1.8	3.5	0	+2.5
7.09–9.85	0.2	+1.2	+2.2	0.2	+1.8	+3.2	1.2	+1.8	+4.2
	2.2	0	+1.4	3.2	0	+2.0	4.2	0	+3.0
9.85–12.41	0.2	+1.2	+2.3	0.2	+2.0	+3.4	1.5	+2.0	+4.7
	2.3	0	+1.4	3.4	0	+2.2	4.7	0	+3.5

11.5.1.1 Straightness

It is a form control that may be used as a surface, derived median line or derived median plane control. If used as a surface control, any deviations from a perfectly straight surface must be contained within the feature's MMC limits of size. Straightness is an individual control. Figure 11.2 illustrates this concept.

Table 11.6 Interference force fits

	FN1			FN2			FN3		
	Limits of interfer-ence	Standard limits		Limits of interfer-ence	Standard limits		Limits of interfer-ence	Standard limits	
Range		H6	n6		H7	s6		H7	t6
0–0.12	0.05	+0.25	+0.5	0.2	+0.4	+0.85			
	0.5	0	+0.3	0.85	0	+0.6			
0.12–0.24z	0.1	+0.3	+0.6	0.2	+0.5	+1.0			
	0.6	0	+0.4	1.0	0	+0.7			
0.24–0.40	0.1	+0.4	+0.75	0.4	+0.6	+1.4			
	0.75	0	+0.5	1.4	0	+1.0			
0.40–0.56	0.1	+0.4	+0.8	0.5	+0.7	+1.6			
	0.8	0	+0.5	1.6	0	+1.2			
0.56–0.71	0.2	+0.4	+0.9	0.5	+0.7	+1.6			
	0.9	0	+0.6	1.6	0	+1.2			
0.71–0.95	0.2	+0.5	+1.1	0.6	+0.8	+1.9			
	1.1	0	+0.7	1.9	0	+1.4			
0.95–1.19	0.3	+0.5	+1.2	0.6	+0.8	+1.9	0.8	+0.8	+2.1
	1.2	0	+0.8	1.9	0	+1.4	2.1	0	+1.6
1.19–1.58	0.3	+0.6	+1.3	0.8	+1.0	+2.4	1.0	+1.0	+2.6
	1.3	0	+0.9	2.4	0	+1.8	2.6	0	+2.0
1.58–1.97	0.4	+0.6	+1.4	0.8	+1.0	+2.4	1.2	+1.0	+2.8
	1.4	0	+1.0	2.4	0	+1.8	2.8	0	+2.2
1.97–2.56	0.6	+0.7	+1.8	0.8	+1.2	+2.7	1.3	+1.2	+3.2
	1.8	0	+1.3	2.7	0	+2.0	3.2	0	+2.5
2.56–3.15	0.7	+0.7	+1.9	1.0	+1.2	+2.9	1.8	+1.2	+3.7
	1.9	0	+1.4	2.9	0	+2.2	3.7	0	+3.8
3.15–3.94	0.9	+0.9	+2.4	1.4	+1.4	+3.7	2.1	+1.4	+4.4
	2.4	0	+1.8	3.7	0	+2.8	4.4	0	+3.5
3.94–4.73	1.1	+0.9	+2.6	1.6	+1.4	+3.9	2.6	+1.4	+4.9
	2.6	0	+2.0	3.9	0	+3.0	4.9	0	+4.0
4.73–5.52	1.2	+1.0	+2.9	1.9	+1.6	+4.5	3.4	+1.6	+6.0
	2.9	0	+2.2	4.5	0	+3.5	6.0	0	+5.0
5.52–6.30	1.5	+1.0	+3.2	2.4	+1.6	+5.0	3.4	+1.6	+6.0
	3.2	0	+2.5	5.0	0	+4.0	6.0	0	+5.0
6.30–7.09	1.8	+1.0	+3.5	2.9	+1.6	+5.5	4.4	+1.6	+7.0
	3.5	0	+2.8	5.5	0	+4.5	7.0	0	+6.0
7.09–7.88	1.8	+1.2	+3.8	3.2	+1.8	+6.2	5.2	+1.8	+8.2
	3.8	0	+3.0	6.2	0	+5.0	8.2	0	+7.0
7.88–8.86	2.3	+1.2	+4.3	3.2	+1.8	+6.2	5.2	+1.8	+8.2
	4.3	0	+3.5	6.2	0	+5.0	8.2	0	+7.0

	FN4			FN5		
	Limits of interference	Standard limits		Limits of interference	Standard limits	
Range		H7	u6		H8	x7
0–0.12	0.3	+0.4	+0.95	0.3	+0.6	+1.3
	0.95	0	+0.7	1.3	0	+0.9
0.12–0.24	0.4	+0.5	+1.2	0.5	+0.7	+1.7
	1.2	0	+0.9	1.7	0	+1.2
0.24–0.40	0.6	+0.6	+1.6	0.5	+0.9	+2.0
	1.6	0	+1.2	2.0	0	+1.4

(continued)

Table 11.6 (continued)

	FN4			FN5		
	Limits of	Standard limits		Limits of	Standard limits	
Range	interference	H7	u6	interference	H8	x7
0.40–0.56	0.7	+0.7	+1.8	0.6	+1.0	+2.3
	1.8	0	+1.4	2.3	0	+1.6
0.56–0.71	0.7	+0.7	+1.8	0.8	+1.0	+2.5
	1.8	0	+1.4	2.5	0	+1.8
0.71–0.95	0.8	+0.8	+2.1	1.0	+1.2	+3.0
	2.1	0	+1.6	3.0	0	+2.2
0.95–1.19	1.0	+0.8	+2.3	1.3	+1.2	+3.3
	2.3	0	+1.8	3.3	0	+2.5
1.19–1.58	1.5	+1.0	+3.1	1.4	+1.6	+4.0
	3.1	0	+2.5	4.0	0	+3.0
1.58–1.97	1.8	+1.0	+3.4	2.4	+1.6	+5.0
	3.4	0	+2.8	5.0	0	+4.0
1.97–2.56	2.3	+1.2	+4.2	3.2	+1.8	+6.2
	4.2	0	+3.5	6.2	0	+5.0
2.56–3.15	2.8	+1.2	+4.7	4.2	+1.8	+7.2
	4.7	0	+4.0	7.2	0	+6.0
3.15–3.94	3.6	+1.4	+5.9	4.8	+2.2	+8.4
	5.9	0	+5.0	8.4	0	+7.0
3.94–4.73	4.6	+1.4	+6.9	5.8	+2.2	+9.4
	6.9	0	+6.0	9.4	0	+8.0
4.73–5.52	5.4	+1.6	+8.0	7.5	+2.5	+11.6
	8.0	0	+7.0	11.6	0	+10.0
5.52–6.30	5.4	+1.6	+8.0	9.5	+2.5	+13.6
	8.0	0	+7.0	13.6	0	+12.0
6.30–7.09	6.4	+1.6	+9.0	9.5	+2.5	+13.6
	9.0	0	+8.0	13.6	0	+12.0
7.09–7.88	7.2	+1.8	+10.2	11.2	+2.8	+15.8
	10.2	0	+9.0	15.8	0	+14.0
7.88–8.86	8.2	+1.8	+11.2	13.2	+2.8	+17.8
	11.2	0	+10.0	17.8	0	+16.0

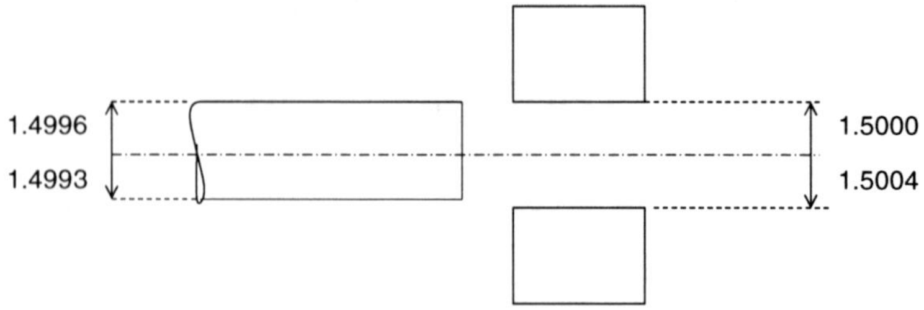

Fig. 11.1 Analysis for RC1

Table 11.7 Symbol modifier

Modifier definitions	Symbol
Maximum material condition (MMC)	Ⓜ
Least material condition (LMC)	Ⓛ
Regardless of feature size (RFS)	Ⓢ
Diameter (DIA)	∅

Table 11.8 GD&T symbols

Feature type	Tolerance type	Characteristics	Symbols
Individual	Form	Straightness	⏤
		Flatness	⏥
		Circularity	○
		Cylindricity	⌭
Individual or related	Profile	Profile of a line	⌒
		Profile of a surface	⌓
Related	Orientation	Angularity	∠
		Perpendicularity	⊥
		Parallelism	//
	Location	Position	⌖
		Concentricity	◎
		Symmetry	⌯
	Runout	Circular	↗
		Total	⌰

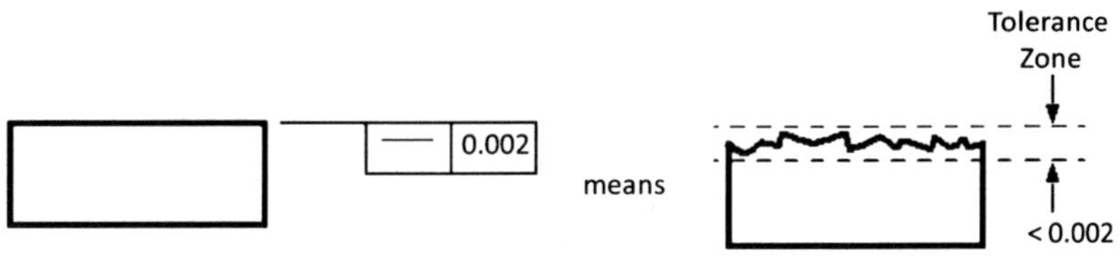

Fig. 11.2 Straightness control

11.5.1.2 Flatness

A perfectly flat surface has all its elements in the same plane. Flatness feature control frames create a tolerance zone not related to any datums. All elements of the produced feature under control must lie with the tolerance zone determined by two parallel planes. Flatness control is generally used on planar surfaces to be used as primary datum features. Figure 11.3 illustrates this concept.

11.5.1.3 Circularity

It is a two-dimensional surface form control. The circularity tolerance zone consists of two concentric circles that are spaced radially at a distance equal to the tolerance in the feature control frame. Each feature is compared to a perfectly round geometric counterpart of itself. Datums are not used. Figure 11.4 illustrates this concept.

11.5.1.4 Cylindricity

Cylindricity is an individual geometric characteristic. It is a surface form control that simultaneously limits the allowed out of roundness, straightness, and taper that a surface may experience. The feature control frame normally consists of only a cylindricity symbol and a geometric tolerance. The tolerance zone consists of two coaxial cylinders different in size radially by the amount shown in the cylindricity control. Figure 11.5 illustrates this concept.

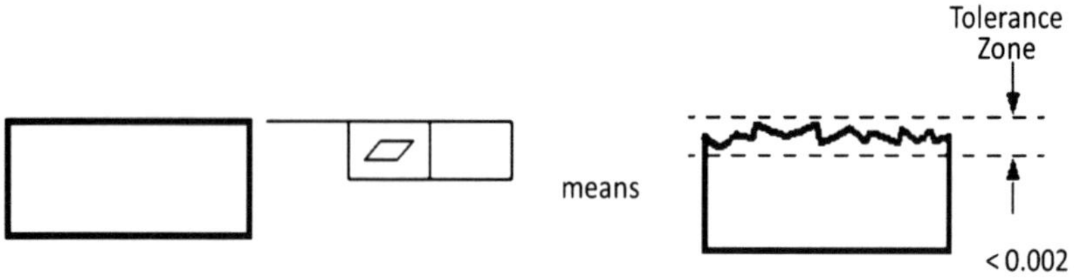

Fig. 11.3 Flatness form control

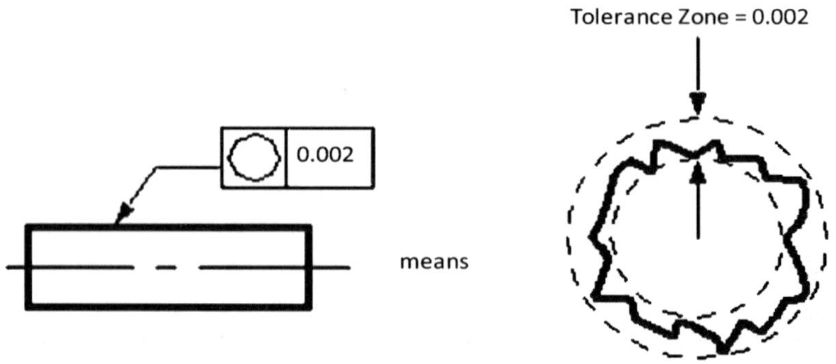

Fig. 11.4 Circularity form control

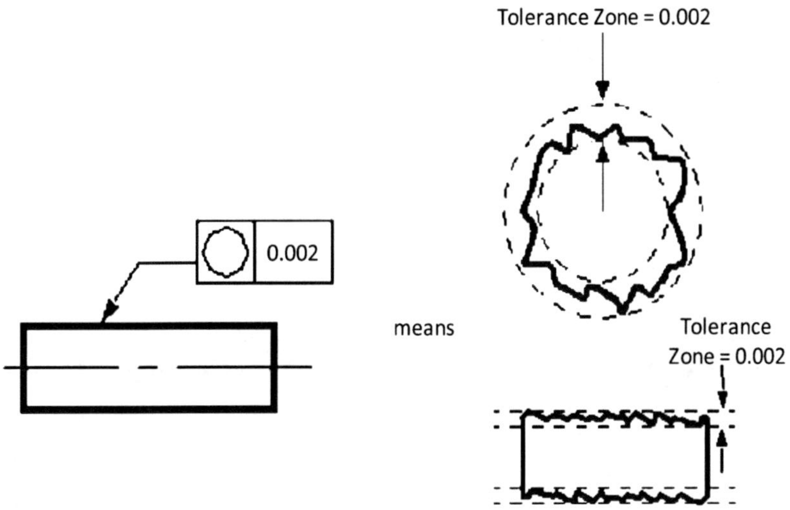

Fig. 11.5 Cylindricity form control

11.5.2 Profile Control

A profile is the outline of an object in a given plane. Profiles are formed by projecting a three-dimensional figure onto a plane or by taking cross-sections through the figure. The elements of a profile are straight lines, arcs, and other curved lines. With profile tolerancing, the true profile may be defined by basic radii, basic angular dimensions, basic co-ordinate dimensions, or formulae.

The profile tolerance specifies a uniform boundary along the true profile within which the elements of the surface must lie. It is used to control form or combinations of size, form, orientation, and location. A profile tolerance may be applied to individual profiles taken at various cross-sections through the part or to an entire surface.

11.5.2.1 Profile of a Line

The tolerance zone established by the profile of a line tolerance is two-dimensional, extending along the length of the considered feature. This applies to the profiles of parts having irregular line patterns, such as the tapered wing of an aircraft. Figure 11.6 illustrates this concept.

11.5.2.2 Profile of a Surface

The tolerance zone established by the profile of a surface tolerance is three-dimensional, extending along the length and width of the considered feature or features. Figure 11.6a, b illustrates this concept.

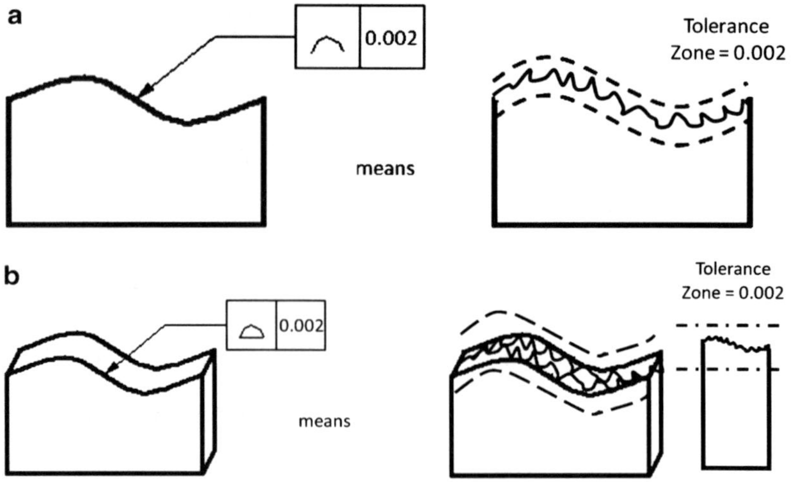

Fig. 11.6 (**a**) Profile of line, (**b**) profile of surface

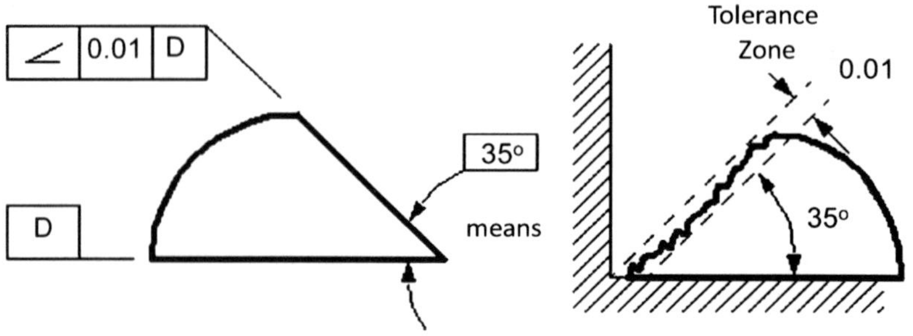

Fig. 11.7 Angularity orientation control

11.5.3 Orientation

There are three orientation controls. These are angularity, perpendicularity, and parallelism. All controlled features are related to datum planes, to datum axes, or to a combination of both.

11.5.3.1 Angularity

Angularity is the condition of a surface, centerplane, or axis at a specified angle from a datum plane, surface, or an axis. It can be used to tolerance all angles other than perpendicular or parallel ones. Figure 11.7 illustrates this concept. An angularity tolerance specifies one of the following [4]:

- A tolerance zone defined by two parallel planes within which the axis of the considered feature exists.
- A cylindrical tolerance zone within which the axis of the considered feature exists.
- A tolerance zone defined by two parallel lines within which the line element of the surface exists.

11.5.3.2 Perpendicularity

It is the condition of a surface, center plane, or axis at a right angle to a datum plane or axis. Figure 11.8 illustrates this concept.

A perpendicularity tolerance specifies one of the following [4]:

- A tolerance zone defined by two parallel planes perpendicular to a datum plane or axis.
- A tolerance zone defined by two parallel planes perpendicular to a datum axis.
- A cylindrical tolerance zone perpendicular to a datum plane.
- A tolerance zone defined by two parallel lines perpendicular to a datum plane or axis.

11.5.3.3 Parallelism

It is the condition of a surface or center plane, equidistant at all points from a datum plane, or an axis, equidistant along its length from one or more datum planes or a datum axis. Figure 11.9 illustrates this concept. A parallelism tolerance specifies one of the following [4]:

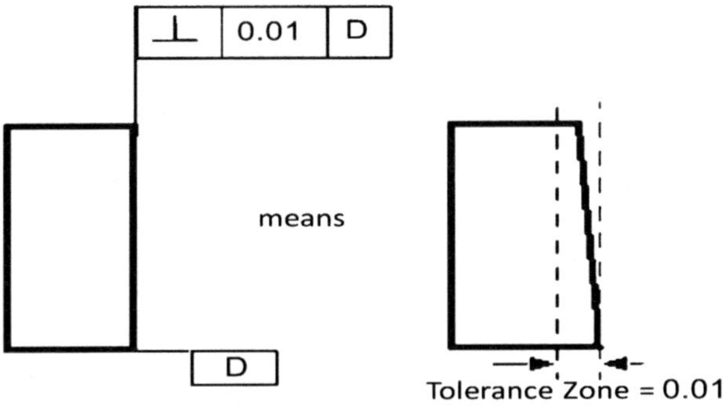

Fig. 11.8 Perpendicularity orientation control

- A tolerance zone defined by two parallel planes parallel to a datum plane or axis.
- A cylindrical tolerance zone parallel to one or more datum planes or a datum axis.
- A tolerance zone defined by two parallel lines parallel to a datum plane or axis.

11.5.4 Location

There are three tolerances of location. These are position, concentricity, and symmetry. A positional tolerance defines a zone within which the center, axis, or center plane of a feature of size is permitted to vary from a true position. Another definition for positional toleranced is the boundary, defined as the virtual condition, located at the theoretically exact position that may not be violated by the surface or surfaces of the defined feature. A positional tolerance is indicated by the position symbol, a tolerance value, applicable material condition modifiers, and appropriate datum references placed in a feature control frame. Figures 11.10 and 11.11 illustrate the boundary and axial conditions.

Concentricity and symmetry are similar controls used on dissimilar features. Concentricity is most commonly used to control median points on a figure of revolution to a datum axis, whereas symmetry is used to control the median points

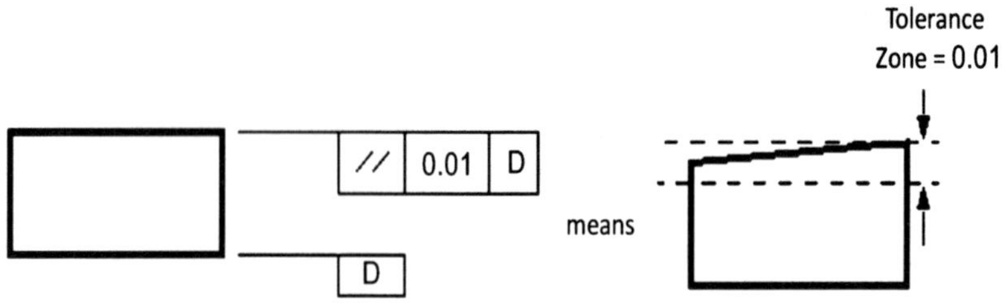

Fig. 11.9 Parallelism orientation control

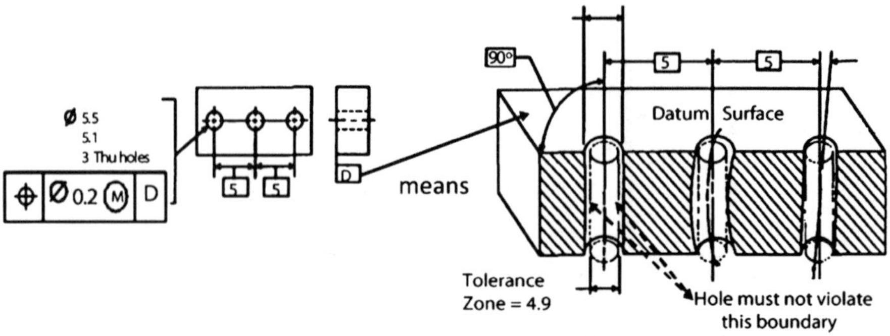

Fig. 11.10 Boundary representation for position control

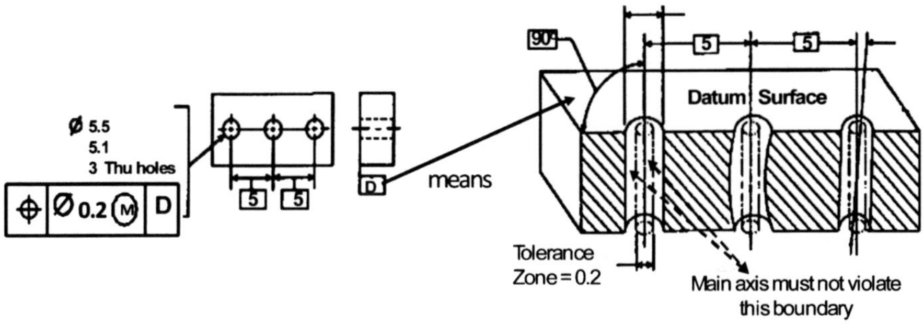

Fig. 11.11 Axis representation for position control

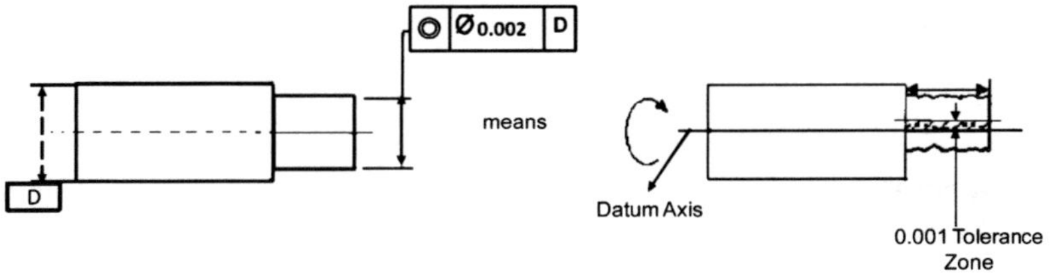

Fig. 11.12 Concentricity control

Fig. 11.13 Symmetry control

of two opposing parallel planar type surfaces to a datum plane. Both require the use of a datum feature(s) and are always controlled on a RFS. Figures 11.12 and 11.13 illustrate these concepts.

11.5.5 Runout

Runout is a composite tolerance used to control the functional relationship of one or more features of a part to datum axis. There are two types of runout control, circular runout and total runout. Circular runout provides control of circular

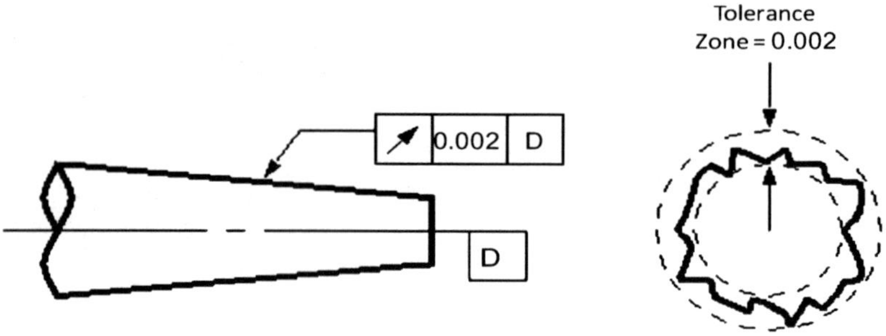

Fig. 11.14 Circular runout

elements of a surface. The tolerance is applied independently at each circular measuring position as the part is rotated 360°. Circular runout may be used to control the cumulative variations of circularity. Total runout provides composite control of all surface elements. The tolerance is applied simultaneously to all circular and profile measuring positions as the part is rotated 360°. Figure 11.14 illustrates the circular runout concept.

11.6 Manufacturing Processes and Tolerances

Depending on the types of manufacturing processes, the part's accuracy could vary. Machining operations are typically accurate due to the high levels of automation used for their control. Although primary processes such as casting are inaccurate. It is important to consider the level of variations associated with each process during the process selection and planning. In many cases, sequences of processes are selected due to the machine performance, required GD&T, and surface finish. The surface finish or roughness is the result of the selected processes impact on the surface quality (tool, mechanical vibrations, cut types, etc.). It is the irregularities on the surface of the part and it is measured as the arithmetic average (R_a) of the highs (peaks) and lows (valley) deviations from the nominal (actual) surface of the feature. Figure 11.15 illustrates this concept. The roughness features includes roughness height and spacing. It is the close space between the high (peaks) and (lows) valleys. Roughness is usually due to the surface forming operations. The waviness consists of the wide irregularities surface pattern. It is typically produced due to the machine vibrations. Lay is the direction of the machining pattern. The arithmetic average of the highs (peaks) and lows (valley) deviations from the nominal (actual) surface of the feature is illustrated in Fig. 11.16.

The R_a is calculated as follows:

$$R_a = \sum \frac{d_i}{n} \quad for \quad i = 1,2 \ldots n \tag{11.1}$$

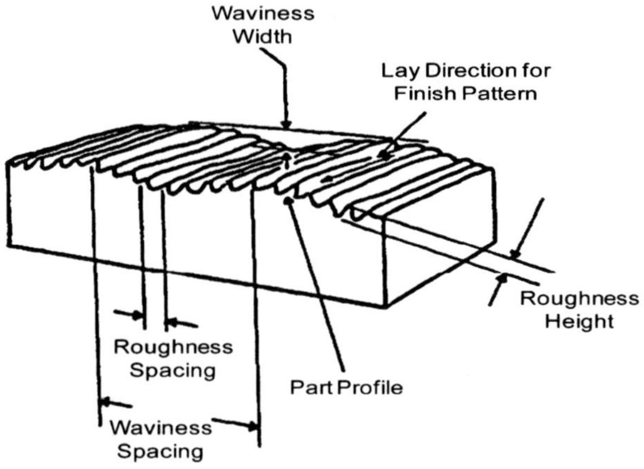

Fig. 11.15 Surface finish profile and characteristics

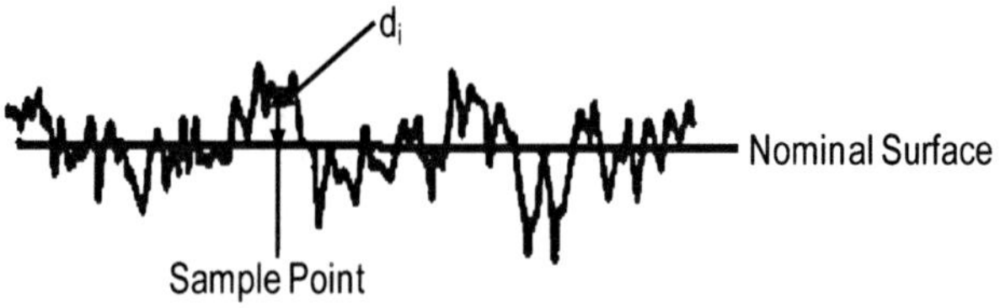

Fig. 11.16 Surface finish profile

where d_i is the distance between the peak and the center line (absolute value) and n is the total number of sample points. Table 11.9 lists an example of a compatibility chart used for selecting manufacturing processes based on the required tolerances and surface finishes [2]. Tables 11.3–11.6 are the required tables for calculating functional tolerances using fit analysis.

11.7 Summary

In this chapter, an introductory discussion is provided to the topics of geometric dimension and tolerancing and its impact on selecting the right manufacturing processes.

Also, improvements in measurements provide a need to better understand that variation is unavoidable in manufacturing. However, acceptable levels of variation will result in a good assembly. It is important to understand the limits of this variation. After the development of GD&T, drawings became the main tool for communication among design, manufacturing, testing, etc.

Table 11.9 Compatibility analysis (units in inches)

Process	Subprocess	GD&T and surface finish specifications			
				Roughing	Finish
Milling	Face	Tolerance		0.2	0.001
		Flatness		0.001	0.001
		Angularity		0.001	0.001
		Parallelism		0.001	0.001
		Surface finish		50	30
	Peripheral	Tolerance		0.002	0.001
		Flatness		0.001	0.001
		Surface finish		50	30
	End	Tolerance		0.004	0.004
		Parallelism		0.0015	0.00155
		Surface finish		60	
Turning	Facing	Surface Finish – 250 to 16			
	Parting	Diameter		Tolerance	
	Boring	to 1.0		0.001	
	Drilling	1–2		0.002	
	Reaming	2–4		0.003	
Boring		Diameter	Tolerance		
		0–3/4	0.001 (0.0002)	Straightness	0.0002
		3/4–1	0.0015 (0.0002)	Roundness	0.0003
		1–2	0.002 (0.0004)	Positional	0.0001
		2–4	0.003 (0.0008)	Surface Finish	8
		4–6	0.004 (0.001)		
		6–12	0.005 (0.002)		
Grinding	Tool type	Diameter	Tolerance	Subprocess	
	Internal	0–1	0.00015 (0.00005)	Internal grinding	
		1–2	0.0002 (0.00005)	Cylindrical grinding	
		2–4	0.0003 (0.0001)	Center-less grinding	
		4–8	0.0005 (0.00013)	External grinding	
		8–16	0.0008 (0.0002)	Surface grinding	
	Center and Centerless	Tolerance	0.05 (0.0001)		
		Parallelism	0.0005 (0.0002)		
		Roundness	0.0005 (0.0001)		
		Surface Finish	8 (2)		
	Flat	Tolerance	0.001 (0.0001)		
		Parallelism	0.001 (0.001)		
		Surface Finish	32 (2)		
Drilling	Twist drill	Diameter	Tolerance	Positional	0.008
	Spade drill	0–3/4	+0.003–0.001	Roundness	0.004
	Center drill	3/4–1	+0.004–0.001	Surface Finish	100
	Combination	1–2	+0.006–0.001		
	Countersink	2–4	+0.008–0.002		
	Counter-bore	4–6	+0.010–0.003		
		6–12	+0.012–0.004		

(continued)

Table 11.9 (continued)

Process	Subprocess	GD&T and surface finish specifications			
Honing		Diameter	Tolerance	Roundness Surface Finish	0.0005
		1	Roughing Finishing		4
		2	0.0005 +0.0001		
		4	0.0008 +0.0005		
			0.001 +0.0008		
Broaching		Tolerance		0.001	
		Surface finish		125–32	
Sawing		Length tolerance	Squareness		Surface finish
		0.01	0.2		200–300
		0.01	0.2		200–300
		0.008	0.2		125
Tapping		Tolerance	0.003		
		Roundness	0.003		
		Surface finish	75		
Lapping		Tolerance	Roughing		Finish
		Flatness	0.000025		0.000015
		Surface finish	0.000025		0.000012
			4–6		1–4

11.8 Review Questions

11.1 What are the importances of GD&T?
11.2 Define GD&T and its standard.
11.3 Discuss functional dimensioning.
11.4 What is the meaning of engineering tolerance?
11.5 Define the following terms: straightness, flatness, circularity perpendicularity, and cylindricity.

References

1. American Society of Mechanical Engineers (1995) Mathematical Definition of Dimensioning and Tolerancing Principles.
2. Chang TC (1990) Expert process planning for manufacturing. Addison-Wesley, Reading.
3. Drake P (1999) Dimensioning and tolerancing handbook. McGraw-Hill, New York, pp 5–17.
4. Foster LW (1994) Geo-metrics III: the application of geometric and tolerancing technique. Addison-Wesley, Reading.

5. Krulikowski A (1992) Geometric dimensioning: a self-study course. SME Publications, Effective training Inc.
6. Meadows DJ (1995) Geometric dimensioning and tolerancing: applications and techniques for use in design, manufacturing, and inspection. Marcel Dekker, New York.
7. Wilson BA (1995) Dimensioning and tolerancing handbook. Genium, New York.
8. Ibrahim Z (2005) Mastering CAD/CAM. McGraw-Hill, New York.

Chapter 12
Rapid Prototyping

Abstract This chapter provides an overview of the rapid systems: stereolithography (SLA), solid ground curing (SGC), laminated object manufacturing (LOM), selective laser sintering (SLS), direct shell production casting (DSPC), and fused deposition modeling (FDM).

Keywords Rapid prototyping • SGC • DSPC • SLS • FDM

12.1 Introduction

Rapid prototyping (RP) is a technique for direct conversion of 3D CAD data into a physical prototype using a number of techniques, mostly based on slicing a 3D object into multiple 2D layers and building them up one layer at a time. Industries have been using RP techniques increasingly to reduce their product development cycle. Having realized the potential of RP for prototyping applications, a large number of processes have been developed allowing the use of various materials ranging from plastics to metals for the development of prototypes. Rapid manufacturing (RM) is the term for applying RP for production of final manufactured parts. Rapid tooling (RT) generally concerns the production of molds and tooling inserts using RP. RP and RT are means for compressing the time to market of products and lowering the cost, considered as competitiveness-enhancing technologies [12].

Rapid prototyping can be used to check for form, fit, and function of prototypes especially in tooling. RP parts are used for visual inspection, ergonomic evaluation, and masters for secondary manufacturing processes in various stages of new product development (NPD) [4]. The process has been used in the automotive, consumer products, casting, and toy industries. An estimated 3.55 million models and prototype parts were build worldwide using RP techniques in 2001.This is an increase of 18.3% from the three million models produced in 2000 [19]. Since at the design stage, two copies of a unique design are built on the average, it is estimated that some 1.72 million distinctive parts were produced in 2001. Research has shown that the total costs for new products can be reduced by as much as 30–60% [8, 20], and lead times can be

A.K. Kamrani and E.A. Nasr, *Engineering Design and Rapid Prototyping*,
DOI 10.1007/978-0-387-95863-7_12, © Springer Science+Business Media, LLC 2010

reduced by 50–60% [14]. RP has a direct fabrication ability that reduces tooling cost. It does not involve tooling and fixturing. RP can also be used for design of very complicated objects with self-similarity feature [15]. Contemporary CAD systems can no longer handle the object when the object has a very complicated shape with self-similarity. Most objects in nature are by design self-similar.

The RP processes are used in reverse engineering (RE) and NPD. RE takes a manual or computer-aided method to duplicate an existing object or system. RE has gained importance to speed up product development process in recent years. This is especially true since the manufacturing process is being outsourced to nations with cheaper labor and 70% of the product cost is committed at the design stage.

The RP can be used for a variety of applications. These applications have been categorized into four main functions: parts for modeling and presentations, functional prototypes, tooling and process parts, production, and manufacturable parts. Model prototypes are nonfunctional parts, although they are developed with high accuracy in shape and dimension [2]. The RP model can be used for concept validation, design verification and form, and fit analysis of assembly parts. Functional prototypes are developed to be analyzed under the same conditions as in actual use. Material selection and use is an important part of this process because the properties of the prototype material should be very similar to the properties of the material that will be used for the actual component. These are physical properties such as stress and strain and thermal properties such as expansion. Process parts are used to create functional prototypes. RP models can be used as patterns for molds in investment and sand casting processes. This process is known as the indirect approach to RT. They can also be used as direct approach in case of the design of the pattern in spray metal operation. Finally, the production parts are directly manufactured using RP equipment. This process is known as RM. Material used in this category of RP application must have all the characteristics of the final product as it is processed using the traditional manufacturing operations [10]. The first recorded rapid manufactured part was developed on a selective laser sintering (SLS) machine in 1992. These parts must have the required durability and functional. Figure 12.1 illustrates the RP Generic process.

12.2 Benefits of Rapid Prototyping Technology

Rapid prototyping has a strong impact on productivity, which means getting a product from concept to prototype to reality in as fast and as an inexpensive method as possible. Conventionally, producing prototypes, testing them, and approving them have been a slow and costly process. Conventional prototyping can take a long period of time from weeks to months, depending on the complication of the design and the method used [8].

The RP can assist radically in reducing the prototype life cycle and the related costs. Moreover, RP computerizes the fabrication of prototypes from 3D CAD designs. A physical prototype provides more information about a product compared with the information that can be obtained from the design drawing. The following are the general benefits of using RP technology in the industries [3]:

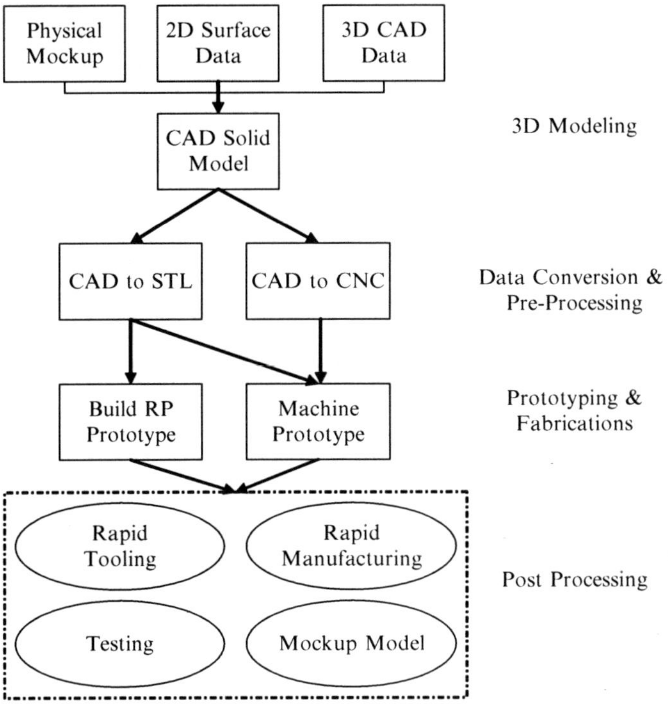

Fig. 12.1 RP generic process

- Quickly try out "new" design concepts
- Troubleshoot existing or new designs
- Obtain ideas for future design changes
- Better communication among engineering teams
- Functional testing of designs
- Marketing
- Production tooling and fixturing can be better planned
- Packaging and shipping methods can be explored

The RP cycle for Rapid Prototyping Design is illustrated in Fig. 12.2.

12.3 Rapid Prototyping Terminology

The RP processes are categorized based on the material types (powder, liquid, solid, etc.) and the building technology (laser, adhesive, extrusion, etc.) The stereolithography file (STL) is generated from the solid model CAD data and then used as input to the RP machines to generate the part. Some of the available RP technologies are fused deposition modeling (FDM), laminated object manufacturing (LOM), stereolithography

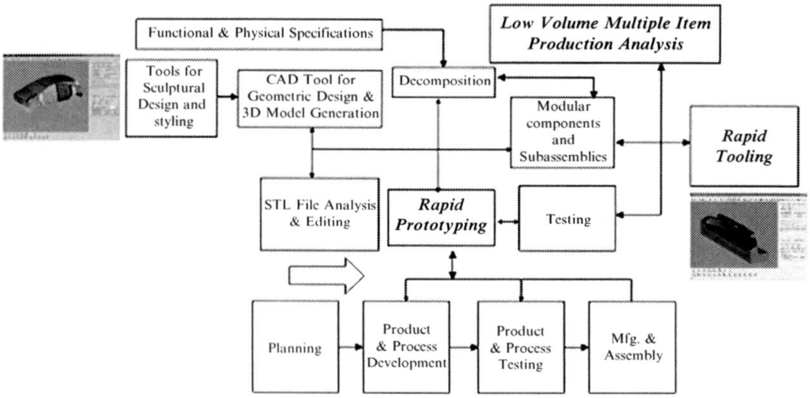

Fig. 12.2 RP cycle for rapid product design

(SLA), and SLS. Most of these processes use layered manufacturing techniques (LMT). RM refers to the direct manufacture of solid 3D parts using LMT techniques. SLS has been used to manufacture parts for the space shuttle and space labs at Boeing. SLS has also been used to make parts at NASA's Jet Propulsion Lab.

The first stage of producing a rapid prototype is a 3D CAD solid model or surface model generated by any CAD software. Because the RP system generates a precise physical duplicate of the solid modeling CAD design, it is important that the CAD design should be excellent. The geometrical solid model CAD data are converted into STL file format by the use of CAD software. The word STL is especially developed for stereolithography RP systems. It has become the standard format/interface between a 3D solid model CAD design and any RP systems [1].

Normally, the 3D solid surface is approximated by a surface consisting of triangular facets. The STL file gives the *X*, *Y*, and *Z* coordinates of the vertices of each triangular facet. Therefore, it is necessary that the generated faceted surfaces should be perfectly defined in the STL file format. Today's RP systems support software that validates the STL file. The software has the capability to recognize the defective regions of the file and fix them. The repair process is automated by the computer software and it is not always guaranteed. So, it is essential to review the corrected file for validation purposes.

Many problems can affect the accuracy of the STL file. Because STL files are made of a group of triangular facets and each triangle consists of straight sides, the RP design might not consist totally of straight sides. For example, the complex curves and designs are basic features of today's design. To overcome this problem, the curved surface using a planar surface should be perfectly defined. If the curved surface is defined by a small number of triangular facets, the STL file will not be compatible with the actual design. Therefore, the generated RP part will not be extremely accurate and the flat surfaces will be noticeable. As a result of defining a curved surface with an enormous number of triangular facets, the curved surface can be represented more accurately. Generating the typical STL file for curved surfaces is an important task in the RP processes; therefore most of the solid CAD softwares support techniques for controlling the number of the generated triangles [17].

The CAD data can be used to create two types of prototypes: soft prototypes and hard prototypes. Soft prototypes can be defined as computer-generated graphics of a design in which visualization tools integrate CAD-based systems and animation to facilitate a detailed observation of the model. The generated soft prototypes by visualization tools cannot be touched; however, it can provide another way of conveying the product design specifications. On the other hand, hard prototypes can be defined as the physical models that can be touched and observed in 3D views.

There are many factors affecting the selection of the RP process. These factors are [11]:

- Part resolution
- Postprocessing operations
- Available RP systems to the designer
- Prototype material characteristics
- Available time

12.4 Rapid Prototyping Systems

The rapid systems covered in this chapter are as follows:

1. SLA
2. SGC
3. LOM
4. SLS
5. DSPC
6. FDM

These systems are discussed in detail in the following subsections.

12.4.1 Stereolithography

Stereolithography is developed via 3D solid modeling systems. It is a popular RP process; approximately 60% of RP operations can use the SLA process. The A in SLA stands for apparatus.

12.4.1.1 Stereolithography Process Preparation

To create a part using the SLA process, the following steps should be done:

1. The STL file is generated by CAD software.
2. The generated file is provided into the RP machine.

3. The support file is generated.
4. The STL file and support file are sliced with the slice program.
5. The BFF file is generated.
6. The BFF file is uploaded into the RP machine.

The material container is an element of the machine and never removed unless to change resin. The platform is mounted onto the elevator of the machine and moves up and down within the resin container.

12.4.1.2 Stereolithography Process

The SLA process develops parts as follows [5]:

1. A directed laser beam reproduces the shape of slice files onto the build material. Supports are built first.
2. The region laser traced onto the build material cures, solidifying the first layer of the laser trace. Support structures are reproduced concurrently as needed.
3. The platform with the first layer of the part and/or support is let down the surface of the build material.
4. The solidified build material is wiped to level it and avoid buildup of material at one end of the container.
5. The platform is lowered one layer below the surface of the build material.
6. The previous steps are repeated until the part and supports are built layer by layer.
7. When the part is completed, the model is detached from the machine and the supports are manually removed from the part.
8. The part is cleaned, postcured in an ultraviolet postcuring apparatus, and finished as required.

12.4.1.3 Advantages and Disadvantages of Stereolithography

Advantages of Stereolithography

The following are the advantages of the SLA process [5]:

1. *Productivity*: Parts can be fabricated in a short period of time (hours/days), depending on the design complexity without the need for machining or tooling.
2. *Unattended operation*: The SLA process can run from build start to build finish without the need of a worker.
3. *Accuracy*: The accuracy of SLA parts is good.
4. *Reliability*: SLA systems are very reliable.
5. *Flexibility*: Several parts can be built in the same build envelope.
6. *Material usage*: Slight waste is created.

7. *Surface finish and appearance*: The surface finish and shape of produced parts are very good.
8. *Part strength*: Part strength is excellent; therefore parts are used for the following tests: stress, airflow, fluids, and photoelastic analysis.

Disadvantages of Stereolithography

Photocurable resins such as those used in SLA are brittle. Use of these resins causes the following problems [5]:

1. *Cracking*: Cracking of SLA-built parts is common when parts are drilled and have fasteners driven into them.
2. *Changing in dimension due to absorption of moisture*: Because the material continues to cure and absorb ambient moisture from the air, parts generated from SLA resins face some problems such as changes in dimension and distorting.
3. *Support creation and subsequent removal*: Numerous parts produced via SLA require the construction of supports. As a result, manual removal and disposal of the supports are essential.
4. *Cleaning and curing*: Some SLA resin-based parts must be cleaned using a solvent such as alcohol.
5. *Long Build Time*: Because SLA run time is proportional to the volume of the designed part, so building of large solid parts may require long time.

12.4.2 Solid Ground Curing

Solid ground curing (SGC) was developed by Cubital. SGC is a RP system that utilizes an ultraviolet light to cure each layer of liquid resin. In contrast with laser-based systems, SGC process cures a whole layer in one exposure, rather than tracing the part with a beam [20].

12.4.2.1 Solid Ground Curing Process Preparation

To create a part using the SGC process, the following steps should be done:

1. The STL file is generated and loaded into the controller of the SGC RP machine.
2. Containers of build material and wax are filled.

12.4.2.2 Solid Ground Curing Process

Solid ground curing process generates parts as follows [5]:

1. A xerographic negative of the initial working slice is printed onto a glass plate. The segment of the glass plate representing the slice to be built is transparent.

2. The build platform is coated with a thin layer of photocurable liquid resin.
3. The build platform moves into a position under the glass plate and both are aligned under an ultraviolet lamp.
4. High-intensity ultraviolet light passes through the plate to cure the resin under the transparent portion of the glass.
5. Uncured resin is vacuumed off the build platform.
6. A layer of liquid support wax is deposited onto the build platform, filling in the spaces among the cured areas, and is hardened by a chill plate.
7. A part of the solidified resin and wax is milled off to give a consistent surface for deposition of resin for the next layer.
8. The build platform moves under a dominant ultraviolet lamp for final curing of the layer.
9. The process continues until the part or parts are built.
10. The solid cube of wax and parts is removed for dewaxing.
11. The wax is dissolved and washed away with hot water or a solution of hot water and citric acid.

12.4.2.3 Advantages and Disadvantages of Solid Ground Curing

Advantages of Solid Ground Curing

The following are the advantages of the SGC process [5]:

1. *Stable build environment*: The build surrounding decreases the possibility of curling and warping. Moreover, it reduces the need for supports and final curing.
2. *Throughput*: Because parts can be nested and stacked, a whole layer can be cured with one 6-s exposure of light.
3. *Build geometry*: Any 3D solid geometric design can be built in any configuration.
4. *Build time*: Parts can be built overnight, with attended operation.
5. *Interruptable operation*: In an urgent situation, the build cycle can be intermittent for a high-priority build. On the other hand, because of the potential for something to go wrong, this is not recommended unless totally needed.

Disadvantages of Solid Ground Curing

There are many limitations of using the SGC process. They include the following:

1. *Attended operation*: The SGC process is labor intensive. The process needs to be continuously supervised.
2. *Dewaxing cycle*: After curing, the wax needs to be washed away with water/citric acid solution.
3. *Solid waste disposal*: A portion of each layer of build material and wax is milled off.

4. *Liquid waste disposal*: Although the working environment is good, the wax and water, or wax/water/citric acid from the dewaxing cycle must be disposed of.

12.4.3 Laminated Object Manufacturing

Laminated object manufacturing is a RP system that utilizes 3D solid model CAD data to build up prototype models from sheet material.

12.4.3.1 Laminated Object Manufacturing Process Preparation

To create a part using the LOM process, the following steps should be done:

1. 3D solid model CAD data is entered into the process controller.
2. The software generates cross-sectional slices.
3. The paper feeder is loaded with the desired material.

12.4.3.2 Laminated Object Manufacturing Process

The LOM process creates parts as follows [16]:

1. A sheet of build material is placed over the LOM machine's elevator platform on which a build plate rests.
2. The elevator moves up until the plate makes contact with the build material.
3. The initial layer of material is laminated to a layer of tape on the elevator platform.
4. A laser beam cuts the cross-sectional outline of the initial layer of material.
5. The laser beam cuts a crosshatch design into the excess material to assist the removal process.
6. The laser beam also cuts a wall that supports the model and crosshatched excess material.
7. The elevator platform raises to detach the traced design from the roll of build material, and then lowers.
8. A second layer of build material is advanced.
9. The elevator platform rises to make contact with the build material.
10. The heated laminating roller presses down the second layer to bond it to the initial layer.
11. The previous steps continue until the complete model is completed.
12. The laminated stack and surrounding wall and plate are removed from the elevator platform.
13. The surrounding wall is detached.
14. Excess external and internal material is removed to reveal the model.

12.4.3.3 Advantages and Disadvantages of Laminated Object Manufacturing

Advantages of Laminated Object Manufacturing

The following are the advantages of the LOM process [5]:

1. *Inexpensive*: Both material investment and the LOM process are very inexpensive.
2. *No toxic materials*: Only nontoxic fumes from laser cutting are generated. The fumes can be vented.
3. *Accuracy*: LOM parts were once prone to dimensional nonstability, particularly in the Z-axis. Improvements in materials and the process of the machine can reduce this problem.
4. *Semi-attended operation*: The LOM process can run unattended.

Disadvantages of Laminated Object Manufacturing

There are many limitations of using the LOM process. They include the following:

1. *Dimensional stability*: Until they are sealed, LOM papers are susceptible to swelling because of humidity. Even though improvements in the process and materials can reduce this problem, LOM parts have some Z-axis accuracy problems. Z height should be supervised during the building process.
2. *Internal cavities*: Some parts require stopping the process to eliminate the waste. Moreover, the part should be split or delaminated to clean out interior cavities, after which the part can be bonded back together.
3. *Postproduction time*: Depending on the complexity of the designed model, postproduction time to eliminate the whole waste can be labor intensive.
4. *Secondary processes*: To generate accurately functional parts, secondary processes are required.

12.4.4 Selective Laser Sintering

Selective laser sintering is a RP system that utilizes a laser to fuse (sinter) powdered material to build the layers of a part.

12.4.4.1 Selective Laser Sintering Process Preparation

To create a part using the SLS process, the following steps should be done:

1. A CAD file in STL format is loaded into the controller of the SLS RP machine.
2. The cartridges containing the required build material are installed.

12.4.4.2 Selective Laser Sintering Process

Selective laser sintering process creates parts as follows [6]:

1. A thin layer of heat-fusible build material is deposited onto the part-building bed within the build chamber.
2. The 3D solid modeling software slices the part files and the heat generating CO_2 laser draws the initial layer of the part to be built onto the powdered build material. The initial layer is supported by unfused material beneath it.
3. The particles of material contacted by the laser fuse together to form a solid mass. The intensity of the laser beam is modulated to fuse the powder only in the areas defined by the design geometry of the slice.
4. The part-building bed lowers a depth of one slice.
5. A roller mechanism deposits the next layer of build material on top of the previous layer.
6. The previous steps are repeated, building the part as each layer fuses to the one beneath it.
7. When the build is accomplished, the part is removed from the build chamber.
8. Loose material falls away or is brushed off in the breakout station.
9. The part may be postprocessed as required.

12.4.4.3 Advantages and Disadvantages of Selective Laser Sintering

Advantages of Selective Laser Sintering

The following are the advantages of the SLS process:

1. *Wide variety of build materials*: A very wide variety of build materials are used in the SLS process. Moreover, materials are cheap.
2. *Unattended operation*: The SLS process can run from build start to build finish without the supervision of a worker.
3. *No support materials required*: Powders perform as natural supports.
4. *Postcuring*: No postcuring is needed, excluding sand and metal. Postcuring of plastic parts is not required in the SLS process.
5. *Multiple part building*: Many parts are built and stacked within the SLS build envelope.
6. *Strength and durability*: Strength and durability are excellent, especially in nylon materials.
7. *Little waste*: Material that is not fused in the SLS process can be recycled and reused.
8. *Accuracy and surface finish*: The accuracy, surface finish, and shape of parts generated by the SLS process are excellent.
9. *Thin walls*: Very thin walls can be accomplished.
10. *Easy cleanup of parts*: Free materials usually fall off the completed part. What remains is easy to brush off.

11. *Metal and sand offer unique applications*: SLS provides metal and sand applications that other RP processes do not support.
12. *Functional prototypes*: The SLS process produces some prototypes that are functional.

Disadvantages of Selective Laser Sintering

There are many limitations of using the SLS process. They include the following:

1. *Temperature sensitivity*: To avoid oxidation of materials during the build process, it is necessary to maintain a nitrogen inert environment. The material fusing temperature should be held to just below its melting point.
2. *Metal builds*: As a consequence of the temperature requirements for the build, metals can be very difficult to use.
3. *Surface sealing requirement*: Cleanup of SLS parts is uncomplicated; however, to improve the rough, powdery surface finish of the part, some form of sealing and sanding is necessary.
4. *Build envelope is limited*: The build envelope is small.
5. *Use of entire platform*: Some materials will not build in a good manner if the whole platform is used.

12.4.5 Direct Shell Production Casting

Direct shell production casting (DSPC) was developed by Soligen. DSPC is based on 3D printing process invented at the Massachusetts Institute of Technology and licensed to Soligen. DSPC is a RP system that is different in two significant areas [13]:

- A process that is different from all other RP processes
- A company that markets services rather than equipment

The DSPC process is basically a 3D printing process that generates ceramic molds for metal castings with interior cores directly from a 3D solid model CAD file, avoiding the requirement of the traditional tooling.

12.4.5.1 Direct Shell Production Casting Process Preparation

To produce a part using the DSPC process:

1. A 3D solid model CAD file is generated in STL format.
2. The CAD file is forwarded to Soligen.

12.4.5.2 Direct Shell Production Casting Process

Direct shell production casting process creates parts as follows [1, 5]:

1. Using a 3D solid model CAD file, Soligen designs a virtual model for net shape casting and the gating system.
2. Soligen then creates a digital model of a ceramic mold, including integral cores to produce the hollow sections. The digital model is transferred to the DSPC RP system.
3. The DSPC RP system automatically spreads the initial layer of build material, a ceramic powder.
4. A liquid binder is "printed" onto the powder layer to define the cross-section of the mold. The initial layer is lowered and the next layer of ceramic powder is spread.
5. The process is repeated until the whole mold is "printed." The mold is then fired to create a rigid ceramic mold surrounded by unbound powder.
6. Unbound ceramic powder is removed from the mold.
7. At Soligen's foundry, the mold is filled with molten metal. After the metal cools, the ceramic and gating metal are removed.

12.4.5.3 Advantages and Disadvantages of Direct Shell Production Casting

Advantages of Direct Shell Production Casting

The following are advantages of the DSPC process:

1. *Patternless process*: DSPC is a patternless process that produces molds directly and eliminates tooling, wax, and shell dipping.
2. *Integral cores*: In the DSPC process, there is no need for core prints or core boxes.
3. *Reduced production time*: Functional metal parts are produced in a short period of time.
4. *Unique casting capability*: Castings are easily produced that cannot be produced by conventional casting processes.

Disadvantages of Direct Shell Production Casting

There are many limitations of using the DSPC process. They include the following:

1. *One part from one mold*: The ceramic mold is destroyed after releasing the cast part. Therefore, only one part can be produced per mold.
2. *Surface finish*: DSPC castings often show the layer steps of the DSPC process. Internal surfaces that show the layer divisions are hard to sand smooth out.

3. *Mass production*: Castings can be produced that cannot be produced by conventional casting processes. This advantage can lead to a disadvantage because the part that is prototyped cannot be mass produced conventionally.

12.4.6 Fused Deposition Modeling

Fused deposition modeling is a process that uses a computer-controlled extrusion head to deposit build material from the STL file. The FDM process is basically the same for all Stratasys models.

12.4.6.1 Fused Deposition Modeling Process

The FDM process builds 3D prototypes from 3D solid model CAD file or surface models [18]

1. Build material is fed to the FDM machine extrusion head from a spool.
2. The build material is melted by the temperature-controlled extrusion head.
3. The semiliquid build material is extruded and deposited in thin layers onto a fixtureless base.
4. The deposition process is repeated layer by layer to build up a 3D model.

12.4.6.2 Fused Deposition Modeling Support System

Because the FDM process uses two extrusion heads, support material may be either the same as or different from the build material. Moreover, the following are some important aspects of this matter:

- To help easy, snap-off removal, perforations are created where the supports join the model.
- No special cleanup is necessary.

12.4.6.3 Advantages and Disadvantages of Fused Deposition Modeling

Advantages of Fused Deposition Modeling

The following are advantages of the FDM process:

1. *A variety of build materials*: As a consequence of the availability of many different build materials, FDM is mainly good for functional prototyping.
2. *Easy removal of support material*: Support material is easily removable. Moreover, except for the support material, no material is wasted.

Disadvantages of Fused Deposition Modeling

There are many limitations of using the FDM process. They include the following:

1. *Delamination*: FDM prototypes built with certain materials exhibit weak interlaminar strength. They tend to delaminate when bent, or to shear along layer boundaries.
2. *Envelope size*: Because FDM uses a gantry, however, potentially the envelope is unlimited.
3. *Speed*: The FDM process is relatively slow.

12.5 Summary

Rapid prototyping is a technology that enables the building of physical models and prototype parts from 3D CAD data. RP builds the physical models and prototype parts from plastic, liquid, powdered, and sheet materials in thin cross-sections based on a computer model. The rapid systems covered in this chapter are SLA, SGC, LOM, SLS, DSPC, and FDM. Each one of these systems had some advantages and some limitations.

12.6 Review Questions

12.1. Define RP and state its importance.
12.2. What are the benefits of RM technology?
12.3. Define the term, STL.
12.4. Differentiate between soft prototypes and hard prototypes.
12.5. Name five RP systems.
12.6. Explain the steps of the SLA process.
12.7. What are the advantages and limitations of SLA?
12.8. Describe SGC process.
12.9. List the advantages and limitations of the SGC process.
12.10. Explain in detail LOM and its advantages and limitations.
12.11. What is SLS process? Explain.
12.12. Describe the DSPC process.
12.13. Explain the FDM process.
12.14. Explain the advantages and limitations of the FDM process.

References

1. Ashley S (1991) Rapid prototyping systems – special report. Mech Eng 113:34–43.
2. Ashley S (1995) Rapid prototyping systems is coming of age. Mech Eng 117(7):62–68.

3. Byun B, Hong S, Kwan H (2006) Determination of optimal build direction in rapid prototyping with variable slicing. Int J Adv Manuf Technol 28(3–4):307–313.
4. Choi SH, Samavedam S (2001) Visualization of rapid prototyping. Rapid Proto J 7(2):99–114.
5. Ford Motor Company (1997) Participant's guide: rapid prototyping, fabrication, and tooling.
6. Helisys (1998) Selective laser sintering product brochure, Austin.
7. Lotta L, Kanerva J (2002) Effects of rapid prototyping on small and medium sized companies' product development, Technical Paper, Society of Manufacturing Engineers, 7–14.
8. Lin, Liang S (2002) Rapid prototyping through scanned point data. Int J Prod Res 40(2):293–310.
9. Lopez SM, Wright PK (2002) The role of rapid prototyping in the product development process. Rapid Proto J 8(2):116–125.
10. Miller J (1996) Fundamental of rapid prototyping and applications in manufacturing. Westec Conference, Los Angeles.
11. Pegna A, Joseph J (2006) Environmental impacts of rapid prototyping: an overview of research to date. Rapid Proto J 12(2):64–71.
12. Pham DT, Dimov SS (2001) Rapid manufacturing. Springer, London.
13. Noorani R (2006) Rapid prototyping: principles and applications. Wiley, New York.
14. Shan Z, Yan Y, Guan L (2003) Rapid manufacture of metal tooling by rapid prototyping. Int J Adv Manuf Technol 21(7):469–475.
15. Soo SC, Yu KM (2003) Rapid prototyping for self-similarity design. J Mater Process Technol 139(1):219–225.
16. Stratasys (1998) LOM products brochure. Helisys, Carson.
17. Tie Y, Ma R, Ye M, Wang D, Wang C (2006) Rapid prototyping fabrication and finite element evaluation of the three-dimensional medical pelvic model. Int J Adv Manuf Technol 28(3–4):302–306.
18. Tomas CL (1996) An introduction to rapid prototyping. SDC, Mission.
19. Wohlers T (2003) Rapid prototyping/manufacturing in industry. Adv Mater Process 161(1):35–44.
20. Wohlers T (2004) Rapid prototyping, tooling, and manufacturing state of the industry. Wohlers Report 2004, Wholer Associates, Fort Collins.

Chapter 13
Data Mining Methodology and Techniques

Abstract The chapter provides the basic concepts of data mining. Data mining deals with the discovery of hidden knowledge, unexpected patterns, and new rules from large databases. It is regarded as the key element of a much more elaborate process called knowledge discovery in databases, or KDD, which is closely linked to data warehousing. Data mining can bring significant gains to organizations, for example, through better-targeted marketing and enhanced internal performance.

Keywords Data mining • Knowledge • KDD

13.1 Introduction

Data mining deals with the discovery of hidden knowledge, unexpected patterns, and new rules from large databases. It is regarded as the key element of a much more elaborate process called knowledge discovery in databases, or KDD, which is closely linked to data warehousing. According to Adriaans and Zantinge [1], data mining can bring significant gains to organizations, for example, through better-targeted marketing and enhanced internal performance. The long-term goal of data mining is to create a self-learning organization that makes optimal use of the information it generates. Recent publications on data mining concentrate on the construction and application of algorithms to extract knowledge from data. Skarmeta et al. [38] developed a data mining algorithm for text categorization. Andrade and Bork [2] used a data mining algorithm to extract valuable information on molecular biology from large amounts of literature. Lin et al. [31] developed an efficient data mining algorithm to measure proximity relationship measures between clusters of data. Delesie and Croes [9] presented a data mining approach to exploit a health insurance database to evaluate the performance of doctors in cardiovascular surgeries nationwide. The emphasis given by most authors and researchers on data mining focuses on the analysis phase of data mining. When a company uses data mining, it is important to also see that there are other activities involved in the process. These activities are usually more time-consuming and have an important influence on the success of the data mining procedure [21].

A.K. Kamrani and E.A. Nasr, *Engineering Design and Rapid Prototyping*,
DOI 10.1007/978-0-387-95863-7_13,

13.2 Data Mining

In the past, data mining has been referred to as knowledge management or knowledge engineering. Until recently, it has been an obscure and exotic technology, discussed more by theoreticians in the artificial intelligence fields. Fayyad et al. [12] defined data mining as a step in the KDD process consisting of applying computational techniques that, under acceptable computational efficiency limitations, produce a particular enumeration of patterns or models over the data. Adriaans and Zantinge [1] gave a more general definition used by many researchers. They stated that data mining is the process of searching through details of data for unknown patterns or trends. They stressed the importance of having an efficient method of searching in large amounts of data until a sequence of patterns emerges, whether complete or only within an allowable probability. Many times, large databases are searched for relationships, trends, and patterns, which prior to the search, are not known to exist nor visible. These relationships or trends are usually assumed to be there by engineers and marketers, but need to be proven by the data itself. The new information or knowledge allows the user community to be better at what it does. Often, a problem that arises is that large databases are searched for very few facts that will give the desired information. Moreover, the algorithm and search criteria used in a single database may change when a new trend or pattern is to be studied. Also, each database may need a different search criterion as well as new algorithms that can adapt to the conditions and problems of the new data.

More often than not, humans find it difficult to understand and visualize large data sets. Furthermore, as Fayyad and Stolorz [11] described, data can grow in two dimensions defined as the number of fields and the number of cases for each one of these fields. As they explained, human analysis and visualization abilities do not scale to high dimensions and massive volumes of data. A second factor that is making data mining a necessity is the fact that the rate of growth of data sets completely exceeds the rates that traditional "manual" analysis techniques can cope with. This means that if a company uses a regular technique for extracting knowledge from a database, vast amounts of data will be left unsearched, as the data growth surpasses the traditional mining procedures. These factors call for a need of a technology that will enable humans to tackle a problem using large amounts of data without disregarding or losing valuable information that may help solve any kind of problem involving large data sets. Yevich [46] stated that: "data mining is asking a processing engine to show answers to questions we do not know how to ask." He explained that instead of asking in normal query language, a direct question about a single occurrence on a database, the purpose of data mining is to find similar patterns that will somehow answer the desired questions proposed by the engineers or marketers. If the questions or the relationships asked to be found on a database are too specific, the process will be harder and will take more time. Moreover, a lot of important relationships will be missed or disregarded.

Data mining is being used widely in the United States, while in Europe, it has been used to a less extent. Large organizations such as American Express and

AT&T are utilizing KDD to analyze their client files. In the United Kingdom, the BBC has applied data mining techniques to analyze viewing figures. However, it has been seen that the use of KDD brings a lot of problems. As much as 80% of KDD is about preparing data, and the remaining 20% is about mining. Part of these 80% is the topic that will be analyzed and discussed in the next section. It is very difficult to introduce data mining into a whole organization due to:

- Lack of long-term vision
- Struggle between departments
- Not all files are up-to-date
- Legal and privacy restrictions
- Poor electric data processing capabilities
- Interpretation problems
- Lack of standardizations

In addition to these common problems, the company needs to have a minimum level of expertise on the data mining processes. Scenarios, in which the area expert does not have a specific question and asks the analyst to come up with some interesting results, are sentenced to fail. The same holds true for situations where the expert provides the data analyst with a set of data and a question, expecting the analyst to return the exact answer to that question. According to Feelders et al. [13], data mining requires knowledge of the processes behind the data, in order to:

- Determine useful questions for analysis;
- Select potentially relevant data to answer these questions;
- Help with the construction of useful features from the raw data; and
- Interpret results of the analysis, suggesting possible courses of action.

Knowing what to ask and what to expect from the information in a database is not enough. Knowledge of the available data from within is also required or at least desired. This will enable the expert and the data analyst to know where the data are and have it readily available depending on the problem being studied. Finally, data analysis expertise is also desired. Hand [24] discussed that phenomena such as population drift and selection bias should be taken into account when analyzing a database. Data mining expertise is required in order to select the appropriate algorithm for the data mining problem and the questions being raised.

13.3 The Methodology

Data mining is an iterative process. As the process progresses, new knowledge and new hypothesis should be generated to adjust to the quality and content of the data. This means that the quality of the data being studied will determine the time and precision of any given data mining algorithm, and if the algorithm is flexible enough, important information about a problem will be found even if the central

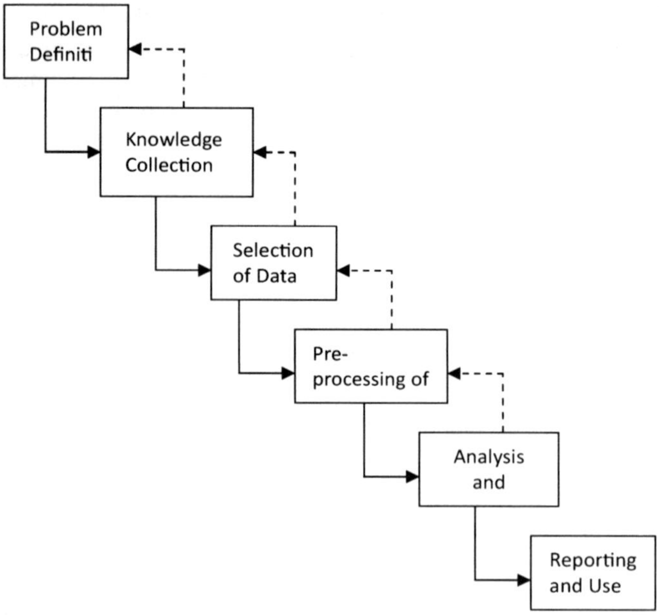

Fig. 13.1 Major steps in the data mining process

question is not fully answered. Fayyad et al. [12] developed a structured methodology outlining the different steps and stages of the data mining process. They outlined a six-step methodology that involved defining the problem to be solved, the acquisition of background knowledge regarding this problem, the selection of useful data, the pre-processing of the data, the analysis and interpretation of the results, and the actual use of these results. In their methodology, they stressed the importance of the constant "jump-backs" between stages. Feelders et al. [13] then used this methodology to explain each step of the data mining process. As Fig. 13.1 illustrates, the mining stage of the process (Analysis and Interpretation) is just one of the basic stages of the data mining methodology. The discussion on the stages of this methodology will show that all phases play a major role in the process, especially those that come before the mining stage.

13.3.1 Problem Definition

The problem definition phase is the first stage of the data mining process. During this stage, the objectives of using data mining on the desired problem are identified. These are questions or assumptions that, as it was mentioned earlier, are known to exist by marketers and engineers, but that need to be proven by the data. Most of the time, the initial question asked in a data mining project should be very vague. This will help the data mining process because it will disregard large amounts of data that are not useful to the problem. This way, the selection of data and the pre-processing of data will work with a set of data that has been initially "dissected" to help solve the initial problem. During the problem definition stage, it is also important to know how the results of the data mining process are going to be used. Glymour et al. [15] discussed the different common uses of the results of a data mining process. Some of these include:

- *Intervention and prediction*: Results can lead to an intervention of the system being studied. Also, they can predict certain behaviors of the system.
- *Description and insight*: Results give an intelligent description an insight about the topic being studied.

Glymour et al. [15] also stressed the fact that one should be cautious about the source of the data. Data may be bias, an issue that will directly affect the results of the data mining process. Biased descriptions and insights will lead to biased and possibly harmful predictions about a system. Another issue that may arise is the problem of causality. Feelders et al. [13] suggested a closer look at the data before assuming the results given by the data mining process. This will ensure that the conclusions drawn by the process are not just the result of chance. The problem definition stage sets the standards and expectations of the data mining process. To an extent, this stage helps the users know the quality of the data being studied. If many iterations are required, and the problem definition ends up being too vague without getting acceptable results, the problem may lie on the quality of the data and not in the definition of the problem.

13.3.2 Acquisition of Background Knowledge

As it was mentioned in the previous stage of the data mining process, possible bias and selection effects of the data being studied should be known. This knowledge will give the development team the possible limitations of the data under consideration. Another important type of knowledge that is important to have before any selection of data is the typical causal relations found on data. Heckerman [25] proposed Bayesian Networks as a solution to this problem. They allow the incorporation of prior knowledge, which may signal possible causality found on a given result from the data. The acquisition of prior knowledge will also prevent the occurrence of "knowledge rediscovery," in essence, the data mining algorithm will tackle a problem with a number of assertions that will prevent it from having to relearn certain patterns that are known to exist. Feelders et al. [13] proposed a method called rule induction, in which "the user would have to guide the analysis in order to take causal relations into account." The acquisition of knowledge plays a critical role in the data mining process. This stage can help directly on the time it takes the process to give positive results. It also prevents the process to learn facts and rules that are already known to be true.

13.3.3 Selection of Data

After the background knowledge is known, the data mining process reaches the important stage of selecting the data that will be used and analyzed to give an answer to the problem under consideration. This selection of the relevant data should be "open-minded" because the purpose of data mining is not for the human to solve the problem, but rather to let the data speak for itself. With the background knowledge in place, the data mining process will prevent the human expert from introducing new unaccounted biases that could harm the conclusions made by the process.

Subramanian et al. [39] and Yevich [46] proposed the use of a data warehouse as an ideal aid for selecting potential relevant data. However, a data warehouse is rarely available for use in the present time, so companies have to go through the process of "creating" one before the selection of data to achieve an acceptable selection of data. If a data warehouse is not readily available at the selection stage, the process will be a long one, and the data mining process will suffer a big delay. The selection of data is a crucial step in the data mining process. Assuming the previous steps are performed properly, data selection narrows down the range of the potential conclusions to be made in the following steps. It also sets the range where these conclusions may be applicable.

13.3.4 Pre-processing of Data

It is often required to pre-process the data before it can be analyzed. This stage also allows the expert the freedom of adding new attributes to the process, which will in a way, help the data mining procedure. These additions constitute certain relations between data that may be difficult for the data mining algorithm to assert. Some algorithms, such as the classification tree algorithm, fail in the assertion of certain relations that may be important in the data mining process [13]. Other algorithms take long amounts of time to make the assertions, so if the knowledge is readily available by the expert, it is better to add it directly as an attribute rather than letting the algorithm make the assertion. Much of the pre-processing in the data mining process is due to the fact that many of the relations between entities in a database are one-to-many. Data mining algorithms often require that all data concerning one instance of the entity should be stored in one record, so that the analysis can be done in one big table that contains all the records regarding the possible instances of a single entity. In order for the pre-processing to be successful, the expert should use domain knowledge and common sense to determine the possible attributes and the creation of records that will enable the data mining process to be successful over time.

13.3.5 Analysis and Interpretation

The analysis phase follows a long process of problem definition, selection of data, and pre-processing of data. At this time of the process, at least 70% of the time used in data mining process has elapsed. During this phase, it is critical for the expert to have experience and knowledge on the subject area being studied, on data analysis, and of course, on data mining. Knowledge on the subject being studied is required primarily to interpret results, and most importantly, to assert which results should be taken into account for further study for possible corrective actions if they are necessary at one point. Data analysis experience is principally required to explain certain strange patterns found on the data, and to give importance to more interesting parts of the data being studied. Finally, data mining experience and expertise are required for the technical interpretation of the results. This technical interpretation is in essence, as Feelders et al. [13] mention in their paper, the translation of the results to the language of the domain and the data expert. The analysis and interpretation stage of the data mining process is where the actual mining takes place. After the data have been selected, pre-processed, and the problem to be solved is known, this stage tries to find certain patterns, similarities, and other interesting relations between the available data. All these patterns are usually translated into rules that are used in the last phase of the data mining process.

13.3.6 Reporting and Use

Results of a data mining process have a wide range of uses. Results can be used in a simple application, such as being input for a decision process, as well as in an important application, like a full integration into an end-user application. The results of a data mining process can also be used in a decision support system or a knowledge-based system. This field of application enables the learning process of the knowledge-base system to be faster and more efficient since it will not have to wait or adjust to certain biases that a human expert may have. The knowledge obtained will be in some cases more accurate, and if the background knowledge obtained during the data mining process is reliable, then the results will have a higher reliability as well. As it was mentioned in an earlier section of this paper, the data mining process has been used as a tool for various purposes. Its results can be used to predict patterns and behaviors, or just to organize, sort, and choose certain amounts of data to prove an assertion made by an expert in the field under consideration. In almost any event, the results of the data mining process should be presented in an organized fashion. This implies the use and development of a well-defined user interface. A good and robust user interface is critical in the data mining process as well as in a wide range of problems. It enables the users to report and interact with the programs effectively, which improves the success chances of the process.

In summary, the data mining process is an iterative process. The selection stage and the pre-processing stage are the most critical steps in the data mining process. The two steps account for almost 80% of the time used in the data mining effort. Thus, special attention should be given to these two steps if the data mining process is to be a success. The last two steps of the process are usually "routine work" if the required data analysis and mining expertise are used properly. Many researchers have discussed these two steps, but the good performance and use of the analysis and the results of the process are completely dependent on the previous stages. Without a good definition of the problem, or, with an inadequate selection of data, the results are not going to be useful. Companies have to realize that the data mining process is a complicated process. Expertise in many fields is required for the process to be successful. Workers have to be patient, as the process may take a long time to be able to come up with useful answers. Management has to support the procedure, and know that analyzing and reporting results are not the only two parts of the data mining process. In fact, they have to understand that these two steps are just the end work of an extensive and complex process.

13.4 Data Mining Techniques

From the data mining methodology explained in the last section, it can be seen that the actual mining is only done at the end of the process, specifically at the analysis and interpretation stage. At this point of the process, various goals have been set and many innovative techniques can be used to reach these goals and improve the data mining stage of the process. These relatively new techniques have been widely used by researchers as a tool for solving data mining problems, as problems have turned harder, databases have grown larger, and the traditional methods for data mining have become obsolete. Each one of these techniques has some advantages, disadvantages, and limitations that should be accounted for, as the use of the techniques is very problem specific.

13.4.1 Traditional Methods of Data Mining

Yevich [46] summarized some of the traditional mining methods and goals and divided them into three categories: association, clustering, and classification. In the associations' method, associations usually refer to searching all details or transactions from the systems for patterns that have a high probability of repetition. An associative algorithm is required to find all the rules that will correlate one set of events with another set of events. This method is very effective whenever the size of the database being studied is relatively small. However, as the database grows, the method becomes unreliable because the searching space is too large for the algorithm to handle. In addition to the information needed in this method (probability of repetition of events), the algorithm will have to assume a lot of information in order to be able to cover all the search space. The clustering method can be used when there is no known way to define a class of data to be analyzed.

In essence, the goal of clustering is to come up with certain relations between data sets, by comparing data and finding relations from the initial information given by the data itself. Clustering algorithms are usually used to discover a previously unknown or suspected relation in a database. This methodology is also known as unsupervised learning and it has been used widely by many researchers in data mining and other areas of research. A problem that may be encountered with this methodology is the task of finding certain relations between data sets. As with associations, a lot of information has to be assumed in order to find relations between data points that seemingly have nothing in common. At this point is where these new techniques, especially those that involve machine learning, come into play.

The classification method identifies the process and must discover roles that define whether an item or event belongs to a particular set of data. Classification can become a very complex process because often many tables and thousands of attributes must be examined to find a relation between components. This creates a complexity problem similar to the one found on the last two methods. As databases grow in size, the time it takes for this method to reach its goal increases rapidly. Classification is also known as supervised learning.

These three methods of data mining and their goals are sometimes very similar in nature. In essence, clustering and classification are pursuing similar goals, and both are grouping together chunks of data in order to arrive at certain conclusions. However, on all three methods a complexity problem is encountered whenever conventional searching techniques are used and when databases grow in size. A number of techniques, old and new, are used to resolve these problems. As it was mentioned earlier, some involve methods such as machine learning while others involve the use of old techniques such as SPC.

13.4.2 Modern Methods of Data Mining

Adriaans and Zantinge [1] summarized the various techniques used for tackling and solving complex data mining problems, stressing that the selection of a technique should be very problem specific. Some of the techniques available are query tools, statistical techniques, visualization, online analytical processing (OLAP), case-based learning, decision trees, neural networks, and genetic algorithms. The complexity of these techniques varies, and a good selection of the technique used is critical to arrive at good and significant results. Table 13.1 shows the techniques available for data mining and their basic advantages and limitations. The simplest technique available is the use of query tools for a rough analysis of the data. Just by applying simple structured query language (SQL) to a data set, one can obtain a lot of information. Before applying more advanced

Table 13.1 Data mining techniques

Data mining technique	Characteristics advantages/disadvantages
Query tools	• Used for extraction of "shallow" knowledge • SQL is used to extract information
Statistical techniques	• SPC tools are used to extract deeper knowledge from databases • Limited but can outline basic relations between data
Visualization	• Used to get rough feeling of the quality of the data being studied
Online analytical processing (OLAP)	• Used for multi-dimensional problems • Cannot acquire new knowledge • Database cannot be updated
Case-based learning	• Uses *k*-nearest neighbor algorithm • A search technique rather than a learning algorithm • Better suited for small problems
Decision trees	• Good for most problem sizes • Gives true insight into nature of the decision process • Hard to create trees from a complex problem
Neural networks	• Mimics human brain • Algorithm has to be trained during encoding phase • Complex methodology
Genetic algorithms	• Use Darwin's evolution theory • Robust and reliable method for data mining • Requires a lot of computing power to achieve anything of significance

techniques, there is a need to know some basic aspects and structures of the data set. This was stressed in the previous section, and it was regarded as background knowledge. SQL is a good technique for discovering knowledge that is in the "surface" of the data, which is the information that is easily accessible from the data set.

Scott and Wilkins [37], Adriaans and Zantinge [1], and Yevich [46] have stated that for the most part, about 80% of the interesting information can be abstracted from a database using SQL. However, as Adriaans and Zantinge [1] stressed in their book, the remaining 20% of hidden information can only be extracted using more advanced techniques, and for most problems, this 20% can prove of vital importance when solving a problem or extracting valuable knowledge for future use.

Statistical techniques can be a good simple start for trying to extract this important 20% of knowledge from the data set. Patterns can be found in the data being studied with the help of histograms, Pareto diagrams, scatter diagrams, check sheets, and other statistical tools. If important relations are not found using SPC tools, at least some information will be learned, and some relations that are definitely not in the data will be known, so that more advanced techniques do not have to look for weak relations in the data being studied. Visualization techniques are very useful for discovering patterns in data sets, and are usually used at the beginning of a data mining process to get a feeling of the quality of the data being studied. This technique uses SPC as well to gain knowledge from databases based on simple but important patterns. Online analytical processing (OLAP) tools are widely used by companies, they support multi-dimensional problems that involve many sorts of information requested by managers and workers at the same point in time. OLAP tools store the data being studied in a special multi-dimensional format, and managers can ask questions from all ranges. A drawback of OLAP tools is that data cannot be updated. Also, as Fayyad and Stolorz [11] mentioned, OLAP is not a learning tool, so no new knowledge can be created, and new solutions cannot be found. Essentially, if the data has the solutions, OLAP will work well, but if the solutions are not there, then OLAP is useless and obsolete. Case-based learning uses the k-nearest neighbor algorithm to assert relations in a database. This algorithm, however, is not really a leaning algorithm. Russell and Norvig [36] used the algorithm as a search method rather than a learning technique. However, as they mention in their book, this search technique proves to be very useful since the data set itself is used as reference. In essence, the search technique only searches the data set space, and thus, it does not get "corrupted" by outside data or knowledge. A problem with the search technique used in case-based learning is its complexity. The algorithm searches and compares every single record or input from the database with each other, in order to find relations between records in a data set, so as the amount of data increases, the algorithm's complexity increases as well. According to Russell and Norvig [36], this search technique leads to a quadratic complexity, which is obviously not desirable when searching large data sets. This technique has been used widely, but only in small problems that have small databases as input for data assertion.

Decision trees are usually used for classification purposes. The database is classified into certain fields that will enable the expert to assert certain behaviors and patterns found in the database. In essence, the data will be divided into categories and to make an assertion or find a pattern in the data, one has to follow a path in the decision tree to arrive at a conclusion. The path taken represents the assertions, facts, and other information used to make the desired conclusion. Many algorithms have been proposed for the creation of such decision trees. Adriaans and Zantinge [1] used the tree induction algorithm to create trees used by car companies to predict customer's behaviors. An advantage of this approach is the complexity of most of the algorithms. Most decision tree algorithms are very effective, and have an $n \log n$ complexity. An important requirement for a decision tree to be successful is to have good knowledge of the problem at hand and of the data available. A lot of assertions and decisions regarding relations between data have to be made by the expert, and the technique relies heavily on these decisions.

Neural networks are an advanced technique that tries to depict the architecture and functionality of the human brain. By modeling neurons, the building block of the brain, neural networks try to mimic the human brain's learning process. This way, the data mining process will imitate the thinking process of a human, and thus, the results of the process will be more reliable than with the other techniques. The architecture of a neural network has been discussed and defined by many researchers, as each researcher defines a neural network for the problem at hand. However, all

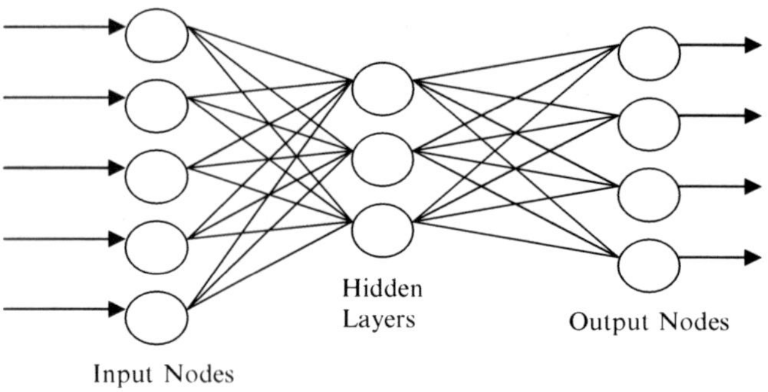

Fig. 13.2 A neural network architecture for data mining

authors have followed a basic architecture that consists of a set of input nodes, output nodes, and a potentially unlimited number of intermediate nodes. Input nodes receive input signals, output nodes give output signals, and the intermediate nodes are essentially a *black box* where the input is converted into output. Figure 13.2 shows an example of a neural network. Note that all intermediate nodes are connected with all input and output nodes. These connections emulate the existing links between neurons in the human brain. Russell and Norvig [36], among other researchers, stress the fact that a neural network is a two-phase process. The first phase is the encoding phase, in which the neural network is trained to perform a certain task. The second phase is the decoding phase, in which the network is used to classify examples, make predictions, or execute any learning task, involved in the data mining process.

13.5 Genetic Algorithm

Genetic algorithms follow the theory of evolution proposed by Darwin. His theory is based on the "natural selection" process of evolution, which essentially states that each species has an over-production of individuals and in a tough struggle for life, only those individuals that are best adapted to the environment survive. The same principle can and has been adapted by many researchers that have used genetic algorithms as a learning tool on various data mining problems.

Holmes et al. [28] developed and used a genetic algorithm to search and learn certain patterns for epidemic surveillance. By searching large databases, the algorithm creates a number of rules regarding possible causes, risks, and solutions for certain problems. Koonce et al. [29] used genetic algorithms in data mining for learning manufacturing systems. Michalski [33] developed the learnable evolution model (LEM), a genetic approach that differs in many aspects to the Darwinian model used by most researchers. LEM uses machine learning to improve the evolutionary process. In data mining, this "new" approach improves the performance and the ability of the algorithm to learn and find interesting patterns. Vila et al. [41]

used genetic algorithms and neural networks to improve the quality of the data used in data mining, as they stressed that finding patterns and solutions on incomplete and/or unreliable data will result in useless conclusions.

Fu [14] compared a greedy search method with a genetic-based approach for two-dimensional problems involving large data sets. He explained that the improvements shown by the GA approach are vast. They include a more robust methodology, a faster method for finding patterns, and more reliable results. Yuanhui et al. [47] combined a neural network with a genetic algorithm to mine classification rules. They showed that the genetic approach generates better rules than the decision tree approach. The advantages and disadvantages of genetic algorithms follow those of natural selection in general. An important drawback is the large overproduction of individuals. Genetic algorithms work with populations of chromosomes, and large amounts of data are used to solve even the easiest problems. Another problem with genetic algorithms is the random character of the searching process. As Adriaans and Zantinge [1] discuss, the end-user of a genetic algorithm does not really see how the algorithm is creating and selecting individuals for finding patterns in large amounts of data. An advantage of generic algorithms is their reliability. One can be sure that if a solution exists in a large database, the genetic algorithm will find it. This is of course, assuming that all the requirements of a genetic algorithm have been addressed and used properly. Another important advantage concerning reliability and robustness is that genetic algorithms do not need to have previous "experience" on the problem at hand. This means that a genetic algorithm used for data mining will eventually find a pattern (if there is any) on the data even if the problem at hand is brand new and no past solutions have been found. As databases expand, researchers have grown more and more dependent in artificial intelligence to solve these complicated data mining problems. AI techniques potentially have low complexities, and more importantly, they resemble the human way of thinking, which may make these techniques the most reliable of all, at least from the human's stand point.

13.6 GA Methodology

Genetic algorithms represent a great alternative for solving these difficult optimization problems. They have shown great power with very promising results from experimentation and practice of many industrial engineering areas. GAs use *meta-heuristics* to solve these problems. These heuristics provide the user great flexibility because he/she does not have to have great problem-specific knowledge in order to arrive to good solutions. Some classic examples of these problems include the knapsack problem, the quadratic assignment problem, the minimum spanning tree problem, the traveling salesman problem, and the film-copy deliverer problem. All these problems have been solved (or partially solved) using genetic algorithms.

13.6.1 Genetic Algorithms

Many optimization problems from the manufacturing systems world are very complex and hard to solve using conventional optimization techniques. The rapid development of computer science, and especially, the rapid development of artificial intelligence techniques has created a great interest among engineers in imitating living beings to solve those kinds of difficult problems. The simulation of the natural evolutionary process of human beings results in stochastic optimization techniques called evolutionary algorithms [32]. Currently, there are three areas of research in the field of evolutionary algorithms: genetic algorithms (GAs), evolutionary programming (EP), and evolution strategies (ESs). GAs are by far, the most widely known type now. Some applications of genetic algorithms in the manufacturing world include scheduling and sequencing, group technology, reliability design, vehicle routing and scheduling, facility layout and location, transportation, among many others.

Genetic algorithms are stochastic search techniques based on the mechanism of natural selection and natural genetics [26, 44]. A genetic algorithm starts with a set of random solutions called population. Each individual or solution in the population is called chromosome. A chromosome is a string of symbols (usually, a binary bit string) [20]. Chromosomes evolve via successive iterations called generations. During each iteration, the chromosomes are evaluated using a fitness function that will eventually decide if a chromosome passes to the next generation or dies. After an iteration of the algorithm, a new generation is created with new chromosomes called offspring. They are formed by either merging two chromosomes from the current generation using the crossover operator or by modifying a single chromosome using the mutation operator. The new generation is then formed by selecting some of the chromosomes according to their fitness value and by rejecting others that do not qualify as valuable individuals. Fitter chromosomes are more likely to live longer. In essence, they have a higher probability of survival. After a few runs (a few generations), Holland [26] stated that the algorithm should converge to the best chromosome, which hopefully, represents the optimal solution to the problem. The use of genetic algorithms in any study has a series of advantages and disadvantages. In particular, genetic algorithms have been receiving considerable attention as a novel optimization technique [34]. Some of the advantages of using genetic algorithms include [32, 44]:

1. *Simple*: The algorithm is easy to develop and to validate.
2. *Efficient*: The algorithm has a "parallel" nature. It works with a set of solutions (population) instead of working with a single chromosome at a time. During evolution, individuals interchange information using the crossover operator. New information is introduced when mutation is used. This implies that the algorithm is efficient. Even if it begins with a very poor population, it will progress rapidly towards good solutions.
3. *Global optimum*: The use of a population of solutions, crossover and mutation, leads the results toward a global optimum instead of being trapped on local optima.

4. *Flexible*: Genetic algorithms can handle any kind of objective functions and any kind of constraints when dealing with an optimization problem. They also provide great flexibility with the use of heuristics to make an efficient implementation for a specific problem.

There are also some disadvantages of using genetic algorithms to solve optimization problems. Some of these disadvantages include [32, 43]:

1. *Size*: When executing a GA model, the population size must have a finite limit. This means that the population is really just a sample of the whole population. A large population is desired to avoid the possible errors produced by finite sampling (the behavior of the genetic algorithm may vary from run to run).
2. *Data format*: Data structure is often string like in format, which is hard to utilize. Pre or post data converting procedures have to be applied to solve a problem.
3. *Cost*: Generating a large number of generations (iterations) may be computationally expensive as well as time-consuming. The cost of waiting for an optimal solution may be too high.

John P. Holland described the basic structure of a genetic algorithm in 1975. His proposed structure follows the same basic process shown in Fig. 13.1. The crossover operation uses the single point approach, and the selection probability of chromosomes for reproduction is determined by the chromosome's fitness value. A good representation of Holland's original algorithm can be found in Ref. [34]. It has a rather simple structure that can "shadow" the algorithm's potential:

Procedure: Genetic Algorithm

1. [Initialization] Generate random population of n chromosomes.
2. [Evaluation] Evaluate the fitness $f(x)$ of each chromosome x in the population.
3. [New population] Create a new population by repeating the following steps until the new population is complete:
   - [Selection] Select two parent chromosomes from a population according to their fitness (the better fitness, the bigger chance to be selected).
   - [Crossover] Given a crossover probability, crossover the parents to form new offspring. If no crossover is performed, offspring is an exact copy of parents.
   - [Mutation] With a mutation probability, mutate new offspring at given positions in a chromosome.
   - [Accepting] Place new offspring in a new population.
4. [Replace] Use new generated population for a further run of algorithm (new generation).
5. [Test] If the end condition is satisfied (usually, required fitness value is obtained), stop, and return the best solution in current population.
6. [Loop] Otherwise go to step 2 (a new generation will be created).

The initialization that creates the first set of solutions (first generation) can follow any technique; but it is usually a random process that creates strings of numbers, each of which represent a possible solution to the problem at hand. The initialization process however, is not perfect. A random string of numbers may have all the basic requirements needed in a chromosome (such as length), but the combination of these numbers may result in the creation of an illegal solution. Many solutions have been developed to solve the initialization problem. They are basically divided into two methods: the penalty method and the decoder method [5]. Gordon and Whitney [22] gave a simple penalty for each infeasible solution [22]. Olsen examined three kinds of penalty methods for the knapsack problem [35]. The idea of the penalty method was that every time an infeasible solution was found a penalty was given to it, and this penalty would prevent the solution from ever being picked as a parent for a new generation. In essence, after the first generation, these infeasible solutions would die.

Gordon and Whitney also developed the decoder method [22], which generates a solution from a chromosome using the greedy approximation heuristic. This heuristic is essentially a blind strategy, because at each step of the way, the heuristic chooses the best alternative possible for that single step, which, in some cases, may lead to a non-optimal solution. The greedy strategy has been used to solve the initialization problems for the knapsack problem with some promising results, but this is a very problem-dependent heuristic since most problems will not find an optimal solution using this kind of strategy. The evaluation process needs to be carefully defined for each problem. The performance of the fitness function is crucial for the algorithm. If the evaluation technique is too general, the algorithm may never reach an optimal solution because the fitness function does not measure the chromosomes properly. On the other hand, if the process is too strict, the algorithm will be unable to find any kind of optimal solution. It will just run forever trying to satisfy some impossible conditions. The initialization and evaluation processes are very important in the performance of any genetic algorithm, however; the three steps that really measure the algorithm's performance are the crossover and mutation steps, and the selection step. These three steps can be divided into two kinds of operations that are done within any genetic algorithm:

1. Genetic operations: crossover and mutation.
2. Evolution operation: selection.

The genetic operations mimic the process of heredity of genes to create new offspring at each generation, while the evolution operation mimics the process of Darwinian evolution to create populations from generation to generation. Genetic algorithms differ from conventional optimization and search techniques in several fundamental ways. Goldberg [16, 20] and Cheng and Gen [5] summarized these important differences and divided them into four different "groups":

1. Genetic algorithms work with a coding solution set, not the solutions themselves.
2. Genetic algorithms search from a population of solutions, not a single solution at a time.
3. Genetic algorithms use "payoff" information (the fitness function), not derivatives or other auxiliary knowledge.
4. Genetic algorithms use probabilistic transition rules, not deterministic rules.

One of the more important benefits of using genetic algorithms when trying to solve a problem is the fact that these algorithms "move" inside the solution space (i.e. all possible alternatives are tested at the same time rather than just evaluating a single solution at a time and stopping when an optimal solution is supposedly found).

13.6.2 Genetic Operations

13.6.2.1 Crossover

Crossover operates on two chromosomes at a time and generates offspring by combining them [5]. A cut-point is randomly chosen (or deterministically chosen, but this choice could create problems) and the segments are combined to create two new chromosomes. The cut-point can be one or many. If the crossover operation follows a one cut-point approach, then both chromosomes are mixed together: the "head" of one is paired with the "tail" of the other one and so, two new chromosomes are created. If there is more than one cut-point, the new chromosome will be an intercalated version of both parents. There are many different approaches when doing crossover. It can be said that each author develops a "new" crossover operation depending on problem requirements and needs. However, the basic approaches are summarized in Table 13.2.

Cheng and Gen [5] studied a simplification of the crossover operator that would work on chromosomes of unequal length. This simplification was called *Injection Crossover*. The original version was developed by Falkenauer and Delchambre [10], and it was originally called *grouping crossover*. Grouping crossover was developed to solve a version of the knapsack problem, an interesting combinatorial optimization problem. Tate and Smith [40] developed a crossover operator for the quadratic assignment problem, another application of the combinatorial problems. In essence, their method took two parent chromosomes and mixed them together to get a single child by finding common features of both parents and passing them to the new chromosome. One of the most interesting problems in combinatorial optimization is the traveling salesman problem (TSP). The proposed solutions to this problem have given a number of crossover techniques that have been

Table 13.2 Basic crossover operations [8]

Single point crossover	Originally proposed by Holland [26]. It is basically a simple cut-and-paste
Multiple point crossover	Two-point crossover is most popular. Works the same way as the single point crossover
Segmented crossover	It is a variant of multi-point crossover that allows the number of crossover points to vary. A switching probability is defined to end a crossover segment at any point
Uniform crossover	Randomly select each gene from one of the two parents. It is most disruptive, but least biased

used by many authors when trying to solve this and other difficult optimization problems. During the past decade, several methods have been proposed, such as the partial-mapped crossover (PMX), order crossover (OX), cycle crossover (CX), position-based crossover, order-based crossover, heuristic crossover, among many others.

- *Partial-mapped crossover* (PMX) was proposed by Goldberg et al. [19] and it can be viewed as an extension of two-point crossover. This algorithm however, has a repairing procedure that resolves the illegitimacy caused by the two-point crossover approach.
- *Order crossover* (OX) was proposed by Davis [7]. It follows the same idea as the PMX but it uses a different repairing procedure for all the unacceptable chromosomes that are drawn from combining two chromosomes together.
- *Position-based crossover* and *order-based crossover* were proposed by Davis [7]. The first method is a kind of uniform crossover with a repairing procedure for all unwanted chromosomes. The second method is just a variation of position-based crossover for the traveling salesman problem.
- *Cycle crossover* (CX) was proposed by Holland et al. [27]. As with the position-based crossover, this crossover operator takes two parent chromosomes to create a new offspring. In this case however, the information from the first parent is passed to the offspring by selecting pieces of information in a deterministic fashion rather than randomly selecting the information that will be passed on to the new chromosome.
- Yamamura et al. [44] proposed a crossover called *Subtour exchange crossover* for the traveling salesman problem. This crossover selects subtours from parents and the combination of these subtours creates offspring, which are then tested for optimality.
- *Heuristic crossover* was first presented by Grefenstette et al. [23] for solving the TSP problem. Their approach followed the nearest neighbor heuristic for finding the best path from one place to another. It is important to note that this heuristic crossover is just an example of a heuristic crossover technique, as each problem and each author can develop his/her own heuristic for the specific problem at hand.

Table 13.3 contains a summary of the crossover operation techniques and their applications during the past decade. Crossover is the most important feature of a genetic algorithm. The performance of the algorithm is closely related to the performance of the crossover operation. A very critical characteristic of the crossover operation is the crossover rate, which controls the number of chromosomes that undergo the crossover operation. This rate should be carefully defined so that an optimal solution can be found in a reasonable amount of time. A high crossover rate allows a more deep exploration of the solution space and reduces the chances of settling for a false optimum; but if the rate is too high, the result is wastage of computation time in exploring bad regions of the solution space.

Table 13.3 Crossover techniques for combinatorial optimization problems

Crossover technique	Researcher	Application problem	Reference
Injection crossover	Cheng and Gen	Combinatorial problem	1997 [5]
Grouping crossover	Falkenauer and Delchambre	Knapsack	1992 [10]
"Common features" crossover	Tate and Smith	Quadratic assignment	1995 [40]
Partial mapped crossover (PMX)	Goldberg and Lingle	Traveling salesman problem	1987 [18]
Order crossover (OX)	Davis	Traveling salesman problem	1985 [7]
Position-based crossover	Davis	Combinatorial optimization	1985 [7]
Order-based crossover	Davis	Traveling salesman problem	1985 [7]
Cycle crossover (CX)	Holland et al.	Traveling salesman problem	1986 [27]
Subtour exchange crossover	Yamamura et al.	Traveling salesman problem	1992 [44]
Heuristic crossover	Grefenstette et al.	Combinatorial optimization	1987 [23]

Table 13.4 Simple mutation operations [8]

Bit-flipping	Flip a bit with certain probability
Random assignment	Mutate a bit with probability by assigning 0 or 1 at random to the position, regardless of what is there

13.6.2.2 Mutation

Mutation is an operation that produces spontaneous random changes in some chromosomes. It serves one of two roles: replacing genes lost from the population during the selection process (so that they can be tried in a new context), or providing the genes that were not present in the initial population. There are two basic ways of doing mutation: bit-flipping and random assignment. Table 13.4 summarizes the basic mutation approaches.

Falkenauer and Delchambre [10] developed a mutation operator that worked together with their grouping crossover [10] for the knapsack problem. After crossover was performed, the mutation operator would delete a number of genes at random, and then it would append those same items in random order to arrive at new chromosomes. Tate and Smith [40] also developed a mutation operator that worked together with their crossover technique used for the quadratic assignment problem. This technique was called *inverse mutation* and it essentially took two genes at random and reversed the order of all sites within the subsequence bounded by the two selected genes [5]. Some other mutation operators have been developed in the past decade for solving the traveling salesman problem. These include: *insertion mutation*, *displacement mutation*, *reciprocal exchange mutation* and *heuristic mutation*.

- *Insertion mutation* selects a gene from a chromosome at random and inserts it in another place within the same chromosome. This operator can be very disruptive if the procedure is used before studying the effects of moving genes around. Moving all genes one place to the right or to the left could give as a result a whole new chromosome that does not resemble the parent or parents from where it was created. However, this procedure was first used to solve the traveling salesman problem, and in that case, shifting all genes one place just gives a new path that follows the exact same structure as the old one but with a location exchange to achieve a better solution.
- *Displacement mutation* is another mutation technique developed to solve the traveling salesman problem. It resembles the subtour exchange crossover developed by Yamamura et al. [44]. In essence, displacement mutation selects a subtour of the salesman at random and inserts it in a random position. Insertion mutation is actually a special case of the displacement mutation operator where the subtour length is equal to 1.
- *Reciprocal exchange mutation* selects two positions within a chromosome at random and then swaps their contents to produce new chromosomes. This procedure can be very disruptive as well, since these new chromosomes may be illegal. A repairing procedure has to be used to correct the chromosomes or just disregard them for the next generation of solutions using the penalty or decoder methods introduced before.
- *Heuristic mutation* for the traveling salesman problem was first proposed by Cheng and Gen [6]. It used the same heuristic proposed for the heuristic crossover operator described before (neighborhood heuristic) in order to produce an improved offspring. Again, these heuristics are problem dependent, as each author develops his/her own heuristic for the problem at hand.

Table 13.5 contains a summary of the mutation operation techniques and their applications during the past decade. All the techniques shown can be utilized when trying to solve any kind of optimization problem. However, it has to be noted as it

Table 13.5 Mutation techniques for combinatorial optimization problems

Mutation technique	Researcher	Application problem	Reference
"Grouping" mutation	Falkenauer and Delchambre	Knapsack and other combinatorial problems	1992 [10]
Inverse mutation	Tate and Smith	Quadratic assignment problem	1995 [40]
Insertion mutation	Used by many researchers	Traveling salesman problem	1997 [5]
Displacement mutation	Used by many researches	Traveling salesman problem	1997 [5]
Reciprocal exchange mutation	Used by many researchers	Traveling salesman problem	1997 [5]
Heuristic mutation	Cheng and Gen	Traveling salesman problem	1996 [6]

was briefly mentioned before, that a repairing procedure similar to that introduced in crossover has to be used to fix illegal chromosomes, since not all chromosomes will be legal after the mutation operator does its job. As for the crossover operator, a mutation rate has to be carefully defined for each problem.

This mutation rate is defined as a percentage of the total number of genes in the population. This rate controls the pace at which new genes are introduced into the population for trial. If the rate is too low, many useful genes will never be tried out; but if the rate is too high, there will be much random perturbation and the offspring will start losing their resemblance to the parents, and so, the algorithm will lose its ability to learn from the history of the search [5].

13.6.3 Searching in Genetic Algorithms

Search is one of the more universal problem-solving methods for such problems where one cannot determine a priori the sequence of steps leading to a solution [5]. Over the years, searching has been performed with either blind strategies or heuristic strategies that eventually should lead an algorithm to find a solution to a problem. Blind strategies are used when there is little or no knowledge about the problem to be solved, while heuristic strategies are used when certain information about the problem is known to be true.

Booker [4] stated that there are two important issues in the discussion of search strategies: exploring the search space and exploiting the best solution [4]. Michalewicz [32] did a comparison of random search, hill-climbing search, and genetic search. He stated that random search is an example of a strategy which explores the search space while ignoring the exploitation of the promising regions of the search space. Hill-climbing on the other hand, is an example of a strategy which exploits the best solution for possible improvement while ignoring the exploration of the search space. Michalewicz finished his discussion stating that genetic algorithms are a class of general purpose search methods combining elements of directed and stochastic search which can make a great balance between exploration and exploitation of the search space [32]. When a genetic search begins, there is a widely random population and the crossover operator tends to perform a widespread search for exploring all solution space. As high fitness solutions develop, the crossover operator provides exploration in the "neighborhood" of each of them.

Genetic algorithms perform a multiple directional search by maintaining a population of potential solutions. The population-to-population approach attempts to make the search escape from local optima. In essence, the population undergoes a simulated evolution: At each generation the relatively good solutions are reproduced, while the bad solutions die. One advantage of applying genetic algorithms to combinatorial optimization problems is that they do not have a lot of mathematical requirements about the optimization problem [5]. Given their evolutionary nature, genetic algorithms search for solutions without regard to the specific inner workings of the problem. Genetic algorithms can handle any kind of objective functions and any kind of constraints and if used in the right way, the algorithms will eventually reach an optimal solution.

13.6.4 Encoding Problems

So far, it has been assumed that a set of solutions is given to a genetic algorithm and that after a while, the algorithm will find a solution to any given problem. However, there is a very important topic that should be discussed. When a person wants to use a genetic algorithm to solve a problem, he/she has to find a way to represent the possible solutions to the problem so that the algorithm can understand and work with these solutions toward finding the optimal solution to a problem.

Encoding a solution of a problem into a chromosome is the key issue for genetic algorithms. The most common and easy way to encode a solution is to use binary strings. However, for many GA applications, especially for the problems from the manufacturing world, the simple GA is difficult to apply because the binary string is not a natural coding. There are several other ways to encode a solution; for example, real number encoding for constrained optimization problems and integer encoding for combinatorial optimization problems. Holland [26] discussed some alternate encoding representations for solving combinatorial optimization problems. The first encoding representation is called *permutation encoding*. This encoding is usually used in ordering problems such as the traveling salesman problem. It uses integers and each digit denotes a number in a sequence of steps to be taken. The second encoding representation discussed by Holland is called *value encoding* and it is used in such problems where the use of real numbers is most recommended. In value encoding, every chromosome is a string of some values. Values can be anything connected to a problem, from numbers, real numbers, or characters to some complicated objects. Another kind of encoding is called *tree encoding*. This kind of encoding is used mainly for evolving programs or expressions [34]. In tree encoding, every chromosome is a tree of some object, such as functions or commands in a programming language. Table 13.6 summarizes these encoding possibilities.

Before testing the performance of a genetic algorithm on any given problem, it is necessary to perform a careful analysis to ensure an appropriate representation of solutions is being used. Using a representation that does not fully represent the wanted solutions may create problems in the performance of the algorithm because the procedure will be working with a set of wrong or inaccurate solutions from the beginning of the run.

13.6.5 Selection

The principle behind genetic algorithms, as it was mentioned earlier, is Darwinian natural selection. Selection deals with the problem of selecting the "valuable" individuals or chromosomes that will survive and pass to the next generation (some will

Table 13.6 Summary of some encoding representations

Encoding type	Notation used	Advantages/disadvantages
Binary encoding	Chromosomes are strings of ones and zeros	Most common, but not natural representation of solutions to problems
Permutation encoding	Chromosomes are usually strings of integers	Less restrictive but problem dependant. Used only for ordering problems
Value encoding	Chromosomes can be string of integers, real numbers, letters, etc.	Very useful but very difficult to manage. Many illegal chromosomes will be created
Tree encoding	Chromosomes are trees representing expressions	Good for evolving programs. Restrictive representation of other problems

survive, others will reproduce to create new individuals). Selection is the driving force in any genetic algorithm, and thus, selection pressure is critical on the performance of the genetic algorithm. Michalewicz [32] stated that usually, low selection pressure is indicated at the start of the genetic algorithm search in favor of a wide exploration of the search space, while high selection pressure is recommended at the end in order to exploit the most promising regions in the search space. The selection procedure may create a new population for the next generation based on all parents and offspring or only on part of them. Selection may replace parent chromosomes by their offspring. When all parents are replaced by their offspring, a *generational replacement* takes place [5]. This occurrence is not recommended, since genetic algorithms are blind in nature, and offspring chromosomes may be worst than their parents. This would lead the genetic algorithm to some unwanted places when looking for an optimal solution.

Over the past few years, some solutions have been given to overcome the problem of having generational replacement. Holland gave a first suggestion in 1975 and it involved a random selection of a parent whenever an offspring was born. This chosen parent was then replaced by the offspring that was just created. De Jong [8] suggested another solution to the generational replacement problem. It is called the crowding strategy, which selects the parent to be replaced depending on its similarities with the just created offspring. (The most similar parent dies and gives space for the new offspring.) If the sampling space is large and includes both parents and offspring, they will fight for survival and the idea of choosing a parent to be replaced by the offspring is not used. Instead, the fittest individuals will survive and will pass on to the next generation. There are different ways of evaluating and comparing chromosomes to decide which ones will survive and which ones will die. The sampling mechanism is concerned with the problem of how to select chromosomes from the sampling space. Cheng and Gen [5] summarized the three basic approaches for sampling chromosomes:

- Stochastic sampling
- Deterministic sampling
- Mixed sampling

The *roulette wheel selection* [26] is the most recognized method for selection among stochastic sampling. The basic idea of this kind of selection is to determine the selection probability for each chromosome proportional to its fitness value. After all probabilities have been found, a roulette with the values is created. The selection process is based on spinning the wheel pop_size times, and each time a chromosome is selected for the new population. The probability that a chromosome will be chosen for survival is directly related to its fitness value. In essence, chromosomes with high fitness values will be selected more times [34]. Among deterministic sampling techniques, *truncation selection* is the most recognized method for selection. This technique, proposed by Goldberg and Thierens [17] rank all chromosomes according to their fitness and selects the best ones as parents for the next population. *Elitist selection* ensures that the best chromosome is passed on to the new generation if it is not selected through another selection process. Mixed sampling contains both random and deterministic features at the same time. A typical example is the tournament selection method proposed by Goldberg et al. [19]. This method chooses a set of chromosomes and picks out the best one from the set for reproduction. The number of chromosomes is called the *tournament size*. The most common size used over the years is two. The selection of the two chromosomes to be involved in the tournament can be stochastic, and then the selection of the best one can be deterministic by calculating their fitness value and choosing the fittest individual.

From the methods discussed, it can be inferred that the selection probability of a chromosome is usually proportional to its fitness. This could be sometimes undesirable. For example, in early generations, there is a tendency for a few super chromosomes to dominate the selection process, and in later generations, competition among chromosomes is less strong and a random search behavior can emerge. Essentially, the power of the genetic algorithms would be lost and the algorithms would become a simple search technique. To solve the selection problem (i.e., probability of survival is proportional to fitness value), De Jong [8] proposed the *scaling* mechanism. Later on, Baker [3] introduced the notion of *ranking* selection to overcome the scaling problems of the direct fitness-based approach. Scaling basically maps raw objective function values to some positive real values and the survival probability for each chromosome is determined according to these values. Ranking, on the other hand, ignores the actual objective function values and uses a ranking of chromosomes to determine survival probability. There is a wide variety of scaling mechanisms. In general, the scaled fitness f_k from the raw fitness f_k for chromosome k can be expressed as $f_k = g(f_k)$ [5]. Function g transforms the raw fitness into scaled fitness. This function g

can follow different forms of scaling methods such as linear scaling, sigma truncation, logarithmic scaling, power law scaling and windowing, among many others. The ranking selection method sorts the population from the best to the worst and assigns the selection probability of each chromosome according to the ranking but not its raw fitness. This method gives the low fitness chromosomes a chance at being selected for reproduction. Although a generation of low fitness individuals is not wanted, it is neither wanted a selection procedure that chooses the same chromosomes every time. This will just lead to having a population of very similar individuals, which will prevent the algorithm to search for solutions elsewhere. In essence, the algorithm will get stuck in local optima most of the time.

13.7 Summary

In this chapter, the basic concepts of data mining are discussed. Data mining deals with the discovery of hidden knowledge, unexpected patterns, and new rules from large databases. It is regarded as the key element of a much more elaborate process called knowledge discovery in databases, or KDD, which is closely linked to data warehousing. Data mining can bring significant gains to organizations, for example, through better-targeted marketing and enhanced internal performance.

13.8 Review Questions

13.1. Define the data mining process.
13.2. What is the methodology used in data mining?
13.3. How a problem is defined in data mining domain?
13.4. Discuss the stage of data selection in data mining process.
13.5. What are data Mining techniques?
13.6. Define collaborative engineering.
13.7. Describe briefly how Genetic Algorithm used in data mining process.
13.8. What are the advantages of using genetic algorithms in data mining?
13.9. What are the disadvantages of using genetic algorithms in data mining?

References

1. Adriaans P, Zantinge D (1996) Data mining. Addison-Wesley, Harlow
2. Andrade M, Bork P (2000) Automated extraction of information in molecular biology. FEBS Lett 476:12–17
3. Baker J (1985) Reducing bias and inefficiency in the selection algorithm. In: Proceedings of the first international conference on genetic algorithms. Lawrence Erlbaum Associates, New Jersey, pp 14–21

4. Booker L (1987) Improving search in genetic algorithms. In: Genetic algorithms and simulated annealing. Morgan Kaufmann Publishers, Los Altos, CA, 61–73
5. Cheng R, Gen M (1997) Genetic algorithms & engineering design. Wiley-Interscience Publication, New York
6. Cheng R, Gen M (1997) Resource-constrained project scheduling problem using genetic algorithms, Intelligent Automation and Soft Computing 3:273–286
7. Davis L (1985) Applying adaptive algorithms to domains. In: Proceedings of the international joint conference on artificial intelligence, Los Angeles, pp 162–164
8. De Jong K (1975) An analysis of the behavior of a class of genetic adaptive systems, Ph.D. thesis, University of Michigan, Ann Arbor
9. Delesie L, Croes L (2000) Operations research and knowledge discovery: a data mining method applied to health care management. Int Trans Oper Res 7:159–170
10. Falkenauer E, Delchambre A (1992) A genetic algorithm for bin packing and line balancing. In: Proceedings of the IEEE international conference on robotics and automation, Nice pp 1186–1193
11. Fayyad U, Stolorz P (1997) Data mining and KDD: promise and challenges. Future Gener Comput Syst 13:99–115
12. Fayyad U, Madigan D, Piatetsky-Shapiro G, Smyth P (1996) From data mining to knowledge discovery in databases. AI Mag 17:37–54
13. Feelders A, Daniels H, Holsheimer M (2000) Methodological and practical aspects of data mining. Inf Manag 37:271–281
14. Fu Z (1999) Dimensionality optimization by heuristic greedy learning vs. genetic algorithms in knowledge discovery and data mining. Intell Data Anal D. Braha (ed.), Ben-Gurion University, Beer-Sheva, Israel, Norwell, MA, USA 3:211–225
15. Glymour C, Madigan D, Pregibon D, Smyth P (1997) Statistical themes and lessons for data mining. Data Mining Knowledge Discovery 1:11–28
16. Goldberg DE (1989) Sizing populations for serial and parallel genetic algorithms. In: Proceedings of the 3rd international conference on genetic algorithms table of contents. Morgan Kaufmann Publishers Inc. San Francisco, CA, USA, pp 70–79
17. Goldberg D, Thierens D (1994) Convergence models of genetic algorithm selection schemes. In: Y. Davidor, H.-P. Schwefel, and R. Männer (eds.) Parallel Problem Solving from Nature: PPSN III. Springer, Berlin, pp 119–129
18. Goldberg D, Lingle R (1987) Alleles, loci and the traveling salesman problem. In: Proceedings of the first international conference on genetic algorithms. Lawrence Erlbaum Associates, New Jersey
19. Goldberg D et al (1989) Messy genetic algorithms: motivation, analysis and first results. Complex Syst 3:493–530
20. Goldberg D (1989) Genetic algorithms in search, optimization and machine learning. Addison-Wesley, Massachusetts
21. Gonzalez R, Kamrani A (2001) A survey of methodologies and techniques for data mining and intelligent data discovery. In: Dan Braha, Ben-Gurion Univ, Beer-Sheva (eds.) Data mining for design and manufacturing. Kluwer Academic Publishing, Israel Norwell, MA, USA pp 41–59
22. Gordon V, Whitney D (1992) Serial and parallel genetic algorithms as functions optimizers. In: Proceedings of the fifth international conference on genetic algorithms, San Mateo, CA: Morgan Kaufmann, pp 177–183
23. Grefenstette J et al (1987) Genetic algorithms for the traveling salesman problem. In: Proceedings of the first international conference on genetic algorithms. Lawrence Erlbaum Associates, New Jersey, pp 160–168
24. Hand DJ (1998) Data mining: statistics and more? Am Statist 52:112–118
25. Heckerman D (1996) Bayesian networks for knowledge discovery. In: T G Dietterich, S Becker and Z Ghahramani (eds.) Advances in Knowledge Discovery and Data Mining. AAAI Press, pp 273–305
26. Holland JH (1992) Adaptation in natural and artificial systems. MIT Press, Boston

27. Holland JH, Smith D, Oliver I (1986) A study of permutation crossover operators on the traveling salesman problem. Eur J Oper Res 160(1):224–230
28. Holmes JH, Durbin DR, Winston FK (2000) The learning classifier system: an evolutionary computation approach to knowledge discovery in epidemiologic surveillance. Artif Intell Med 19:53–74
29. Koonce DA, Fang C, Tsai S (1997) A data mining tool for learning from manufacturing systems. Comput Indust Eng 33:27–30
30. Kusiak A (1999) Computational intelligence in design and manufacturing. Wiley-Interscience Publications, New York, pp 498–526
31. Lin X, Zhou X, Liu C (2000) Efficient computation of a proximity matching in spatial databases. Data Knowledge Eng 33:85–102
32. Michalewicz Z (1992) Genetic algorithms + data structures = evolution programs. Springer, Berlin, p 17
33. Michalski RS (2000) Learnable evolution model: evolutionary processes guided by machine learning. Mach Learn 38:9–40
34. Obitko M (1998) Introduction to genetic algorithms. Czech Technical University. http://cs.felk.cvut.cz/~xobitko/ga/18
35. Olsen A (1994) Penalty functions and the knapsack problem. In: Proceedings of the first IEEE conference on evolutionary computation. IEEE Press, Orlando, Florida, pp 554–558
36. Russell S, Norvig P (1995) Artificial intelligence: a modern approach. Prentice Hall, New Jersey
37. Scott PD, Wilkins E (1999) Evaluating data mining procedures: techniques for generation artificial data sets. Inf Softw Technol 41:579–587
38. Skarmeta A, Bensaid A, Tazi N (2000) Data mining for text categorization with semi-supervised agglomerative hierarchical clustering. Int J Intell Syst 15:633–646
39. Subramanian A, Smith LD, Nelson AC, Campbell JF, Bird DA (1997) Strategic planning for data warehousing. Inf Manag 33:99–113
40. Tate D, Smith A (1995) A genetic approach to the quadratic assignment problem. Comput Oper Res 22:73–83
41. Vila MA, Cubero JC, Medina JM, Pons O (1997) Soft computing: a new perspective for some data mining problems. Vistas Astron 41:379–386
42. Wall M (1995) Intro to genetic algorithms. Massachusetts Institute of Technology, Boston, MA. http://lancet.mit.edu/~mbwall/presentations/IntroToGAs/
43. Wang R (1998) A genetic algorithm based knowledge acquisition system for equipment diagnosis. University of Michigan-Dearborn, Natural and Artificial Systems. University of Michigan Press
44. Yamamura M, Ono T, Kobayashi S (1992) Character-preserving genetic algorithms for the traveling salesman problem. J Japan Soc Artif Intell 7:1049–1059
45. Yao X (1998) Modern heuristic techniques 4, Lecture from Internet. School of Computer Science, University College, UNSW, ADFA. http://www.cs.adfa.oz.au/~xin/course_notes/ModernHeuristics/lecture06/lecture06.html
46. Yevich R (1997) Data mining. In: Wrembel R. and Konicilla C. (eds.) Data warehouse: practical advice from the experts. Prentice Hall, pp 309–321
47. Yuanhui Z, Yuchang L, Chunyi S (1998) In: Wrembel R. and Koncilla C. (eds.) Mining classification rules in multi-strategy learning approach. Intell Data Anal 2:165–185

Chapter 14
Group Technology and Applications

Abstract The chapter provides the concepts of group technology and cellular manufacturing. Grouping objects into groups based on the object features has been done using group technology (GT) approaches. Similar components can be grouped into design families, and new designs can be created by modifying an existing component design from the same family. The philosophy of group technology is an important concept in the design of advanced integrated manufacturing systems. Group technology is not an answer to all manufacturing problems, but it is a good management technique to standardize efforts and eliminate duplication. A well-designed classification and coding system may result in several benefits for the manufacturing plant.

Keywords Group technology • Integrated manufacturing • Management technique

14.1 Introduction

The evolution of manufacturing from stand-alone automation to computer-integrated manufacturing has occurred since the 1960s. During the 1960s, stand-alone automation received the most attention. This type of automation, also known as point automation, consists of workstations that may vary from one numerically controlled machine to a number of numerically controlled machines operated by a skilled operator. In stand-alone automation, the material handling is normally performed manually. The concept of flexible manufacturing and manufacturing cells production was introduced in the early 1980s. A flexible manufacturing system (FMS) is a computer controlled configuration of automated machine tools interlinked by automated workpiece-flow systems to enable simultaneous machining of different part types, which pass through the system along different routes. The fast development of this production philosophy was the result of an increasing variety in the production-part spectrum as well as a worldwide trend toward decreasing batch sizes, minimizing production time, minimizing raw material, minimizing work-in-process, and maximizing machine utilities.

A.K. Kamrani and E.A. Nasr, *Engineering Design and Rapid Prototyping*,
DOI 10.1007/978-0-387-95863-7_14,

The 1990s were considered the era of computer-integrated manufacturing systems (CIMS). Computer-integrated manufacturing is the most automated form of production. It is the integration of the total manufacturing enterprise through the use of integrated systems and data communications coupled with new managerial philosophies that improve organizational and personnel efficiencies. The objective in implementing CIM is to automate and integrate an entire firm's activities, including business, design, manufacturing, planning, and manufacturing control functions. The technology of CIM uses all forms of computer methods, mainly data processing methods, various computer network applications, decision support systems, and database management techniques. An efficient and well-designed CIM system combines manufacturing flexibility with efficient production management assistance to enhance the overall productivity of a manufacturing organization.

14.2 Traditional Manufacturing Systems: An Overview

Traditional manufacturing systems are grouped into three main categories. These are job shop, flow shop, and project shop [6]. The oldest and the earliest of these systems is the job shop. It is a transformation process in which units for different orders follow individual routes or sequences through processes or machines. The characteristics of this kind of production system are its flexibility, wide variety of product design, and need for many highly skilled operators. Loading, unloading, setup, and adjusting are performed manually. General-purpose machines are grouped by function and adapted to the special requirements of different orders. The process for this flexibility involves long in-process times, large in-process inventories, and the difficult task of scheduling different orders within centers in which machines and people must be shared. The flow shop manufacturing system is used for products where the basic designs are more stable over a longer period of time and are produced in larger quantities. The flow shop is a transformation process in which successive units of output undergo the same sequence of operations with more specialized and dedicated equipment. This form is also known as mass production. The flow shop is viewed as either continuous or intermittent. In a continuous flow shop, the process produces exactly the same type of output in great volume. An example of this process is an oil refinery. The intermittent-flow shop is where the process is interrupted in order to modify the setups to execute different specifications of the same basic design. The most automated form of this system is called the transfer line. The third kind of traditional manufacturing system is known as the project shop. In this system, people, material, and machines all come to the project site for product assembly and processing. This type of system is used for a product that is either very large or one-of-a-kind. The tasks should be well defined and must be accomplished in a specific sequence. Figure 14.1 illustrates the layout of the three manufacturing systems described.

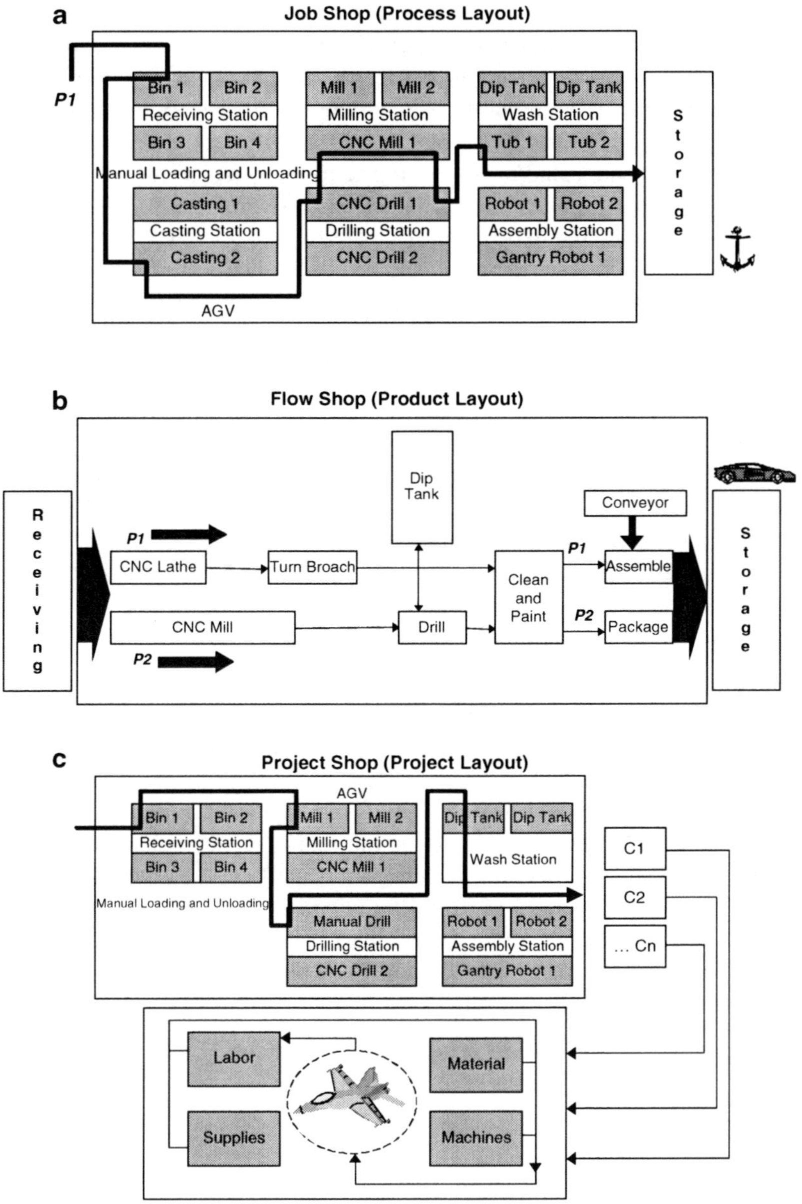

Fig. 14.1 The three kinds of traditional manufacturing systems

14.3 Group Technology

Grouping objects (i.e., components, parts, or systems) into groups based on the object features has been done using *group technology* (GT) approaches [2, 33, 34]. Similar components can be grouped into design families, and new designs can be created by modifying an existing component design from the same family. The philosophy of GT is an important concept in the design of advanced integrated manufacturing systems. GT is not an answer to all manufacturing problems, but it is a good management technique to standardize efforts and eliminate duplication. GT classifies and codes parts by assigning them to different part families based on their similarities in shape and/or processing sequence. Grouping parts into families is a tedious task that requires careful planning and consideration. The method of grouping that is considered to be the most powerful and reliable is classification and coding. In this method, each part is inspected individually by means of its design and processing features. A well-designed classification and coding system may result in several benefits for the manufacturing plant. These benefits include:

- It facilitates the formation of the part families.
- It allows for quick retrieval of designs, drawings, and process plans.
- Design duplication is minimized.
- It facilitates the accurate estimation of machine-tool requirements and logical machine loading.
- It provides reliable workpiece statistics.
- It aids production planning and scheduling procedures.
- It improves cost estimation and facilitates cost.
- It promotes variant design philosophy.

Classification is defined as a process of grouping parts into families based on some set of principles. This approach is further categorized into the visual method (ocular) and coding procedure. Grouping based on the ocular method is a process of identifying part families, visually inspecting parts and assigning them to families and the production cells to which they belong. This approach is limited to parts with large physical geometries, and it is not an optimal approach because it lacks accuracy and sophistication. This approach becomes inefficient as the number of parts increases. The coding method of grouping is considered to be the most powerful and reliable method. In this method, each part is inspected individually by means of its design and processing features. Coding can be defined as a process of tagging parts with a set of symbols that reflect the part's characteristics. A part's code can consist of a numerical, alphabetical, or alphanumerical string. Three types of coding structures exist.

14.3.1 Hierarchical (Monocode) Structure

In this structure, each digit is a further expansion of the meaning of a digit, dependent on the meaning of the previous digit in the code's string. The advantage of this method is the amount of information that the code can represent in a relatively small number of digits. However, a coding system based on this structure is complicated and very difficult to implement. Figure 14.2 illustrates the general structure of this method of coding.

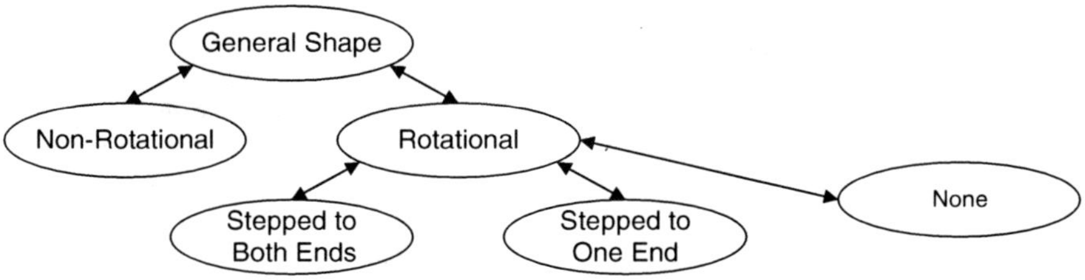

Fig. 14.2 Monocode structure

Digits	Feature	Code				
		1	2	3	4	...
1	External Shape	Symmetrical	Non-Symmetrical	Contour	Complex	...
2	Number of Holes	0	3 to 5	6 to 9	More than 10	...
3	Holes' Diameter	< 0.5	0.5 < hd < 1	1 < hd < 3	3 < hd < 8	...
...	...	...	...	...	...	...

Fig. 14.3 Polycode structure

14.3.2 Chain (Attribute or Polycode) Structure

In this structure, the meaning of each digit is independent of any other digit within the code string. In this approach, each attribute of a part is tagged with a specific position in the code. This structure is simple to implement, but a large number of digits may be required to represent the characteristics of a part. Figure 14.3 illustrates the general layout of a code based on this structure.

14.3.3 Hybrid Structure

Most of the coding systems available are implemented using the hybrid structure. A hybrid coding system is a combination of both the monocode and polycode structures, taking advantage of the best characteristics of the two previously described structures. Figure 14.4 illustrates the general structure of this coding approach.

Clusteringis another efficient method that is used for grouping similar objects and has been practiced, consciously or unconsciously, for many years. The development of computer technology has resulted in cluster analysis, which is given to techniques that allow the production of a range of different types of classification by purely automatic means. This method requires the calculation of a clustering factor known as a similarity or a dissimilarity coefficient by assigning a clustering criterion as an objective to optimize system performance. Similarity and dissimilarity coefficients are calculated values that represent the relationships between parts. Most research has been based on the fact that these coefficients range from 0 to 1.

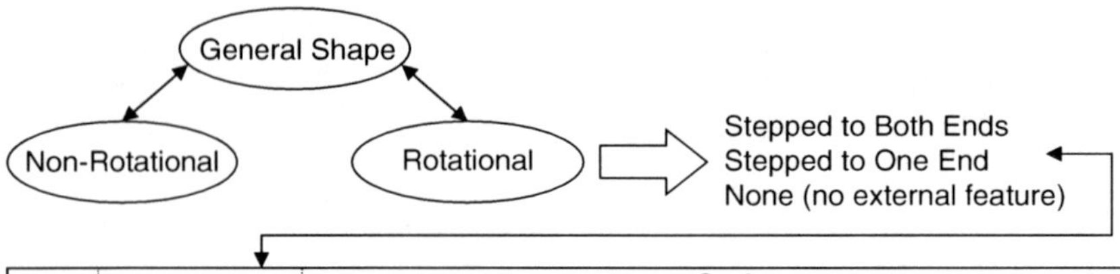

Digits	Feature	Code 1	2	3	4
1	External Shape	Symmetrical	Non-Symmetrical	Contour	Complex
2	Number of Holes	0	3 to 5	6 to 9	More than 10
3	Holes' Diameter	< 0.5	0.5 < hd < 1	1 < hd < 3	3 < hd < 8

Fig. 14.4 Hybrid structure

This indicates that dissimilarity =1.0 – similarity or vice versa. Two methods widely used are hierarchical clustering and non-hierarchical clustering. The hierarchical method results in a graph known as a dendrogram, which illustrates the data grouped into smaller clusters based on their similarity/dissimilarity. The hierarchical method is accomplished in two forms: agglomerative and divisive. In agglomerative hierarchical procedures one starts with $m$objects that are to be classified and at each step the two most similar objects are merged into one single cluster. After $m-1$ such steps, all objects belong to one large cluster. There are many such methods, differing in the criteria used to decide which individual elements or clusters should be merged together and in the way in which the similarity between a newly obtained cluster and other clusters or objects is defined. In the hierarchical method, the structure of the set of objects can be obtained by dividing the set into two or more subsets and continuing the division until all objects have been completely separated. This hierarchical clustering is known as the divisive method. The divisive method has been studied and used much less than agglomerative procedures.

The non-hierarchical method uses partitioning clustering algorithms to search for a division of a set of objects into a number, K, of clusters in such a way that the elements of the same cluster are close to each other and the different clusters are well separated. Because the K clusters are generated simultaneously, the resulting classification is non-hierarchical. The number of clusters, K, can be either given or determined by the algorithm. When K is unknown, the algorithm can be repeated for several values of K, and as a by-product the algorithm will often supply a criterion value for each K. In this way it is possible to evaluate the different clustering and sometimes select an optimal one [17, 21]. Some of these techniques are presented in the following section.

14.4 Sorting Techniques

Sorting techniques begin by constructing an incidence matrix that describes the relationship between parts and machines as shown in Fig. 14.5. The part–machine incidence matrix a_{ij} consists of 0, 1 entries. An entry of 1 indicates that machine

Fig. 14.5 Part–machine incidence matrix

Machine \ Part	1	2	3	4	5
1		1		1	1
2	1		1		
3		1		1	
4	1		1		

i is used to process part *j*, while an entry of 0 indicates that machine *i* is not used to process part *j*. After constructing the matrix, the techniques will attempt to identify groups (clusters) existing in the matrix by rearranging the machines and parts into a more structured form. Most sorting techniques are based on the idea of arranging the matrix entries into a block diagonal form.

Following are some sorting-based clustering algorithms.

14.4.1 Rank Order Clustering Algorithm

The ROC algorithm [18] was developed to group the entries of a part–machine incidence matrix around the diagonal by repeatedly arranging the columns and rows according to the sum of binary values. The ROC was designed to solve the clustering problem of matrices with binary entries. The idea is to cluster the positive entries (1s) into groups and place them along the diagonal of the matrix. The part–machine incidence matrix is read as a binary word. Then the algorithm converts these binary words for each row (column) into decimal equivalents. After that, the algorithm rearranges the rows (columns) in the order of descending values; this step is repeated until there is no change. Following is a step by step example of applying ROC for clustering.

Step 1. Assign binary weights to rows.

	Part Number 1	2	3	4	5	6
Binary Weights	2^5	2^4	2^3	2^2	2^1	2^0
Machine Number 1			1	1		1
2	1	1			1	
3			1	1		1
4	1	1			1	
5	1	1				

Step 2. Calculate decimal equivalent.

Machine Number \ Part Number	1	2	3	4	5	6	Decimal Equivalent
1			1	1		1	13
2	1	1			1		50
3			1	1		1	13
4	1	1			1		50
5	1	1					48

Step 3. Sort the decimal equivalent in descending order.

Machine Number \ Part Number	1	2	3	4	5	6	Decimal Equivalent
2	1	1			1		50
4	1	1			1		50
5	1	1					48
1			1	1		1	13
3			1	1		1	13

Step 4. Assign binary weights and calculate decimal equivalent for columns.

Machine Number \ Part Number	1	2	3	4	5	6	Binary Weights
2	1	1			1		2^4
4	1	1			1		2^3
5	1	1					2^2
1			1	1		1	2^1
3			1	1		1	2^0
Decimal Equivalent	28	28	3	3	24	3	

Step 5. Sort the decimal equivalent in descending order.

Machine Number \ Part Number	1	2	5	3	4	6
2	1	1	1			
4	1	1	1			
5	1	1				
1				1	1	1
3				1	1	1
Decimal Equivalent	28	28	24	3	3	3

Step 6. If no separate groups are visible, repeat the previous steps; otherwise stop.

		Part Number					
		1	2	5	3	4	6
Machine Number	2	1	1	1			
	4	1	1	1			
	5	1	1				
	1				1	1	1
	3				1	1	1

The ROC provides a simple analytical technique that can be easily computerized. The ROC has fast convergence and relatively low computation time depending on the matrix size. The fact that ROC uses binary values will impose restrictions on the size of the matrix that can be solved. Most computers have a maximum of $2^{48}-1$ for integer representation, which means that the maximum number of rows and columns is limited to 47. The algorithm results depend on the initial matrix arrangement. The algorithm collects positive entries (1s) in the top left-hand corner, leaving the rest of the matrix disorganized.

14.4.2 Modified Rank Order Clustering Algorithm

The rank ordering clustering algorithm [33] was improved by incorporating the "block and slice" method. The modified algorithm identified that the ROC has a tendency to collect the 1s in the top left-hand corner. This block of columns was removed from the matrix and then the ROC was applied again. The algorithm will collect another set of 1s in the top left-hand corner. This procedure is repeated again until no elements are left in the matrix. The application of this method will identify mutually exclusive part families, but it may contain some overlapping machines.

14.4.3 Bond Energy Algorithm

The BEA [27] aims at identifying and displaying the natural variable groups and clusters that occur in complex data arrays. The BEA seeks to form a block diagonal form by maximizing the measure of effectiveness (ME). The ME is the summed bond energy over all row and column permutations. The ME is defined as follows:

$$\text{ME} = \frac{1}{2}\sum_{i=1}^{m}\sum_{j=1}^{n} a_{ij}(a_{i,j-1} + a_{i,j+1} + a_{i-1,j} + a_{i+1,j}). \quad (14.1)$$

The procedure for BEA is as follows:

Step 1. Set j=1. Select one of the columns arbitrarily.

Step 2. Place each of the remaining $n-j$ columns, one at a time, in each of $j+1$ positions, and compute each column's contribution to the ME.

Place the column that gives the largest incremental contribution to the ME in its best location.

Increase j by 1 and repeat the preceding steps until $j=n$.

Step 3. When all columns have been placed, repeat the procedure for the rows. For example, the BEA is applied to the following matrix:

	Part Number 1	2	3	4	5
Machine Number 1		*		*	*
2	*		*		
3		*		*	
4	*		*		

Step 1. Set j=1. Select column 2.

Step 2. Place each of the remaining columns in each of j+1 positions, and compute each column's contribution to the ME.

Position	j = 1	j + 1	ME value
Column Number	2	1	0
	2	3	0
	2	4	2
	2	5	1

Column 4 is placed in the j+1position.

Repeating the same procedure for the remaining columns leads to the following column order: 2, 4, 5, 1, 3.

Step 3. Repeating the preceding steps for rows results in the following matrix:

	Part Number 2	4	5	1	3
Machine Number 1	*	*	*		
3	*	*			
2				*	*
4				*	*

14.4.4 Cluster Identification Algorithm

The CIA [24] was designed to identify disconnected blocks if they exist. If there are no disconnected blocks (the matrix is not mutually separable), the entire matrix will form one block. The algorithm begins by masking all columns that have an entry of 1 in any row (selected randomly), then all rows that have an entry of 1 in these columns are masked. This process is repeated until all intersecting columns and rows are masked. These masked columns and rows are placed in a block and removed from the matrix. Then the process of masking starts again and another block is identified. This will continue until all the entries in the matrix are assigned in blocks and all separable blocks are identified. For example, consider the following machine–part incidence matrix:

Part Number

Machine Number	1	2	3	4	5	6	7	8
1				1				1
2			1		1			
3		1				1	1	
4			1		1			
5	1			1				1
6		1				1	1	
7	1			1				1

Step 1. Select any row i of the incidence matrix and draw a horizontal line h_i through it.

Row 3 of the matrix is selected randomly and a horizontal line h_3 is drawn.

Step 2. For each entry of 1 crossed by the horizontal line h_i draw a vertical line v_j. Three vertical lines v_2, v_6, and v_7 are drawn:

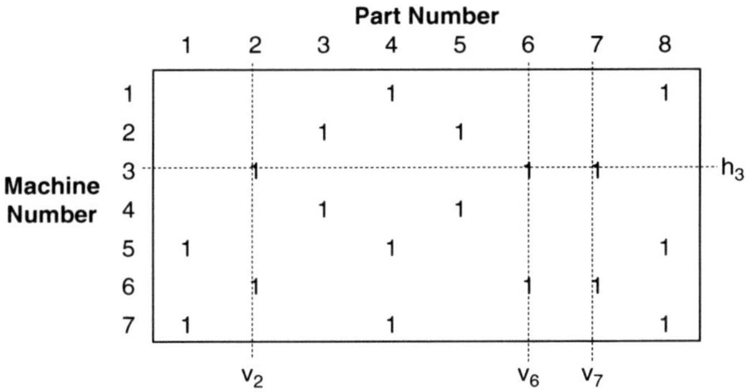

Step 3. For each entry of 1 crossed once by a vertical line v_j draw a horizontal line. A horizontal line h_6 is drawn through all the crossed-once entries of the matrix:

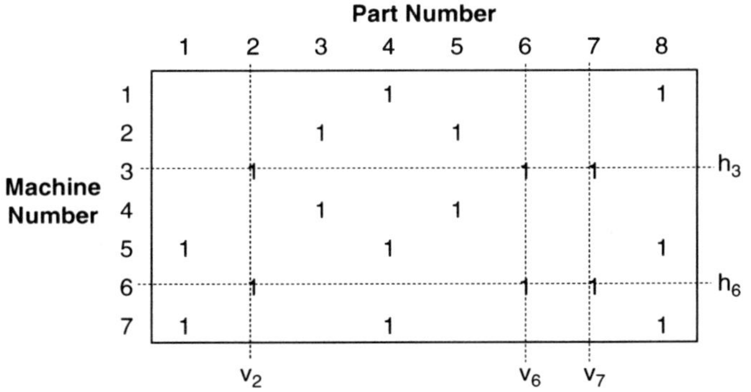

Step 4. The preceding steps are repeated until no more crossed-once entries are left. All the crossed-twice entries are grouped in a block and removed from the matrix. Parts 2, 6, 7 and machines 3, 6 are grouped in one block.

Machine Number \ Part Number	2	6	7	1	3	4	5	8
3	1	1	1					
6	1	1	1					
1						1		1
2					1		1	
4					1		1	
5				1		1		1
7				1		1		1

The grouped parts and machines are removed from the matrix:

	2	6	7
3	1	1	1
6	1	1	1

Step 5. The above procedure is repeated for the remaining matrix entries until all entries are grouped.

	1	3	4	5	8
1			1		1
2		1		1	
4		1		1	
5	1		1		1
7	1		1		1

The resulting matrix is as follows:

	2	6	7	8	1	4	3	5
3	1	1	1					
6	1	1	1					
5				1	1	1		
1				1		1		
7				1	1	1		
2							1	1
4							1	1

14.4.5 Extended Cluster Identification Algorithm

The CIA was extended [24] to deal with overlapping parts (parts that must be processed by machines located at two separate groups or clusters). The extended algorithm combines the expertise of the production engineer and the efficiency of the CIA approach. When overlapping occurs, the production engineer will decide how to resolve this problem by locating the part in the more appropriate block. This decision is usually made based on previous experience and knowledge of the processes involved.

14.4.6 Similarity Coefficient-Based Clustering

McAuley [26] introduced a similarity coefficient between two machines based on the number of parts that visit each machine. The similarity coefficient is defined as the number of components visiting both machines divided by the number of components visiting either of the two machines. The idea is to group similar pairs of machines (or a machine and a machine group, or two machine groups) together. When two machines are merged into one cluster their combined similarity relative to other machines is determined according to the clustering technique used as listed in Table 14.1. This grouping is continued until all the machines have been grouped or the desired number of machine groups is reached.

The use of the similarity coefficient-based clustering is illustrated as follows [33]:

Step 1. Compute the similarity coefficient S_{mn} for machines.

	1	2	3	4	5	6
1	0	0	0.67	0.17	0	0.4
2		0	0.25	0.4	0.75	0.17
3			0	0.12	0.12	0.5
				5	5	
4				0	0.5	0
5					0	0
6						0

Table 14.1 Machines combined similarity methods

Method	Similarity
Single linkage clustering (SLC)	$S_{tv} = \underset{\substack{m\in t\\ n\in v}}{Max}\{S_{mn}\}$
Complete linkage clustering (CLC)	$S_{tv} = \underset{\substack{m\in t\\ n\in v}}{Min}\{S_{mn}\}$
Average linkage clustering (ALC)	$AS_{tv} = \frac{\sum_{m\in t}\sum_{m\in t} S_{mn}}{N_t \leftrightarrow N_v}$

Step 2. Find the maximum value in the matrix. Then join the two machine groups.
Step 3. Compute the new similarity coefficient between the machines using one of the combined similarity methods (only SLC is considered here).

	1	2, 5	3	4	6
1	0	0	0.67	0.17	0.4
2, 5		0	0.25	0.5	0.17
3			0	0.125	0.5
4				0	0
6					0

Step 4. Repeat step 2 until all machines are grouped.

	1	2, 5	4	6
1	0	0.25	0.17	0.5
2, 5		0	0.5	0.17
4			0	0
6				0

	1, 3, 6	2, 5, 4	4
1, 3, 6	0	0.25	0.17
2, 5		0	0.5
4			0

	1, 3, 6	2, 5, 4
1, 3, 6	0	0.25
2, 5 4		0

14.4.7 Mathematical Programming-Based Clustering

A distance measure d_{ij} is considered in most mathematical programming models used in GT. The distance measure is a real-valued symmetric function based on the three fundamental axioms [33]:

Reflexivity, $d_{ii} = 0$
Symmetry, $d_{ij} = d_{ji}$
Triangle inequality, $d_{ij} \leq d_{is} + d_{sj}$, where *s*is any part other than parts *i* and *j*.

There are several distance measures for a part–machine incidence matrix (a_{ij}) that can be used in formulating the mathematical model; the most widely used are [24, 33]:

Minkowski Distance Metric

$$d_{ij} = \left[\sum_{k=1}^{m} \left| a_{ki} - a_{kj} \right|^{r} \right]^{1/r}, \tag{14.2}$$

where r is a positive integer and m the number of machines.

Two special cases are widely used, Absolute distance (for $r = 1$) and Euclidean distance (for $r = 2$).

Weighted Minkowski Distance

$$d_{ij} = \left[\sum_{k=1}^{m} w_k \left| a_{ki} - a_{kj} \right|^{r} \right]^{1/r}, \tag{14.3}$$

where r is a positive integer and m the number of machines.

Two special cases are widely used, Weighted absolute distance (for $r = 1$) and Weighted Euclidean distance (for $r = 2$).

Hamming Distance Measure

$$d_{ij} = \sum_{k=1}^{m} \delta(a_{ki}, a_{kj}) \tag{14.4}$$

where

$$\delta = (a_{ki}, a_{kj}) = \begin{cases} 1 & \text{if } a_{ki} \neq a_{kj}, \\ 0 & \text{otherwise.} \end{cases}$$

Some mathematical models are discussed below.

14.4.7.1 The p-Median Model

The p-median model is used to group n components into p clusters. The p-median technique is a non-hierarchical clustering method, in which the objective is to determine the representative elements such that the sum of the distances or dissimilarities of each element from its representative will be as small as possible. The following definitions are used for general notations:

- Elements are denoted by index i, and ranges from 1 to p.
- The distance between two elements i and j are denoted by d_{ij}.

- x_{ij} =1, if element i is representative of element j, and 0 otherwise.
- X_i =1, if element i is selected as representative element, and 0 otherwise.
- P is the selected number of clusters.

In formulating a clustering problem, the following relationships between the variables should also be considered:

(a) Since each element j belongs to a cluster and is associated with one of the representative elements, therefore

$$\sum_j X_{ij} = 1, \quad j = 1,2,\ldots,m.$$

(b) The relationship between variable x_{ij} and X_i indicates that if an element i is selected as a representative of element j, it must be one of the representative elements; hence

$$x_{ij} < X_i, \quad i,j = 1,2,\ldots,m.$$

(c) Since K elements are selected as representative elements, then

$$\sum_i X_i = K, \quad i = 1,2,\ldots,m.$$

Finally, the mathematical model for the p-median problem is represented as

$$\min \sum_{ij} \sum d_{ij} * x_{ij}, \quad i,j = 1,2,3\ldots m. \tag{14.5}$$

The constraints of the objective functions are the relationships described by notes a, b, and c above.

14.4.7.2 Quadratic Programming Model

The quadratic programming model was developed by Kusiak [24]. The model accounts for the total family interaction and allows the number of part families and part family sizes to be restricted.

The formulation can be represented as follows:

p = the required number of families
n = the total number of parts
m_j = the number of parts in part family j

$$\min \sum_{i=1}^{n-1} \sum_{j=i+1}^{n} \sum_{l=1}^{p} d_{ij} x_{il} x_{jl}. \tag{14.6}$$

Subject to:

$$\sum_{j=1}^{p} x_{ij} = 1 \quad \text{for all } i = 1,\ldots,n \quad \text{[Each part belongs to exactly one part family]}$$

$$\sum_{i=1}^{n} x_{ij} = m_j \quad \text{for all } j = 1,\ldots,p \text{ [Part family } j \text{ contains exactly } m_j \text{ parts]}$$

$$x_{ij} = 0,1 \quad \text{for all } i = 1,\ldots,n, j = 1,\ldots,n. \text{ [Integrality Condition]}$$

14.5 Cellular Manufacturing Systems and Design

In recent years a new manufacturing system has emerged known as cellular manufacturing (CM), which utilizes the philosophy of GT. Cellular manufacturing, also known as group production, family programming, or flexible manufacturing, is a manufacturing process that produces families of parts within a single line or cell of machines, operated by a machinist or by robots that operate only within the line or cell. Some of the benefits of the cellular manufacturing system are [4, 7, 10, 28, 39]:

- Elimination or reduction of setup time.
- Fixed costs (setup costs) in small-lot production are reduced.
- Greater manufacturing flexibility.
- Reduction of work-in-process and less inventory (Just-In-Time).
- Less floor space around the machines.
- Reduction of warehousing of raw materials.
- Reduction of the cost of goods produced.
- Capability of using high-investment machinery in the production process.
- Reduction in direct labor cost and increase of productivity.

Manufacturing cells are further categorized into two groups: (1) manned cells and (2) unmanned cells (Fig. 14.6).

In manned cells, operators have been trained to operate more than one machine, each of which is conventional and programmable. Unmanned cells are those composed of programmable, automated machines and robots. These unmanned cells may be considered as part of a trend toward FMSs and computer-integrated manufacturing systems. Table 14.2 summarizes and compares the characteristics of traditional manufacturing systems and cellular manufacturing systems. Comparing the traditional manufacturing system to cellular manufacturing indicates that the use of a cellular manufacturing philosophy can improve the overall productivity of a manufacturing firm that was previously using any one of the traditional systems.

One of the simplest methods used in designing the machine cell is referred to as the eye-balling method. This method uses an experienced operator to examine the information and assign the proper cell formation. The method is highly dependent on human experience and tends to become infeasible as the number of the machines grows. Classification and codingis another method used to design manufacturing cells. In this method, parts that are similar in their design features and

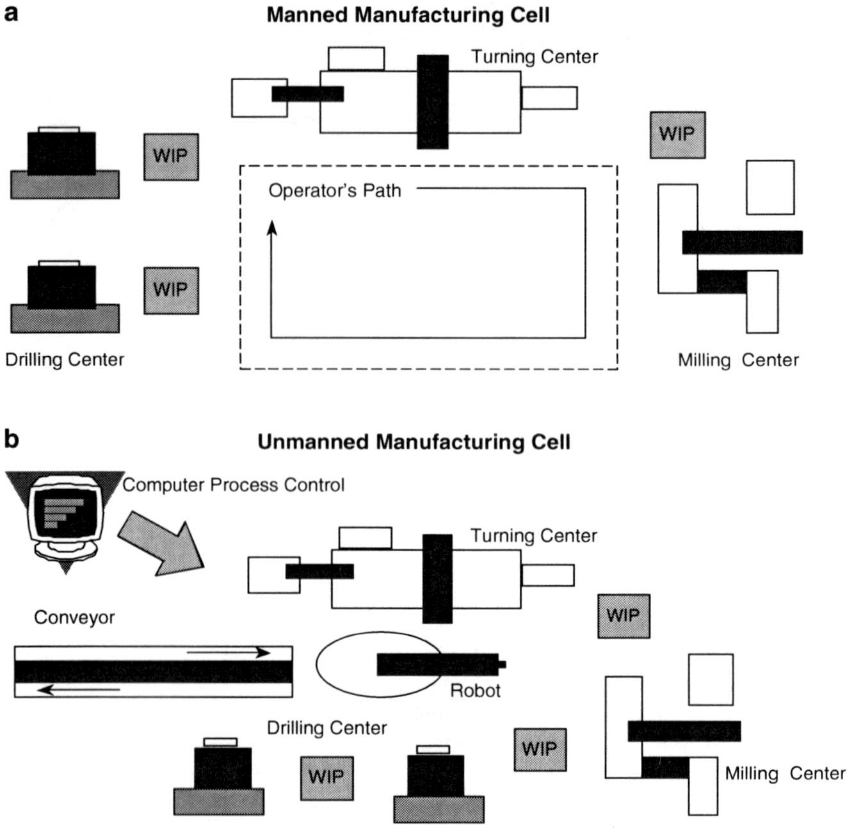

Fig. 14.6 Layouts of manned and unmanned cells

Table 14.2 Characteristics of cellular vs. traditional manufacturing systems

Cellular production	Job shop production	Flow shop production	Project shop production
Flexible and modular machines	Flexible and modular machines	Special-purpose machines	General-purpose machines
Programmable machines	General-purpose machines	Single-function machines	Mobile machines
Mechanized/manual handling	Mechanized/manual handling	Transfer lines	Mechanized/manual handling
Single setup	Long/variable setup	Long/variable setup	Variable setup
Small/medium lot size	Small/medium lot size	Large lot size	Small lot size
Low inventory	Large inventory	Large inventory	Variable inventory
100% Quality assurance	Random sampling	Random sampling	100% Quality Assurance

manufacturing processes are grouped into one family. This method is considered to be both efficient and effective.

Clustering involves the grouping of similar objects based on the degree of similarities between the components within the cells [19, 20, 22, 23, 35, 36]. Four classes of similarity coefficients are often discussed. They are distance coefficients, association coefficients, correlation coefficients, and probabilistic coefficients. These coefficients may be shown in a matrix form. Among the similarity coefficients, distance coefficients are the most widely used. Examples of techniques used for determining these coefficients are Minkowski metric, weighted Minkowski metric [5], and Hamming metric [25]. The association coefficients are widely used for binary and non-binary variables. Examples of these coefficients are [37]:

Simple Matching Coefficient

$$(E_{KL} + E_{OO}) / (E_{KL} + E_{KO} + E_{LO} + E_{OO}), \tag{14.7}$$

Dice Coefficient

$$2E_{KL} / (2E_{KL} + E_{KO} + E_{LO}), \tag{14.8}$$

Tanimoto Coefficient

$$E_{KL} / (E_{KL} + E_{KO} + E_{LO}), \tag{14.9}$$

where

E_{KL}: Number of attributes in common
E_{OO}: Number of attributes in neither set K nor set L
E_{KO}: Number of attributes occurring in set K but not in set L
E_{LO}: Number of attributes occurring in set L but not in set K.

Pearson product-moment correlation is of the third type of similarity coefficient. This method has been used for statistical analysis [9, 12, 27]. Finally, there is the probabilistic coefficient method, which uses the distribution of the frequencies of variables and their data set [11].

14.5.1 The Methodology for Forming Machine Cells

A comprehensive methodology to form part families and machine cells using dissimilarities between design and manufacturing attributes of parts is developed. The methodology forms manufacturing cells by considering several relevant operational costs. The proposed methodology employs a multi-criteria optimization model and allows decision-makers to prioritize their objectives.

In Phase I of the methodology, the KAMKODE coding system is proposed. The attributes selected for this coding system can be easily retrieved from companies' databases. Phase I uses the dissimilarity measures of design and manufacturing attributes in order to determine the families to which the parts are assigned. Since the attributes are of different variable types, several equations are proposed and developed to measure the dissimilarities. Finally, the overall dissimilarity measure between the design attributes is evaluated. A modified version of Jaccard's coefficient [26] is used for measuring the dissimilarity between the sequences of operations. The dissimilarities between tools and fixtures are measured using Dutta's equation [8]. The Hamming metric is used for the determination of dissimilarity between the machine types. Each attribute is ranked, and the weighted dissimilarity measure between parts is then determined using Anderberg's equation. A 0–1 integer program is developed, the objective of which is to minimize the sum of the dissimilarities in order to assign parts into part families. Phase II of the methodology will consider several relevant operational costs. These include:

- Machine investment cost
- Tool investment cost
- Fixture investment cost
- Intracell material handling cost
- Inspection cost
- Machine production cost
- Setup cost

A mixed-integer program is developed to optimize the sum of these costs and determine the assignment of families to cells. It also evaluates the optimum number of machines, tools, and fixtures.

14.5.1.1 Phase I: Part Families Formation

In this phase, a coding system (KAMKODE) is used to categorize parts. The coding system consists of 18 digits. Sample attributes of this coding system are listed in Table 14.3. The data for each attribute could be easily expanded. A sample part coded using the proposed structure is illustrated in Table 14.4. Three strategies are available [1] for measuring the degree of association between objects having mixed variables types. This is the case for the proposed coding system. One such strategy involves the use of disagreement indices. Generally, four types of variables can be identified. These types include binary, nominal, ordinal and continuous.

The linear disagreement indexbetween parts *i*and *j*for attribute *k*, which is of either a binaryor a nominalvariable type, is measured by the following

$$d_{ijk} = \begin{cases} 1, & \text{if } R_{ik} \neq R_{jk}, \\ 0, & \text{otherwise,} \end{cases} \tag{14.10}$$

Table 14.3 KAMKODE coding structure

Design attributes

Attribute 1: *General shape of the part*

(1) Rotational
- (CA1-1) R-Bar
- (CA1-2) R-Tube
- (CA1-3) R-Hexagonal Bar, etc.

(2) Non-rotational
- (CA1-4) NR-Plate
- (CA1-5) NR-Square Bar
- (CA1-6) NR-Sheet Plate
- (CA-7) NR-Rectangular Bar etc.

Attribute 2: *Material*
- (CA2-1) Aluminum Alloys
- (CA2-2) Copper-Zinc Alloys
- (CA2-3) Steels
- (CA24) Cast Irons
- (CA2-5) Plastics etc.

Attribute 3: *Maximum diameter*
- (CA3-1) $D \leq 0.75$ in.
- (CA3-2) $0.75 < D \leq 1.50$ in.
- (CA3-3) $1.50 < D \leq 4.00$ in.
- (CA3-4) $D > 4.00$ in.
- (CA3-5) N/A (Non-rotational Part) etc.

Attribute 4: *Overall length*
- (CA4-1) $L \leq 6$ in.
- (CA4-2) $6 < L \leq 18$ in.
- (CA4-3) $18 < L \leq 60$ in.
- (CA4-4) $L > 60$ in. etc.

Attribute 5: *Diameter of Inside Hole*
- (CA5-1) $d \leq 0.5$ in.
- (CA5-2) $0.5 < d \leq 1.0$ in.
- (CA5-3) $1.0 < d \leq 5.0$ in.
- (CA5-4) $d > 5.0$ in.
- (CA5-5) no hole etc.

Attribute 6: *Product type*
- (CA6-1) Commercial/service
- (CA6-2) Electrical
- (CA6-3) Industrial/mechanical
- (CA6-4) Military
- (CA6-5) Special
- (CA6-7) Other

Manufacturing attributes

Attribute 7: *Number of processing steps*

Attribute 8: *Processing type and sequence*
- (CA8-1) Turning
- (CA8-2) Drilling
- (CA8-3) Reaming
- (CA8-4) Boring
- (CA8-5) Tapping
- (CA8-6) Milling
- (CA8-7) Grinding
- (CA8-8) Broaching
- (CA8-9) Cutting etc.

Attribute 9: *Minimum number of machines required for processing*

Attribute 10: *Processing machine type*
- (CA10-D CNC turning
- (CA10-2) CNC Drilling/tapping
- (CA10-3) Vertical/horizontal CNC Milling
- (CA10-4) External/internal grinding
- (CA10-5) Broaching
- (CA10-6) Band/circular sawing, etc.

Attribute 11: *Number of tool types*

Attribute 12: *Tool type and sequence*
- (CA12-1) Insert
- (CA12-2) Twist drill
- (CA12-3) Adjustable reamer
- (CA12-4) Adjustable boring bar
- (CA12-6) Milling cutter
- (CA12-7) Grinding wheel
- (CA12-8) Broach
- (CA12-9) Band/circular saw blade etc.

Attribute 13: *Number of fixture/jig types*

Attribute 14: *Fixture/jig type*
- (CA14-1) Special fixture
- (CA14-2) Multipurpose fixture
- (CA14-3) Adjustable fixture
- (CA14-4) Rotational adjustable fixture
- (CA14-5) Non-rotational adjustable fixture
- (CA14-6) Special jig
- (CA14-7) Multipurpose jig
- (CA14-8) Adjustable Jig etc.

Attribute 15: *Number of end operations*

Attribute 16: *End operation type and sequence*
- (CA16–1) Clean
- (CA16-2) Polish
- (CA16-3) Buff
- (CA16-4) Coat
- (CA16-5) Paint
- (CA16-6) Assemble
- (CA16-7) Packaging
- (CA16-8) Inspect, etc.

Attribute 17: *Number of devices required for end operations*

Attribute 18: *Devices used for end operations*
- (CA18-1) Dip tank
- (CA18-2) Disk grinder
- (CA18-3) Process robot
- (CA1 8-4) Material handling robot
- (CA18-5) Painting robot
- (CA18-6) Assembly robot
- (CA18-7) Packaging machine
- (CA18-8) Vision system, etc.

Table 14.4 Coding of a sample part with KAMCODE

Part's attributes	KAMKODE
NR-square bar	5
Steel	3
N/A	5
$18<L\leq60$ in.	3
$d\leq0.5$ in.	1
Commercial	1
Two processing steps	2
Processing types: drilling, tapping	2–9
One processing machine	1
Machine type: CNC drilling/tapping	2
Two tool types	2
Twist drill, tap	25
One fixture type	1
Adjustable fixture	3
Three end operations	3
End operation types: clean, paint, and pack	1–5–7
Three end-operation devices	3
Device types: dip tank, process robot, and handling robot	1–3–4

where

d_{ijk} = Disagreement index between parts *i* and *j* for attribute *k*
R_{ik} = Rank of part *i* for attribute *k*
R_{jk} = Rank of part *j* for attribute *k*.

The linear disagreement index for an ordinal variable is determined by the following equation:

$$d_{ijk} = \left|R_{ik} - R_{jk}\right| / (m-1), \tag{14.11}$$

where:

m = Number of classes for attribute *k*
$m-1$ = Maximum rank difference between parts *i* and *j*.

The linear disagreement index for a continuous variable is determined by the following equation:

$$d_{ijk} = \left|R_{ik} - R_{jk}\right| / X_k, \tag{14.12}$$

where X_k is the range of values for the variable.

The linear disagreement index for attributes 1, 2, and 6 is calculated using a binary /nominal equation form, and the linear disagreement index for attributes 3, 4, and 5 is calculated using an ordinal equation form.

The linear disagreement index for the process and end operation sequence is calculated using equation [27, 32]

$$d_{ij} = 1 - \sum_{o} (q_{io} * q_{jo}) \Big/ \sum_{o} (q_{io} + q_{jo} - q_{io} * q_{jo}), \tag{14.13}$$

where

$$q_{io} = \begin{cases} 1, & \text{if part } i \text{ requires operation } o, \\ 0, & \text{otherwise.} \end{cases}$$

The linear disagreement index for tools and fixtures is calculated by [8, 31]

$$d_{ij} = \left(\mathrm{NT}_i + \mathrm{NT}_j - 2\mathrm{NT}_{ij}\right) / \left(\mathrm{NT}_i + \mathrm{NT}_j\right), \tag{14.14}$$

where

NT_i = Number of tools required by part *I*
NT_{ij} = Number of tools common to both parts *i* and *j*.

The linear disagreement index for process and end operation machines is calculated using the Hamming metric as follows [13]:

$$d_{ij} = \sum_{m} \delta\left(X_{im}, X_{jm}\right) \tag{14.15}$$

where

$$X_{im} = \begin{cases} 1, & \text{if part } i \text{ uses machine } m \\ 0, & \text{otherwise} \end{cases}$$

and

$$\delta\left(X_{im}, X_{jm}\right) = \begin{cases} 1, & \text{if } X_{im} \neq X_{jm} \\ 0, & \text{otherwise} \end{cases}$$

After the evaluation of these parameters, the analyst can assign weights to represent his or her subjective evaluation of variables and group parts based on their assigned priorities. The weights can be categorized in any ordinal form. For example, the following structure, Table 14.5, could be applied:

Finally, the weighted dissimilarity measure (*DISij*) between parts *i* and *j* can be determined by the following [1]:

$$\mathrm{DIS}_{ij} = \sum_{k} \left(w_k * d_{ijk}\right) \Big/ \sum_{k} w_k, \tag{14.16}$$

where

w_k = Weight assigned to attribute *k*
d_{ijk} = Disagreement index between parts *i* and *j* for attribute *k*
DIS_{ij} = Weighted dissimilarity coefficient between parts *i* and *j*.

In order to illustrate the calculation of this coefficient, codes are assigned to two sample parts using KAMKODE as presented in Table 14.6. The disagreement

Table 14.5 Example weight categories

Analysis level	Weight
Critical	1
Important	0.75
Required	0.5
Necessary	0.25
Not considered	0.0

Table 14.6 Sample parts used for dissimilarity analysis

	Values	
Coding system attribute	Part I	Part II
Design		
General shape	NR-square bar: 5	NR-square bar: 5
Material type	Steel: 3	Steel: 3
Maximum diameter	N/A: 5	N/A: 5
Overall length	$18 < L \leq 60$ in.:3	$L > 60$ in.:4
Inside hole diameter	$d \leq 0.5$ in.:1	$1.0 < d \leq 5.0$ in.:3
Product type	Commercial: 1	Special: 5
Manufacturing		
Processing sequence	Drilling, tapping: 25	Drilling, reaming: 23
Processing machine	CNC drilling/tapping: 2	CNC drilling/tapping: 2
Tool types	Twist drill, tap: 25	Twist drill, reamer: 23
Fixture type	Adjustable fixture: 3	Non-rotating adjustable: 5
E. operation sequence	Clean, paint, and pack: 157	Clean and pack: 17
E. operation devices	Dip tank, process robot, and handling robot: 134	Dip tank and handling robot: 14

measure between these two parts is then calculated using the proposed formulation. This is presented in Table 14.7.

The dissimilarity of these two parts is measured as

$$DIS_{ij} = (0+0+0+0.33+0.5+1+0.67+1+0.5+1+0.67+0.67)/(12)$$

or 0.528 for all attributes considered critical.

The *p*-medium technique is used to solve the grouping and part family formation problem. The formulation is as follows:

Coefficients:

$p =$ Number of parts
$K =$ Required number of part families
$DIS_{ij} =$ Dissimilarity measure between parts i and j, ($DIS_{ij} = DIS_{ji}$)
$x_{ij} =$ 1, if part i belongs to group j and 0 otherwise

Table 14.7 Dissimilarity measures for two parts

Attributes	Attribute coding values Part I	Part II	Dissimilarity index, d_{ijk}
Design			
General shape	5	5	0 (match)
Material type	3	3	0 (match)
Maximum diameter	5	5	0 (match)
Overall length	3	4	0.33
Inside hole diameter	1	3	0.5
Product type	1	5	1 (mismatch)
Manufacturing			
Processing sequence	25	23	0.67
Processing machine	3	2	1
Tool types	25	23	0.5
Fixture type	3	5	1
E. operation sequence	157	17	0.67
E. operation devices	134	14	0.67

Model:

$$Minimize \sum_i \sum_j DIS_{ij} * x_{ij} \forall i, j = 1, 2, 3 \ldots p \qquad (14.17)$$

subject to:

$$\sum x_{ij} = 1, \forall i, j$$

$$\sum x_{ii} = K, \forall i$$

$$x_{ij} \leq x_{jj} \forall i, j$$

The first constraint assures that each part belongs to only one family. The required number of part families is set in the second constraint. Parts are assigned to part families only if that part family has already been created. This is the last and a conditional constraint to the problem.

The model is a 0–1 integer linear program. The model is solved using a branch-and-bound technique [29]. The branch-and-bound technique is a partial enumeration method in which the set of solutions to a problem is examined by dividing this set into smaller subsets. It can be shown mathematically that some of these subsets do not contain the optimal solution. One important way in which a subset can be examined is to determine a bound. A bound of a subset is a value that is less than or equal to the value of all solutions contained in the subset. By comparing the bound of a subset with an already-found solution, it is sometimes possible to eliminate the subset from further consideration.

14.5.1.2 Phase II: Machine Cell Formation

In phase II of the methodology, the main objective is to determine the number of machines, tools, and fixtures and the number of cells, including the assignment of the proper families to the proper cells. This phase is more of an economic issue than a design issue. The methodology takes into account several relevant operational costs and intends to minimize the total sum of these costs. These include:

- machine investment cost,
- tool investment cost,
- fixture investment cost,
- machine production cost,
- inspection costm,
- setup cost, and
- intra-cell material handling cost.

A mixed-integer mathematical program is developed and used in this part of the methodology for machine cell development. The assumptions associated with developing this mathematical model include:

- Operation time for each part is known.
- The demand for each part is known.
- Machine types can be placed in any selected cell.
- Each part has a fixed routing.
- A minimum of one machine, tool, and fixture type is available for operation, if required for operation.
- At a given time, only one batch is processed on a particular machine type.
- No intercellular material handling is allowed.
- The inspection time for each part is known, and each part is inspected after being processed on a machine type.

The model is as follows:

Index Sets

o = Operation, $o = 1, 2,\ldots,O$
p = Part, $p = 1, 2,\ldots,P$
m = Machine type, $m = 1, 2,\ldots,M$
g = Part group, $g = 1, 2,\ldots,G$
c = Cell, $(C \leq G)$, $c = 1, 2,\ldots,C$
t = Tool type, $t = 1, 2,\ldots,T$
f = Fixture type, $f = 1, 2,\ldots,F$

Objective Function

Machine investment cost: The machine investment cost is defined as the sum of investment costs for the required machines in all cells. The machine investment cost associated with each cell is the product of annual investment and maintenance cost for each machine type and the number of machines required for production. The annual investment and maintenance costs are inputs to the model, and the required number of machines are the decision variables and outputs from the model. This cost computation is summarized by the following expression:

$$\sum_{m}\sum_{c} CM_m * NM_{mc} \tag{14.18}$$

where

CM_m = Annual $ investment and maintenance cost for machine type *m*
NM_{mc} = Number of machines of type *m* assigned to cell *c*.

Tool investment cost: The tool investment cost is defined as the sum of investment costs for the required tools for each machine in each cell. The tool investment cost associated with each cell is the product of annual investment and maintenance cost for each tool type and the number of tools assigned to the process machines required for production. The annual investment and maintenance costs are inputs to the model, and the required number of tools are the decision variables and outputs from the model. This cost computation is summarized by the following expression:

$$\sum_{t}\sum_{m}\sum_{c} CT_t * NT_{tmc} * \sigma'_{tm} \tag{14.19}$$

where

CT_t = Annual $ investment and maintenance cost for tool type *t*
NT_{tmc} = Number of tools of type *t* assigned to machine type *m* in cell *c*.

$$\sigma'_{tm} = \begin{matrix} 1, & \text{if tool type } t \text{ is required by machine type } m \\ 0, & \text{otherwise} \end{matrix}$$

Fixture investment cost: The fixture investment cost associated with each cell is the product of annual investment and maintenance costs for each fixture type and the number of fixtures assigned to the process machines required for production. The annual investment and maintenance costs are input to the model, and number of fixtures are decision variables and outputs from the model. This cost computation is summarized by the following expression:

$$\sum_{f}\sum_{m}\sum_{c} CF_f * NF_{fmc} * \sigma_{fm} \tag{14.20}$$

where

CF_f = Annual $ investment and maintenance cost for fixture type *f*
NF_{fmc} = Number of fixtures of type *f* assigned to machine type *m* in cell *c*

$$\sigma_{fm} = \begin{matrix} 1, & \text{if fixture type } f \text{ is required machine type } m \\ 0, & \text{otherwise} \end{matrix}$$

Machine production cost: In order to calculate the machine production cost, the processing time required for each part type on all machine types required for its production should be determined. This time is dependent on the type of operation and its processing time and on the machine performance rate for that particular operation, since a manufacturing operation can be performed by more than one machine type. By multiplying the total demand of a particular part, the total machining time required to meet the annual demand of a part is calculated. Namely

TMD_{pm} = Total time required to meet annual demand for part p on machine type m

$$\sum_{o} PRF_{om} * TO_{op} * d_p * \mu_{opm}, \forall m, p \tag{14.21}$$

where

$$\mu_{opm} = \begin{cases} 1, & \text{if operation } o \text{ of part } p \text{ requires machine type } m \\ 0, & \text{otherwise} \end{cases}$$

TO_{op} = Time to complete operation o on part p

d_p = Total number of part p demanded annually

PRF_{om} = Machine performance rate for operation o on machine type m

Since it is assumed that each part is inspected after each operation, the cost of rework and machine reliability should be included in the equation. This cost computation is summarized by the following expression:

$$\sum_{p}\sum_{m}\sum_{c}\left[CP_{pm} * R_m + CRW_m * (1 - R_m)\right]\beta_{pm} * TMD_{pm} * Y_{pc} \tag{14.22}$$

where CP_{pm} is the cost of processing part p on machine type m, CRW_m is the rework cost on machine type m, R_m the reliability of machine type m

$$\beta_{pm} = \begin{cases} 1, & \text{if part } p \text{ is processed on machine type } m \\ 0, & \text{otherwise} \end{cases}$$

$$Y_{pc} = \begin{cases} 1, & \text{if part } p \text{ is assigned to cell } c \\ 0, & \text{otherwise} \end{cases}$$

Y_{pc} is the decision variable that identifies optimum selection of cells for parts and their families.

Inspection cost: Since each part will be inspected after being processed on machines, this cost is proportional to the number of machines required for the part production. Inspection cost is the product of unit inspection time and unit inspection cost and the total demand of the parts. The cost computation is summarized by the following expression:

$$\sum_{p}\sum_{m}\sum_{o} CI_{pm} * TI_{pm} * d_p * \mu_{omp} * NM_{mc} \tag{14.23}$$

where TI_{pm} is the inspection time required to inspect part p after being processed on machine type m, CI_{pm} is the inspection cost of part p after being processed on machine type m, d_p the total number of part p demanded annually

NM_{mc} = Number of machines of type m assigned to cell c

$$\mu_{omp} = \begin{cases} 1, & \text{if operation } o \text{ of part } p \text{ requires machine type } m \\ 0, & \text{otherwise} \end{cases}$$

Setup cost: The setup cost for each machine used for group production in each cell is determined by the product of average setup time of a machine and the average cost associated with each setup. The average value includes the setups for machines, tools, fixtures, and inspections. The outputs from the model are the assignment of families to proper cells. Average setup times and costs are inputs to the model. This cost computation is as follows:

$$\sum_m \sum_g \sum_c \mathrm{TS}_{mgc} * \mathrm{CS}_{mgc} * \alpha'_{mg} * X_{gc} \tag{14.24}$$

where CS_{mgc} is the average setup cost for machine type m for group g in cell c, TS_{mgc} is the Average setup time for machine type m for group g in cell c,

$$\alpha'_{mg} = \begin{cases} 1, & \text{if machine type } m \text{ is used for group } g \text{ production} \\ 0, & \text{otherwise} \end{cases}$$

$$X_{gc} = \begin{cases} 1, & \text{if group } g \text{ is assigned to cell } c \\ 0, & \text{otherwise} \end{cases}$$

Intracell material handling cost: The intracell material handling cost is composed of a fixed cost for each part. This includes required system hardware such as fixtures and tools. The next factor included in this cost is the part's intracell transportation cost. Therefore, the unit-handling cost in each cell is represented by average value. The intracell transportation cost is proportional to the number of units demanded and moved within the cell annually. This cost computation is summarized by the following expression:

$$\sum_p \sum_c \left(\mathrm{NMV}_p - 1\right) * d_p * \mathrm{CH}_{pc} * Y_{pc} \tag{14.25}$$

where NMV_p is the number of moves for part p,

CH_{pc} is the average material handling cost of part p in cell c,
d_p is the total number of part p demanded annually,

$$Y_{pc} = \begin{cases} 1, & \text{if part } p \text{ is assigned to cell } c \\ 0, & \text{otherwise} \end{cases}$$

The objective is to minimize the total sum of these costs. The constraints developed for phase II of the methodology are as follows:

Constraints

Budget: The following constraints will restrict the amount of capital expenditure to the annual budgets for machine, tool, fixture, inspection, and material handling. Management sets the budget values. These equations are summarized as follows:

$$\sum_{m}\sum_{c} \mathrm{CM}_m * \mathrm{NM}_{mc} \le \mathrm{BM} \tag{14.26}$$

$$\sum_{t}\sum_{m}\sum_{c} \mathrm{CT}_t * \mathrm{NT}_{tmc} * \sigma'_{tm} \le \mathrm{BT} \tag{14.27}$$

$$\sum_{f}\sum_{m}\sum_{c} \mathrm{CF}_f * \mathrm{NF}_{fmc} * \sigma_{fm} \le \mathrm{BF} \tag{14.28}$$

$$\sum_{p}\sum_{m}\sum_{o} \mathrm{CI}_{pm} * \mathrm{TI}_{pm} * d_p * \mu_{omp} * \mathrm{NM}_{mc} \le \mathrm{BI} \tag{14.29}$$

$$\sum_{p}\sum_{c} \left(\mathrm{NMV}_p - 1\right) * \mathrm{CH}_{pc} * Y_{pc} \le \mathrm{BMH} \tag{14.30}$$

where BM is the Budget available to purchase and maintain machines of all types, BF is the Budget available to purchase and maintain fixtures of all types, BT is the Budget available to purchase and maintain tools of all types, BI is the Budget available for inspection of all parts, and BMH is the Budget available for material handling of all parts in all cells.

Machine capacity: Equations developed in this section will ensure that the capacity of each machine type in each cell is not violated. If so, a number of duplicates will be calculated and proposed in order that the shop meets the annual demand for parts. This constraint is formulated as follows:

$$\beta_{pm} * \mathrm{TMD}_{pm} * Y_{pc} \le \mathrm{TP}_m * \mathrm{NM}_{mc} \quad \forall m, c \tag{14.31}$$

where TP_m is the total annual processing time available on machine type m, NM_{mc} is the number of machines of type m assigned to cell c, TMD_{pm} is the Total time required to meet annual demand for part p on machine type m,

$$\beta_{pm} = \begin{cases} 1, & \text{if part } p \text{ is processed on machine type } m \\ 0, & \text{otherwise} \end{cases}$$

$$Y_{pc} = \begin{cases} 1, & \text{if part } p \text{ is assigned to cell } c \\ 0, & \text{otherwise} \end{cases}$$

Machine–fixture balance: Since each process machine will require at least one fixture for production, this constraint will assure the minimum number required.

In case of a duplicate number of machines, the same number of duplicates for the fixtures is proposed. This constraint is formulated as follows:

$$\mathrm{NF}_{fmc} * \sigma_{fm} \geq \mathrm{NM}_{mc} \quad \forall f, m, c \tag{14.32}$$

where NF_{fmc} is the number of fixtures of type f assigned to machine type m in cell c, NM_{mc} is the number of machines of type m is assigned to cell c,

$$\sigma_{fm} = \begin{matrix} 1, & \text{if fixture type } f \text{ is required by machine type } m \\ 0, & \text{otherwise} \end{matrix}$$

Tool life: Equations developed in this section will ensure that the tool life of each tool type on each machine assigned in each cell is not violated. If so, a number of duplicates will be calculated in order to meet the annual demand. This constraint is formulated as follows:

$$\sum_p \beta_{pm} * \beta'_{tp} * \sigma_{tm} * TM_{dpm} * Y_{pc} \leq TL_t * NT_{tm}c \quad \forall m, c, t \tag{14.33}$$

where TMD_{pm} is the Total time required to meet annual demand for part p on machine type m, TL_t is the total lifetime of tool type t, NT_{tmc} is the number of tools of type t assigned to machine type m in cell c,

$$\beta_{pm} = \begin{matrix} 1, & \text{if part } p \text{ is processed on machine type } m \\ 0, & \text{otherwise} \end{matrix}$$

$$\beta'_{tp} = \begin{matrix} 1, & \text{if tool type } t \text{ is required by part } p \\ 0, & \text{otherwise} \end{matrix}$$

$$\sigma_{tm} = \begin{matrix} 1, & \text{if tool type } t \text{ is required by machine type } m \\ 0, & \text{otherwise} \end{matrix}$$

$$Y_{pc} = \begin{matrix} 1, & \text{if part } p \text{ is assigned to cell } c \\ 0, & \text{otherwise} \end{matrix}$$

Cell capacity: In order to have a high degree of flexibility in each cell, a limit is required to be set to the total number of parts assigned to each cell. This constraint is formulated as follows:

$$\sum_p d_p * Y_{pc} \leq IC_c \quad \forall c \tag{14.34}$$

where IC_c is the maximum number of parts allowed in cell c, d_p is the total number of part p demanded annually,

$$Y_{pc} = \begin{matrix} 1, & \text{if part } p \text{ is assigned to cell } c \\ 0, & \text{otherwise} \end{matrix}$$

Part group assignment: To reassure the assignment of each part family to only one cell and the assignment of the parts in these families to a cell, the following constraints are proposed:

$$\sum_{gc} X_{gc} = 1 \tag{14.35}$$

$$\alpha_{pg} * Y_{pc} = X_{gc} \tag{14.36}$$

where

$$X_{gc} = \begin{matrix} 1, & \text{if group } g \text{ is assigned to cell } c \\ 0, & \text{otherwise} \end{matrix}$$

$$Y_{pc} = \begin{matrix} 1, & \text{if part } p \text{ is assigned to cell } c \\ 0, & \text{otherwise} \end{matrix}$$

$$\alpha_{pg} = \begin{matrix} 1, & \text{if part } p \text{ is a member of group } g \\ 0, & \text{otherwise} \end{matrix}$$

Decision variable binary and integerality: Since the cost equation is a mixed-integer expression, the decision variables should be promptly identified for the solution method used (branch-and-bound). Therefore:

$X_{gc} \in (0,1) \forall gc$

$Y_{pc} \in (0,1) \forall p,c$

$\mathrm{NM}_{mc} \geq 0$ and integer $\forall mc$

$\mathrm{NF}_{fmc} \geq 0$ and integer $\forall f,m,c$

$NT_{tmc} \geq 0$ and integer $\forall t,m,c$

A situation in which the production of 15 parts is required is under analysis. These parts require 15 operations (nine process operations and six end operations). There are eight process machines and five end-operation devices available. A rating factor of 1 is assumed for all machines performing process operations. This assumption indicates that the machines selected are the most suitable ones for performing the process operations. Nine types of tools are available for the process operation. Table 14.8 illustrates the values of the disagreement measure between parts for the proposed method.

Using the proposed formulation and by setting p(required number of part families) to be four, parts and their associated families are as $G1(1,3,5)$, $G2(2,4,6,8,11,14)$, $G3(7,9,10,15)$, and $G4(12,13)$. The dissimilarity values are also used to set up the dendrogramof this example, shown in Fig. 14.7.

Table 14.8 Disagreement measures between all parts

	vs. Part														
Part	1	2	3	4	5	6	7	8	9	10	11	12	13	14	15
1	0.00	1.00	0.60	0.98	0.60	1.00	0.68	0.86	0.73	0.96	0.85	0.88	0.89	1.00	0.74
2	1.00	0.00	0.90	0.45	1.00	0.63	0.94	0.66	0.92	0.78	0.51	0.65	0.71	0.36	0.88
3	0.60	0.90	0.00	0.74	0.62	1.00	0.65	0.69	0.84	0.92	1.00	0.80	0.85	0.93	0.77
4	0.98	0.45	0.74	0.00	0.84	0.60	0.94	0.51	1.00	0.76	0.43	0.55	0.70	0.44	1.00
5	0.60	1.00	0.62	0.84	0.00	0.97	0.76	0.71	0.76	0.77	1.00	0.88	0.89	1.00	0.82
6	1.00	0.63	1.00	0.60	0.97	0.00	1.00	0.59	1.00	0.67	0.63	0.78	0.71	0.51	1.00
7	0.68	0.94	0.65	0.94	0.76	1.00	0.00	0.82	0.75	0.94	0.88	0.76	0.99	0.97	0.56
8	0.86	0.66	0.69	0.51	0.71	0.59	0.82	0.00	1.00	0.75	0.75	0.65	0.87	0.63	0.78
9	0.73	0.92	0.84	1.00	0.76	1.00	0.75	1.00	0.00	0.68	1.00	0.65	0.91	1.00	0.49
10	0.96	0.78	0.92	0.76	0.77	0.67	0.94	0.75	0.68	0.00	0.92	0.72	0.81	0.75	0.59
11	0.85	0.51	1.00	0.43	1.00	0.63	0.88	0.75	1.00	0.92	0.00	0.73	0.74	0.54	1.00
12	0.88	0.65	0.80	0.55	0.88	0.78	0.76	0.65	0.65	0.72	0.73	0.00	0.55	0.77	0.66
13	0.89	0.71	0.85	0.70	0.89	0.71	0.99	0.87	0.91	0.81	0.74	0.55	0.00	0.86	0.97
14	1.00	0.36	0.93	0.44	1.00	0.51	0.97	0.63	1.00	0.75	0.54	0.77	0.86	0.00	0.93
15	0.74	0.88	0.77	1.00	0.82	1.00	0.56	0.78	0.49	0.59	1.00	0.66	0.97	0.93	0.00

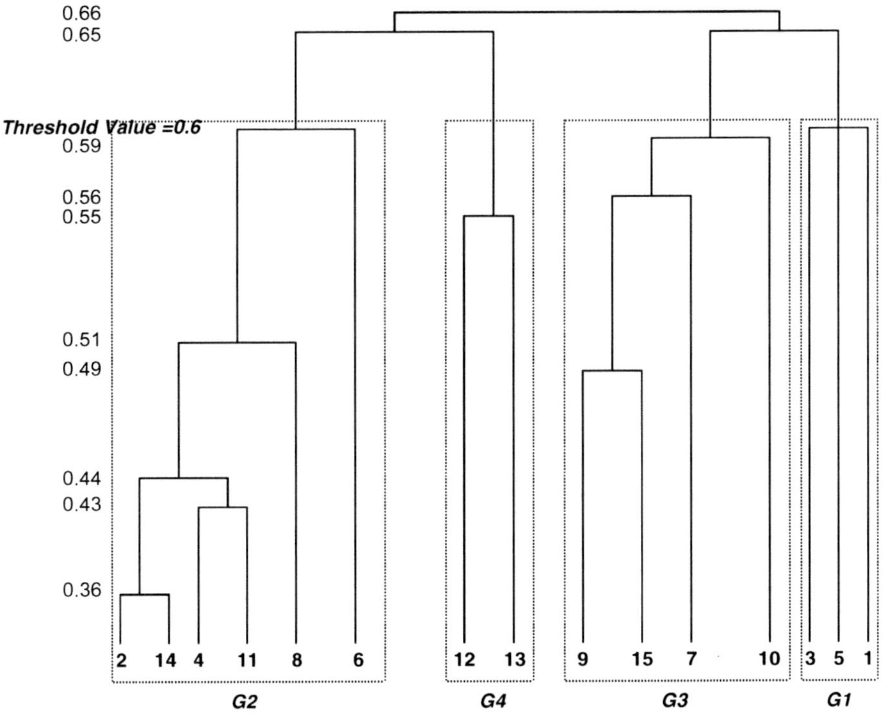

Fig. 14.7 The dendrogram constructed for sample parts

Table 14.9 Machine investment costs, annual available machine time, tool investment cost, and tool life

Machine type	CM_m (\$)	TP_m (min)	Tool type	CM_t	TL_t (min)
1	30,000	102,000			
2	42,000	139,000			
3	24,000	111,000	1	2,538	490
4	35,000	114,000	2	2,576	444
5	41,000	140,000	3	2,526	430
6	22,000	100,000	4	2,436	427
7	20,750	162,000	5	2,562	488
8	30,000	156,000	6	2,334	413
9	22,460	130,000	7	2,454	412
10	27,000	155,000	8	2,154	414
11	32,000	90,000	9	2,244	442
12	26,000	99,000			
13	30,490	89,000			

Table 14.10 Annual demand for various parts (d′)

Part type	d'
1	1,728
2	2,000
3	2,145
4	1,729
5	1,948
6	2,263
7	2,226
8	2,236
9	2,160
10	1,777
11	1,758
12	1,824
13	2,089
14	1,929
15	2,308

For the threshold assignment of 0.60, the four families are as follows: $G1$(2,4,6,8,11,14), $G2$(7,9,10,15), $G3$(1,3,5), and $G4$(12,13). Table 14.9 lists the annual investment and maintenance associated with each machine and tool. It also contains the annual available machining time on each machine and the tool life associated with each tool. The annual demand for each part is given in Table 14.10.

Machine reliability is illustrated in Table 14.11. The inspection time and cost are assumed to be similar for all parts, and all machines have the same rework cost. Setup cost and time are assumed to be similar for all machines. The intracellular material handling cost associated with parts is similar in all cells. The model is solved and results are listed in Tables 14.12–14.14.

Table 14.11 Machine reliability (R)

Machine type	R'' (%)
1	93
2	95
3	94
4	89
5	75
6	85
7	95
8	94
9	97
10	85
11	82
12	85
13	91

Table 14.12 Cell configuration

Part number	Family number	Cell number
1	1, 4	1
3		
5		
12		
13		
2	2	2
4		
6		
8		
11		
14		
7	3	3
9		
10		
15		

Table 14.13 Number of machine types and their assignments

Machine type	Cell number	Number of duplicates
1	1	1
2		2
3		1
8		1
9		1
10		1
11		3
12		2
5	2	3
8		1

(continued)

Table 14.13 (continued)

Machine type	Cell number	Number of duplicates
9		2
10		2
11		1
12		4
13		1
1	3	1
3		2
4		1
8		1
9		1
10		1
11		3
12		1
13		2

Table 14.14 Number of tool types and their assignments

Tool type	Machine type	Cell number	Number of duplicates
1	1	1	176
2	1		194
2	2		404
2	3		91
3	2		263
5	2		137
5	3		83
9	8		49
2	5	2	196
3	5		189
5	5		136
6	5		763
9	8		51
1	1	3	71
2	1		78
2	3		318
3	3		121
4	3		117
4	4		73
9	8		227

14.6 Process Planning and Computer-Aided Process Planning Systems

Process planning is simply the set of activities associated with the manufacturing of a part, presented in a detailed and structured format [38]. A more descriptive definition for process planningis "*the systematic determination of the methods by which a product is to be manufactured, economically and competitively.*" Process planning is the link between design and manufacturing. With the interpretation being made on the product design data, material and process activities are selected and sequenced. Further resources such as machine, tools, and fixtures are also selected. Time for processes is evaluated and documented. Steps associated with the development of a process plan include [3, 14, 16]:

- Design analysis and interpretation
- Raw material analysis
- GD&T and surface finish analysis
- Process selection
- Selection based on design interpretation
- Machine selection
- Capability and availability
- Feasibility
- Multi-tasking
- Selected or designed
- Tolerance stack and charting
- Tolerance and datum surface data analysis
- Complex and tedious sequence of operations
- Operation sequencing
- Logical sequence of the raw material removal
- Layer removal based on GD&T and surface finish
- Tooling and fixturing
- Selected or designed
- Cutting parameters analysis
- Optimum calculation of speed, feed, and depth of cut
- Cycle-time analysis and capacity planning
- Machining operations
- Non-machining operations
- Documentation
- Route sheet development
- In-process text and graphics

The process planning task is experience-based and manual. As a result, modern manufacturing industries are facing a series of problems. (e.g., the departure of the skilled labor force that can produce the part). Manual development of process plans is subjective and may result in the development of plans that are feasible but not optimal. Computer technology has had a great impact on the role of design and manufacturing engineers during the development and implementation life cycle of products. Computer-aided process planning (CAPP) is one such tool that has freed the process planner from making many routine decisions. The CAPP system reduces process planning time and manufacturing cost and enables overall improvement of productivity. CAPP is based on three approaches: variant, generative, and hybrid [30]. The variant approach uses the computer as a design-supporting tool for information retrieval and text editing. Order-specific process plans are stored in the

Table 14.15 Overview of V-CAPP and G-CAPP characteristics

Variant process planning	Generative process planning
Group technology concept	Consistent process plan
Partially automated procedures	Fully automated procedures
Experienced planner required	AI applications
Low cost	High cost
Simple to develop	Complex architecture
Computer-assisted	Computer-aided

process plan database and a standard plan for a similar part is retrieved and modified. Still, the skill and the experience of the process planner play a major role in this approach. The generative approach develops a new process plan for every new product designed. It takes advantage of information-management technologies and capabilities. Company-specific knowledge can be structured in a standardized process planningprocedure. Table 14.15 lists the characteristics of each technique.

14.6.1 Critical Issues in the Design of CAPP Systems

Similar to the design of most complex software systems, development of a CAPP system that meets the requirements of the company is a tedious and difficult task. First, the primary objective for using the system must be identified. A list of technical objectives that should be considered is given below [30]:

- Capture the manufacturing engineering knowledge and make it available to the process planner.
- Minimize activities associated with the development of the process plan (documentation, calculation, etc.).
- Provide editing capabilities for changes required to the developed process plan.
- Develop optimum and consistent process plans for part families.
- Provide an engineering database of manufacturing knowledge, its feasibility, and its availability to the process planner.
- Provide information and data management capabilities in text and graphics.
- CAD integration capabilities.
- Capability for operation by a non-expert.
- Provide alternatives for cost analysis and attempt to select the most economical process plan by realization of all the cost associated with the production of the parts.

Figure 14.8 illustrates an overview of a generic CAPP system's characteristics. Second, the language and method of implementation should be identified. Table 14.16 lists the different programming techniques available for the development of CAPP systems.

The programming technique used characterizes the degree of flexibility in adapting a system to application-specific facts [18]. Traditional programming

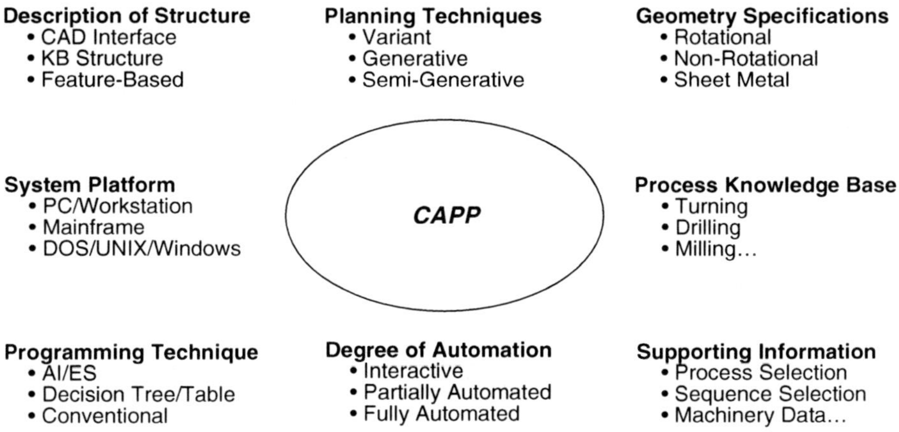

Fig. 14.8 CAPP characteristics

Table 14.16 CAPP systems development techniques

Programming technique	Advantages	Disadvantages
Conventional Programming	Company-specific solution Simple integration	Large development task/time Fixed automation Difficult to transfer
Application Programming	Wide planning knowledge Short response time	Limited field of application Requires extensive work for adoption/extraction Need of programming skills
Decision tables	Wide field of applications No programming shell Updating by user	Long response time Fixed structure of KB Lack of clarity of solutions
Rule-based Systems	Company-specific solution Flexible structure of KB	Extensive programs Qualified personnel

techniques lack the degree of flexibility that is required for the development of company-specific planning procedures toward the automation of process planning. Application of decision tree (DT) and artificial intelligence (AI) systems in the development of CAPP systems can provide the necessary flexibility required. The required flexibility in the system and its development environment is due to the constant changes and improvements in manufacturing technology. The major difficulty with the application of rule-based systems in CAPP is the complexity of acquiring rules and knowledge. This task requires a great amount of effort and time. Also, rule-based systems are not very efficient in the calculation of mathematical functions and equations. By developing a hybrid structure of AI, DT, and conventional programming, some of the problems associated with complexity and efficiency can be resolved.

14.6.2 Structure for a Template-Based System

Figure 14.9 illustrates the overview structure of a generic variant and integrated design and process-planning system. The proposed system is designed using the following three-phase methodology. These three steps include:

- *Product and process strawman generation*

Identify products that could be produced independently.

Develop template for parametric representations and design of the product families.

Determine the process templates for all classified groups.

Define logical process orders in which parts and associate families could be processed from start to finish.

- *Product and process associativity analysis*

Identify the critical design features associated the products and the classified groups.

Associate design features and process templates in order to define logical processing sequence.

Determine the machines and tooling feasible for all process templates.

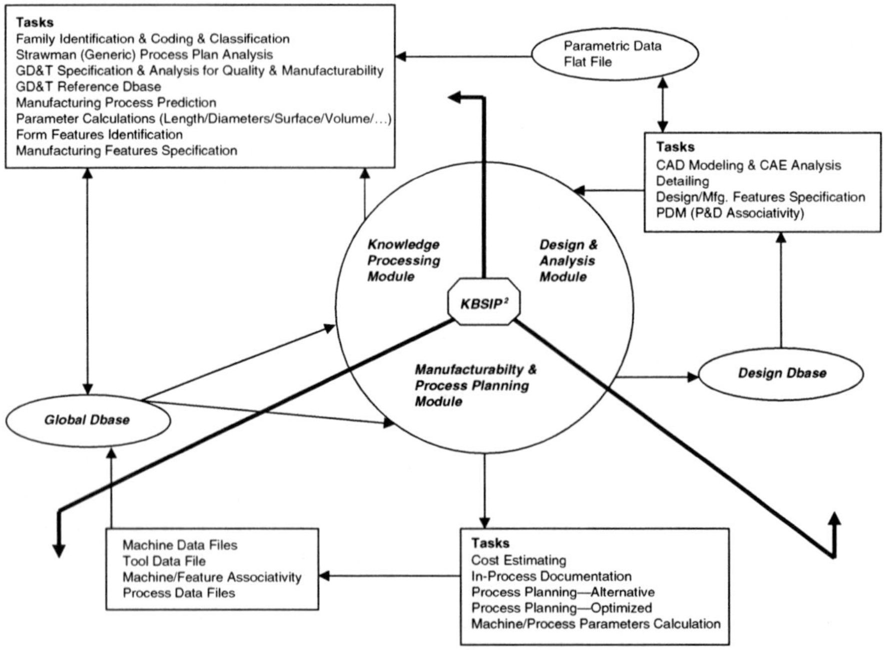

Fig. 14.9 Integrated product design and process planning

Calculate manufacturing cycle time using machine and process based on the identified critical features.

- *Manufacturing systems analysis*

Determine the number of required machines and tools for the manufacture of the product based on the required production volume.

Calculate the production cycle time and determine the cost of machines and required tooling.

The system consists of the following sub-systems.

14.6.2.1 Knowledge Acquisition Modules

In order to build the system, design and process knowledge should be collected and structured in the form of rules. Rules can be collected through interviews of domain experts (design, manufacturing, operator, customers, etc.), design manuals, manufacturing manuals, production information from shop floor data, and technical reports. It is the transfer of expertise into the computational model.

14.6.2.2 Global Database Systems

There are three major database systems required for the development of this environment. These are material, design, and manufacturing. The material database system contains the information for materials and their physical, functional, and manufacturing capabilities. This information is critical since it will have impact on all aspects of design and manufacturing. The material of any product selects the primary process(es) necessary for the production of the components. The primary process should achieve as many of the design and manufacturing features as possible. This concept is known as the near-net-shape process. The design database consists of previous design components, classes of features and their designs, and parametric design data files for parametric modeling; inputs to the system are in the form of either features or parametric data files. It also consists of GD&T files. GT is used for the classification of data. The manufacturing database contains necessary information for production of the components. It includes data for machines, processes, tool, fixtures, machinery, cost, capabilities, and standard practices. These data are global and accessible by all engineering sections.

14.6.2.3 Design for Manufacture Module

Using design and manufacturing modules and global database systems, the DFM module is structured. It includes interface and evaluation modules. The interface module is the link between the design and manufacturing modules. A template-based philosophy is used to develop this interface. The module is developed based on the logic that all manufacturing tasks require data. For automating and integrating these tasks, there is a need to standardize data-manipulating functions.

Typically, these data consist of (1) product data describing the item to be designed and fabricated, which may include geometric data and physical attributes such as material, tolerances, finishes, and heat treatment; (2) process data describing the manufacturing capabilities available and necessary to fabricate the part, including processes, machines, performance, capacity, and tools; and (3) rules data to control design and manufacturing associativities, which may include procedures, specifications, analysis programs, and producibility criteria.

14.6.2.4 Computer-Aided Process Planning Module

The CAPP module is developed based on the templates selected using the DFM module. The report generated from this module is the link between design and manufacturing and is used to produce the component. For each class of parts or features, a composite process plan template is developed. Based on the DFM analysis, segments of this composite plan are retrieved and detailed for the actual process planning task. All necessary calculations such as cost, tooling, and fixturing can be performed in this module. The proposed system will use a parametric data file. Figure 14.10 illustrates a sample of the parametric file template for an engine crankshaft and its family. This file contains the parametric data used for the development of the crankshaft's solid model. It also includes the technological data (GD&T) associated with the design. In case of a new design, the parameters are updated and, if necessary, new constraints are added. The system should allow for direct input of the design data parameters into the process planning module. A decision-tree structure could be used for the implementation of the template-based system. Inputs to the system are parametric files and GD&T data files. The output is the process flow chart, process plan and the initial layout the manufacturing system for

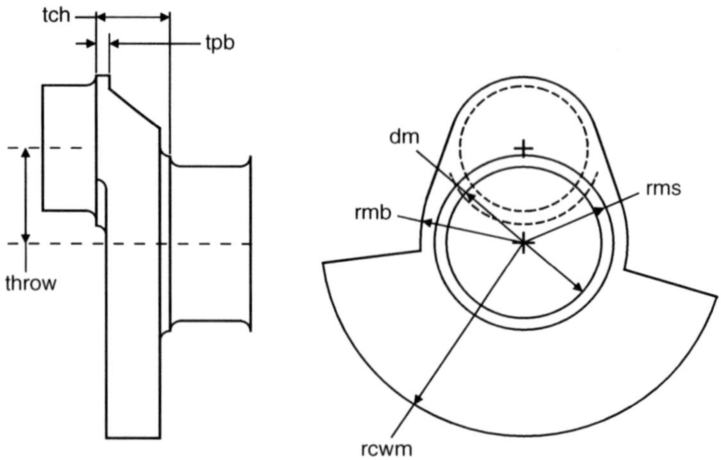

Fig. 14.10 Sample parametric file format

the new crankshaft. A consistency check is required for data-integrity analysis. This routine ensures that all required data are available. In the case of missing data, the user is prompted with the name of the missing variable.

This task is done by comparing the new file to a template, which consists of all possible parameters and constraints. For any design, only a certain set of information is necessary, and the system is designed to examine the family that the new design belongs to. A consistency check based on the selected template from that family is then performed. GD&T data are retrieved either manually or automatically. An integrity check is again performed, and the user is prompted for input in the case of missing data. This module is used for manufacturability analysis, process prediction, and searching the part family for retrieval of a similar operation plan template. A series of calculations must be performed to determine the removal volume(s) from the working surfaces. The result from this calculation is used for selecting the number of surfaces that are to be machined simultaneously. The selection is based on the machine's capabilities and the technical data available. The necessary data required to perform these calculations are also extracted from the parametric data file. These calculations utilize all aspects of the design information, which include the parametric data and information regarding the tolerance and surface finish. For example, the removal volume of the counterweights for the crankshaft is calculated as

CWTMVOL(XCT)0.4=(0.8+TMS)*SA/360*3.14*((RCWM**2)-((DM/2**2))

where XCT is the number of working surfaces), CWTMVOL is the volume removal from the main side of the counterweights, TMS is the thickness of the main journal shoulder, SA is the swept angle *A*, RCWM is the radius of the counterweight from centerline of main, DM IS the main diameter.

Appendix A lists the rest of required calculations. Figure 14.11illustrates sample working surfaces that are selected as CANDIDATES for simultaneous machining. This information is imported into the process planning routine for costing analysis.

The new process plan is developed using the retrieved template files [15]. Template files are selected based on the family that the new design is assigned to. Templates are used for setting the logical sequence of the process and the resources required for processing the design. Figure 14.12 illustrates the crankshaft process template file. A coding structure is used for this task. As information is being retrieved from the parametric files, features associated with the part are identified

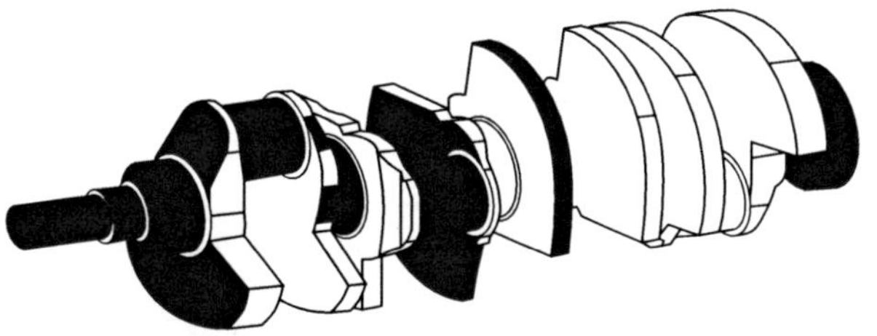

Fig. 14.11 Surfaces that require machining

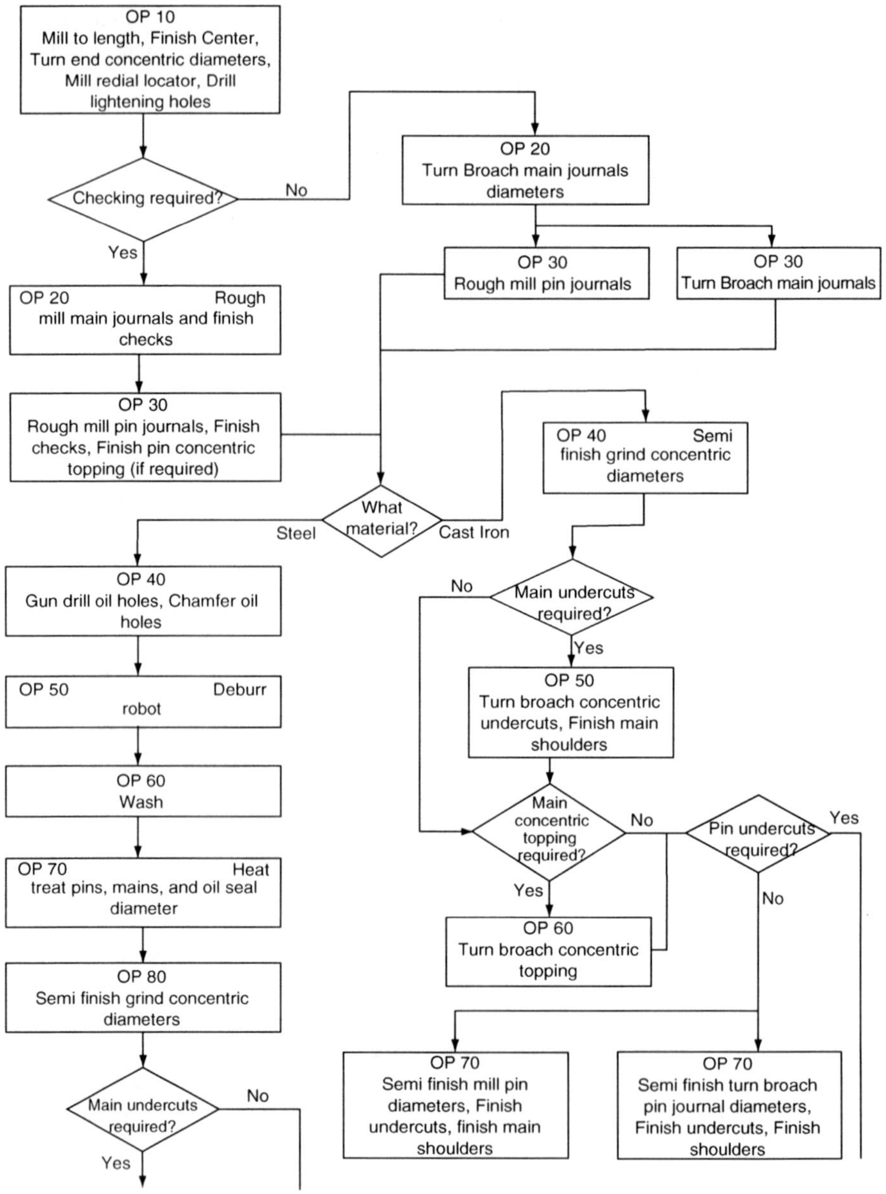

Fig. 14.12 Crankshaft process template file

and a code is set based on the design and manufacturing features. These codes are then sequenced, and required processes from the template database are retrieved. The calculated variables are also exported into this process-planning environment for detailing the decision-tree program. The Facility Planner uses the process plan

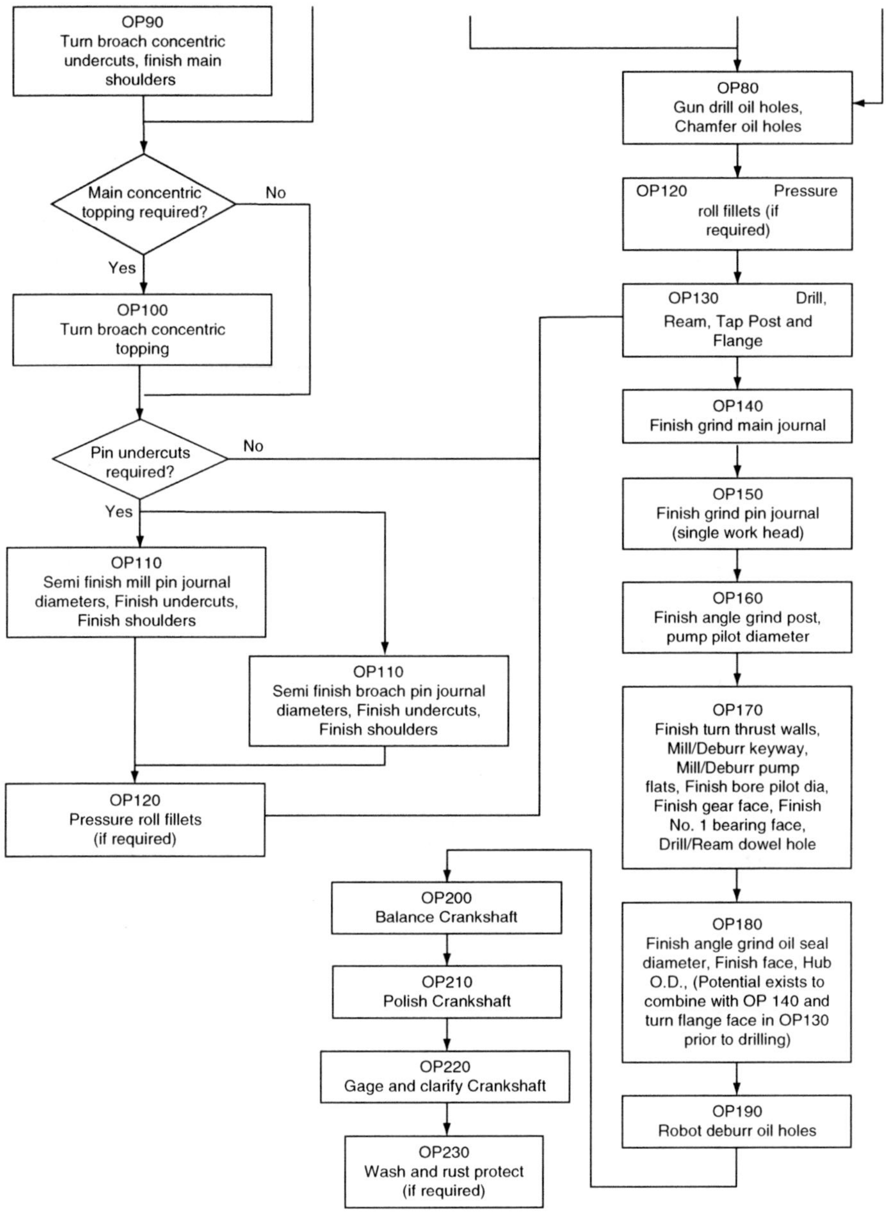

Fig. 14.12 (continued)

to construct the process flow chart. Figure 14.13 illustrates the sample flow chart generated from the template-based process plan. Information provided in this chart are the graphical sequence of the operations, types of the machines, number of the machines, number of stations and the hourly production of each machine.

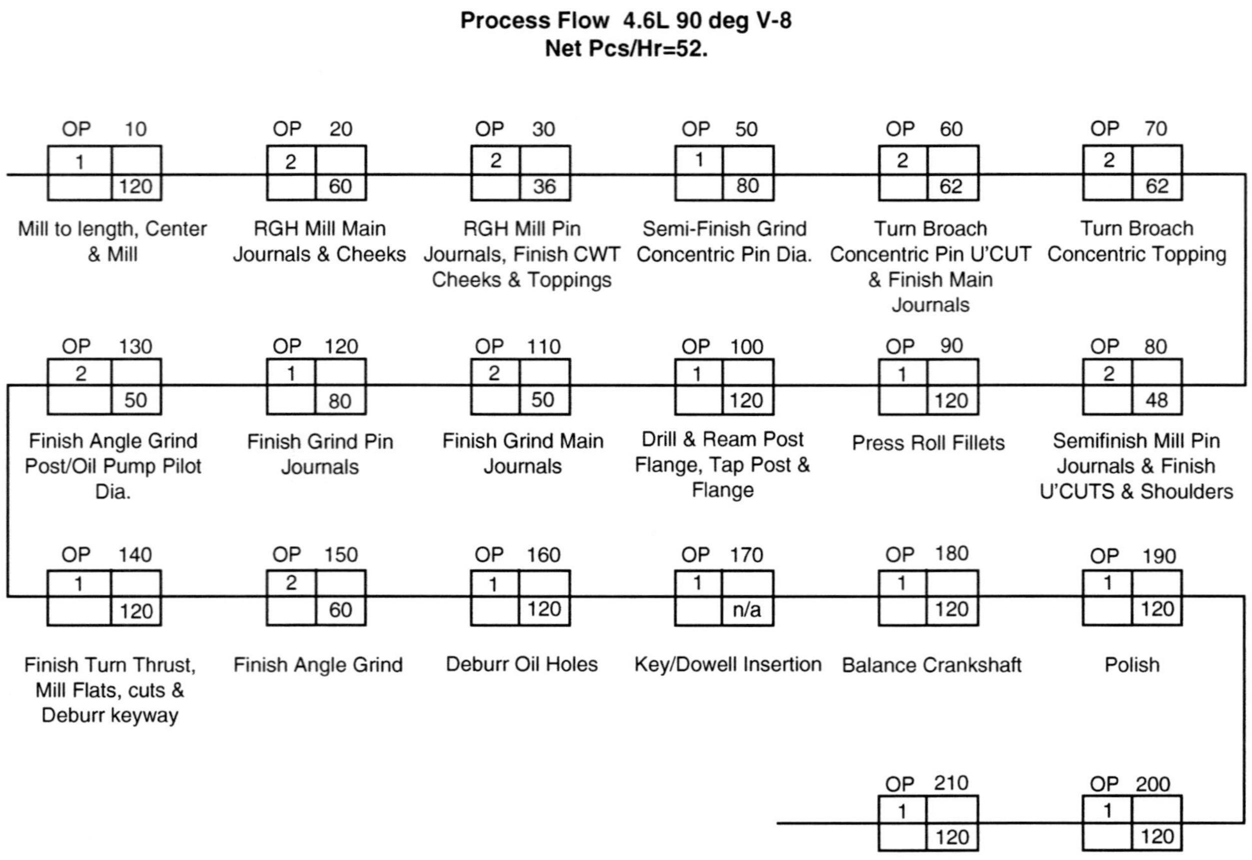

Fig. 14.13 Process flow chart generated from the process plan

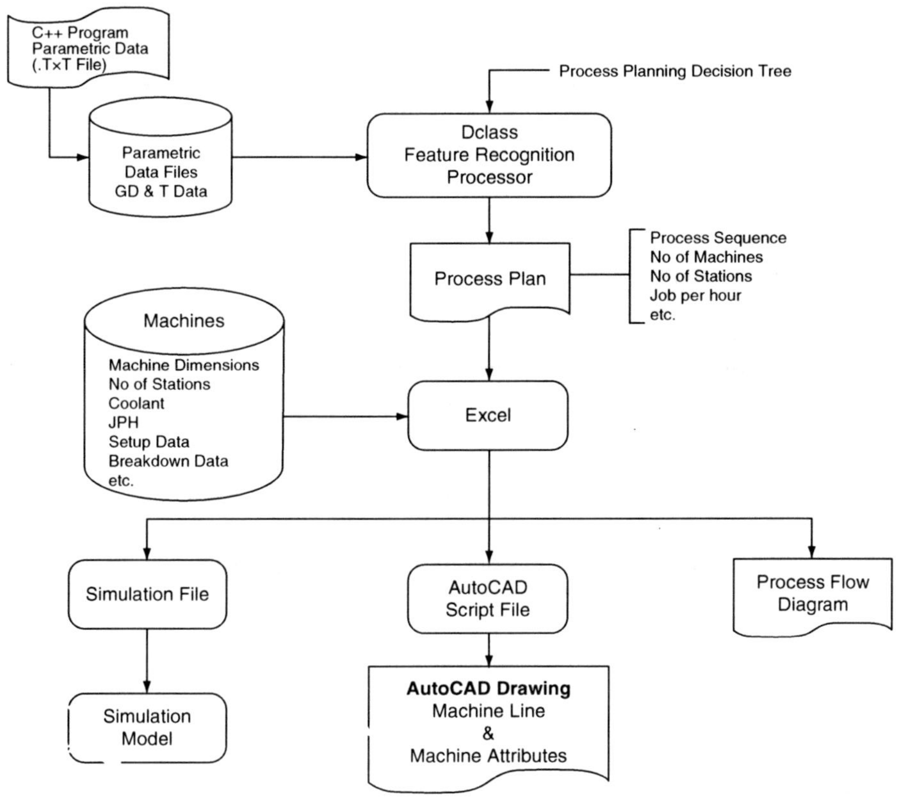

Fig. 14.14 Facility planning flowchart

These data are also presented in textual form in more detail. Figure 14.14 illustrates the steps in the development of the initial layout. The system is designed to respond to the changes made by the user. If there is change to the production volume or any other data necessary for setting the production, the process plan and the layout are quickly updated and a new is generated. The prototype is implemented using Excel®, DT tools, and AutoCAD®. This system provides: (1) the capability to import ASCII design files into the integrated design and analysis environment, (2) the rules associated with the analysis of the design data and GD&T data analysis, (3) the routine required for the material removal calculation, (4) the rules associated with process prediction and associativity analysis, (5) the GT-database structure for supporting the developed rule-based system and (6) the design structure (straw-man) necessary for the development of the process plan and the facility layout. The quality and functionality of a product during design is specified by means of surface finish, maximum allowable tolerance, geometric dimensioning, and geometric tolerancing. These specifications in an engineering design provide the technological details of a product, which are then interpreted into process specifications and used for production, production planning, and other pertinent operations. The integrated variant process-planning system will allow the designer to associate the design's technological information (e.g., GD&T, parametric data) with manufacturing features (e.g., machining, tool) using a template-based structure. The system is implemented using a decision-tree structure. This associativity allows for the analysis of manufacturability and functionality, which is performed automatically and interactively. This system is used for the implementation of route sheets by directly incorporating the parametric and GD&T data. The process thus eliminates the time-consuming and subjective manual process-planning method that is currently employed. The design for manufacturing and assembly (DFMA) process considers all aspects of the design life cycle. These may include development, total part counts, manufacturability, assembly time, modularity, etc. DFMA, through simplification, reduces cost. DFMA promotes the modularity concept. DFMA stresses the importance of part-count reduction through the identification of essential components, while modular design incorporates a function into a single component that "plugs" into the overall assembly. Modularity allows the operator to assemble a module, which can be a collection of several components providing a specific function. The functionality would otherwise have to be manually established by the assembly of components piece by piece. Appendix B illustrates an example of the process plan for the crankshaft.

14.7 Summary

In this chapter, the concepts of GT and cellular manufacturing are presented. Grouping objects into groups based on the object features has been done using GT approaches. Similar components can be grouped into design families, and new designs can be created by modifying an existing component design from the same family. The philosophy of GT is an important concept in the design of advanced integrated manufacturing systems. GT is not an answer to all manufacturing problems, but it is a good management technique to standardize efforts and eliminate duplication. A well-designed classification and coding system will result in many benefits for the manufacturing plant.

14.8 Review Questions

14.1. Define Group Technology.
14.2. What are the traditional manufacturing systems?
14.3. Discuss briefly the three types of coding structures.
14.4. Explain Rank Order Clustering (ROC) Algorithm.
14.5. Explain Cluster Identification Algorithm (CIA)
14.6. Discuss briefly similarity coefficient-based clustering
14.7. What are the benefits of the cellular manufacturing system?

14.8. Formulate machine cells formulation problem by using the learned concepts in this chapter.

14.9. Discuss briefly process planning and computer-aided process planning systems

Appendix A: Formulation Used for Material Removal of Crankshaft

Main Material Removal

DMRVOL(XCT)=LM(XCT)*3.1416*((((DM+3)/2)**2)–(((DM+FIN+GRIND)/2)**2))
DMFVOL(XCT)=LM(XCT)*3.14159*((((DM+FIN)/2)**2)–(((DM+GRIND)/2)**2))
DMGVOL(XCT)=LM(XCT)*3.1416*((((DM+GRIND)/2)**2)–((DM/2)**2))

Pin Diameter Material Removal

DPGVOL(XCT)=LP*3.1416*(((((DP+GRIND)/2)**2)-((DP/2)**2)))
DPFVOL(XCT)=LP*3.1416*((((DP+FIN+GRIND)/2)**2)–(((DP+GRIND)/2)**2))
DPRVOL(XCT)=LP*3.1416*((((DP+3)/2)**2)-(((DP+GRIND+FIN)/2)**2))

Counterweight Material Removal

CWTMVOL(XCT)(0.4)=(0.8+TMS)*SA/360*3.14159*((RCWM**2)–((DM/2)**2))
CWTTVOL(XCT)(0.4)=TCH(XCT)*3.1416*SA/360*(((RCWM(XCT)+3)*2)–RCWM(XCT)*2)
CWTPVOL(XCT)=((0.8+TPS)*3.1416)*(((SA/360)*RCWM**2)+((RMB**2)–(RPS**2)))

Post End Material Removal

PEGVOL=3.1416*(((DPO)/2)**2)*GRIND
PEFVOL=3.1416*(((DPO)/2)**2)*FIN
PERVOL=3.1416*(((DPO)/2)**2)*(2-FIN-GRIND)

Post Material Removal

DPOGVOL=LPO*3.1416*(((((DPO+GRIND)/2)**2)–((DPO/2)**2)))
DPOFVOL=LPO*3.1416*((((DPO+FIN+GRIND)/2)**2)–(((DPO+GRIND)/2)**2))
DPORVOL=LPO*3.1416*((((DPO+3)/2)**2)–(((DPO+GRIND+FIN)/2)**2))

Front Seal Material Removal

DSGVOL=LS*3.1416*(((((DS+GRIND)/2)**2)–((DP/2)**2)))
DSFVOL=LS*3.1416*((((DS+FIN+GRIND)/2)**2)–(((DS+GRIND)/2)**2))
DSRVOL=LS*3.1416*((((DS+3)/2)**2)–(((DS+GRIND+FIN)/2)**2))

Gear Material Removal

DGGVOL=LG*3.1416*(((((DG+GRIND)/2)**2)–((DG/2)**2)))
DGFVOL=LG*3.1416*((((DG+FIN+GRIND)/2)**2)–(((DG+GRIND)/2)**2))
DGRVOL=LG*3.1416*((((DG+3)/2)**2)–(((DG+GRIND+FIN)/2)**2))

Flange Diameter Material Removal

DFGVOL=LF*3.1416*(((((DF+GRIND)/2)**2)–((DF/2)**2)))
DFFVOL=LF*3.1416*((((DF+FIN+GRIND)/2)**2)–(((DF+GRIND)/2)**2))
DFRVOL=LF*3.1416*((((DF+3)/2)**2)–(((DF+GRIND+FIN)/2)**2))

Hub Material Removal

HUBGVOL=3.1416*(((DHUB)/2)**2)*GRIND
HUBFVOL=3.1416*(((DHUB)/2)**2)*FIN
HUBRVOL=3.1416*(((DHUB)/2)**2)*(2-FIN-GRIND)

Flange Main Side Material Removal

MFLGVOL=3.1416*(((((DF)/2)**2)–((DM/2)**2)))*GRIND
MFLFVOL=3.1416*((((DF)/2)**2)–(((DM)/2)**2))*FIN
MFLRVOL=3.1416*((((DF)/2)**2)–(((DM)/2)**2))*(2-GRIND-FIN)

Flange End/Hub Material Removal

FLHGVOL=3.1416*(((DF/2)**2)–((DHUB)/2)**2)*GRIND
FLHFVOL=3.1416*(((DF/2)**2)–((DHUB)/2)**2)*FIN
FLHRVOL=3.1416*(((DF/2)**2)–((DHUB/2)**2))*(2-FIN-GRIND)

Appendix B: Sample Process plan

Part ID: 5.4L/truck/95 90
Part Name: 5.4L 90 deg V-8 Date 12/09/2009
Planner Name: Ali Kamrani Time 12:55pm
Annual Volume=500,000 @ 48 weeks/year & 105h/week & 0.8 efficiency
Net Pcs/Hr = 99.
Gross Pcs/Hr = 124
Max Pcs/Yr = 625,000
Material = C38mod
Cheeking : NO ; Topping : YES ; Undercuts : YES
OPER No. Process Description

10 Step 1. MILL CRANK TO LENGTH ROUGH 569.86 mm.
Post End material removal volume = 1227.9 cu.mm.
Post End material removal volume/min = 2537.7 cu.mm.
Hub End material removal volume = 2403.1 cu.mm.

Hub End material removal volume/min = 4966.4 cu.mm.
Rough cut for post end and hub end = 2.55 mm.

Step 2. DRILL CENTER (14 gage ball)
Step 3. MILL RADIAL LOCATOR
Step 4. LIGHTENING HOLES ARE REQUIRED
Diameter of holes = 25.0 mm
Length of Holes = 55.5 mm
Drill Angle of Holes = 18 degrees
Note: Crank Rough diameter = 198.60 mm

Step 1. TURN BROACH MAIN JOURNAL TO SEMI-FINISH DIAMETER
Material removal volume = 38874.23 cu.mm for all Mains
Material removal volume/min = 8,0340 cu.mm
length of cut = 2.95 mm.
Main diameter after cut = 67.55 mm.

Step 2. TURN BROACH POST END TO SEMI-FINISH DIAMETER
Material removal volume = 8,455.320 cu.mm.
Material removal volume/min = 17474 cu.mm.
length of cut = 2.95 mm.

Step 3. TURN BROACH FLANGE TO SEMI-FINISH DIAMETER
Material removal volume = 12143.76 cu.mm.
Material removal volume/min = 25097 cu.mm.
length of cut = 2.95 mm.

Step 4. TURN BROACH MAIN COUNTERWEIGHTS
Material removal volume = 98596 cu.mm.
Material removal volume/min = 203766 cu.mm.
length of cut = 2.95 mm.

Step 5. TURN BROACH COUNTERWEIGHT TOPS
Material removal volume = 11,193.33 cu.mm.
length of cut = 2.95 mm.

30 Step 1. ROUGH MILL PIN JOURNALS
Material removal volume is 42,087.00 cu.mm for all Pins
Material removal volume/min = 86979 cu.mm.
Pin diameter after cut = 53.450 mm.
length of cut = 2.55 mm.

40 Step 1. GUNDRILL OIL HOLES
Diameter of all oil holes is = 5.5 mm.

Step 2. CHAMFER OIL HOLES
Chamfer on all oil holes; are not required mm.

50 Step 1. DEBURR ROBOT

60 Step 1. IN PROCESS WASH
70 Step 1. HEAT TREAT PINS, MAINS, AND OIL SEALS
80 Step 1. SEMI-FINISH GRIND CONCENTRIC PIN DIAMETERS
Material removal volume = 799.56 cu.mm for all Pins
Material removal volume/min = 1,652.42 cu.mm
Finished diameter = 53.00 mm
Grind length of cut = 0.05 mm

90 Step 1. TURN BROACH CONCENTRIC PIN UNDERCUTS
Undercut of pin fillet = 0.5815 mm

Step 2. FINISH MAIN SHOULDERS
Finish diameter = 0.050 mm.

100 Step 1. TURN BROACH CONCENTRIC TOPPING
Material removal volume = 11,193.33 cu.mm.
length of cut = 1.95 mm.

110 Step 1. SEMIFINISH MILL PIN JOURNALS
length of cut = 0.40 mm.

Step 2. FINISH UNDERCUTS and SHOULDERS

120 Step 1. GUNDRILL OIL HOLES
Diameter of all oil holes is = 5.5 mm.

Step 2. CHAMFER OIL HOLES; not required mm.

130 Step 1. PRESSURE ROLL FILLETS
140 Step 1. DRILL & REAM POST FLANGE

Step 2. TAP POST AND FLANGE

150 Step 1. FINISH GRIND MAIN JOURNALS
Material removal volume = 655.51 cu.mm for all Mains.
length of cut = 0.05 mm.
Finished diameter = 67.5 mm
Tolerance = ±0.011 mm.
Surface finish = 0.15/0.25
Straightness = 0.008 mm.
Circularity = 0.0075/0.004@45° mm.
160 Step 1. FINISH GRIND PIN JOURNALS
Material removal volume = 799.56 cu.mm for all Pins.
length of cut = 0.05 mm.
Finished diameter = 0.01 mm.
Tolerance = ±0.01 mm.
Surface finish = 0.15/2.5
Straightness = 0.008 mm.
Circularity = 0.0075 mm.

170 Step 1. FINISH ANGLE GRIND POST
Grind material removal volume = 136.84 cu.mm for Post.
Grind length of cut = 0.05 mm.
Finished Diameter = 31.76 mm.

Step 2. FINISH OIL PUMP PILOT DIAMETER
Grind material removal volume = 54.02 cu.mm for Oil Pin
Grind length of cut = 0.05 mm.

180 Step 1. FINISH TURN THRUST WALLS

Step 2. MILL/DEBURR KEYWAY
Step 3. MILL/DEBURR PUMP FLATS
Step 4. FINISH BORE PILOT DIAMETER
Step 5. FINSIH GEAR FACE
Step 6. FINISH #1 BEARING FACE
Step 7. DRILL/REAM DOWEL HOLE

190 Step 1. FINISH ANGLE GRIND OIL SEAL DIAMETER
Material removal volume = 60.25 cu.mm for Oil Seal.
Grind length of cut = 0.05 mm.

Step 2. FINISH ANGLE GRIND FLANGE FACE
Material removal volume = 255.36 cu.mm.
Grind length of cut = .05 mm.

Step 3. FINISH ANGLE GRIND HUB OUTSIDE DIAMETER
Material removal volume = 77.51 cu.mm.
Grind length of cut = 0.05 mm.

200 Step 1. ROBOT DEBURR OIL HOLES
210 Step 1. KEY/DOWELL INSERTION
220 Step 1. BALANCE AND POLISH CRANKSHAFT
230 Step 1. GAGE AND CLASSIFY CRANKSHAFT
240 Step 1. WASH AND RUST PROTECT

References

1. Anderberg MR (1973) Cluster analysis for applications. Academic Press, New York.
2. Arn EA (1975) Group technology: An integrated planning and implementation concept for small and medium batch production. Springer, Berlin.
3. Bedworth DD, Henderson MR, Wolfe PM (1991) Computer-integrated design and manufacturing, McGraw Hill, New York.
4. Black JT (1983) Cellular manufacturing systems reduce setup-time, make small lot production economical. Industrial Engineering 36–48.
5. Burbidge JL (1971) Production flow analysis. Prod Eng 50.
6. Degarmo EP, Black JT, Kohser RA (1984) Material and processes in manufacturing, 6th edn. Macmillan, New York.

7. Dumalien WJ, Santernm WP (1983) Cellular manufacturing becomes philosophy of management at components facility. Ind Eng 1:72–76.
8. Dutta SP, Lashkari RS, Nadoli G, Ravi T (1986) A heuristic procedure for determining manufacturing families from design-based grouping for flexible manufacturing systems. Comp Ind Eng 10 (3):193–201.
9. Eades DC (1965) The inappropriateness of the correlation coefficient as a measure of taxonomic resemblance. Syst Zool 14:98–100.
10. Gettelman K (1979) GT and MRP: Productivity. Mod Machine Shop 51(9):90–97.
11. Goodall DW (1966) A new similarity index based on probability. Biometrics 22:882–907.
12. GordonAD (1983) Classification. Chapman & Hall, London.
13. Han C, Ham I (1986) Multi-objective cluster analysis for part family functions. J Manuf Syst 5(4):223–230.
14. Kamrani AK(1995) GD&T classification impact on integrated product and process design. In: Proceedings of the 4th Industrial Engineering Research Conference, Nashville, Tennessee.
15. Kamrani AK, Sferro PR (1998) A group technology knowledge-based system for a rapid response manufacturing environment. In: Dong J (ed) Rapid response manufacturing: contemporary methodologies, tools and technologies, Chapman and Hall, London, pp 120–40.
16. Kamrani AK, Sferro PR (1995) Critical issues in design and evaluation of computer aided process planning systems. In: Proceedings of the 17th International Conference on C&IE, Miami, Florida.
17. Kamrani A (1991) A methodology for manufacturing cell design in a computer integrated manufacturing Environment. Published Ph.D. dissertation, University of Louisville.
18. Kusiak A (1990) Intelligent manufacturing systems. International Series in Industrial and Systems Engineering, Prentice Hall, Englewood Cliffs.
19. Kusiak A (1987) The generalized group technology concept. Int J Prod Res 25(4):561–569.
20. Kusiak A (1985) The part families problem in flexible manufacturing systems. Ann Oper Res 3:279–300.
21. Kusiak A, Vannelli A, Kumar KR (1986) Clustering analysis: models and algorithms. Control Cybernetics 15(2):139–154.
22. Kusiak A, Chow WS (1987) An efficient cluster identification algorithm. IEEE Transactions on Systems, Man And Cybernetics SMC-17 (4):696–699.
23. Kusiak A, Chow WS (1987) Efficient solving of the group technology problem. J Manuf Syst 6(2):117–124.
24. Kusiak A (1992) Intelligent design and manufacturing. Wiley, New York.
25. Lee R (1981) Clustering analysis and its applications. In: Tour (ed.) Advances in Information System Science, Plenum Press, New York.
26. McAuleyJ (1972) Machine grouping for efficient production. Production Engineer, pp. 51:53–57.
27. McCormick WT, Schweitzer PJ, White TW (1982) Problem decomposition and data reorganization by cluster technique. Oper Res 20(5):993–1009.
28. McManus J (1980) Some lessons from a decade of group technology. Prod Eng 59(11):40–42.
29. Ravindran A, Phillips DT, Solbert JJ (1987) Operations research: principles and practice. Wiley, New York.
30. Requicha AAG, Vandenbrande J (1988) Automated systems for process planning and part programming. In: Kusiak A (ed) Artificial intelligence: implications for CIM. IFS Publications, Kempston, pp 301–326.
31. Shafer SC, Rogers DF (1989) Similarity and distance measures for cellular manufacturing: a survey, critique, and extension. Working Paper, University of Miami, August.
32. Shah JJ (1992) Features in design and manufacturing. In: Kusiak A (ed) Intelligent design and manufacturing, Wiley, New York.
33. Singh N, Rajamani D (1996) Cellular Manufacturing Systems: Design, Planning, and Control, Chapman & Hall, London.

34. Snead CS (1989) Group technology: foundation for competitive manufacturing, Van Nostrand Reinhold, New York.
35. Tarsuslugil M, Bloor J (1979) The use of similarity coefficients and cluster analysis in production flow analysis. Proceedings of the International Machine Tool Design and Research Conference, pp 525–531.
36. Vannelli A, Ravikumar K (1986) A method for finding minimal bottle-neck cells for grouping Part–machine families. Int J Prod Res 24(2):387–400.
37. Willett P (1987) Similarity and clustering in chemical information systems, Wiley, New York.
38. Zhang HC, Alting L (1994) Computerized manufacturing process planning systems, Chapman and Hall, London.
39. Zisk BI (1983) Flexibility is key to automated material transfer for manufacturing cells. Ind Eng, pp. 58–64, November.

Index

Lightning Source UK Ltd.
Milton Keynes UK
16 September 2010

159918UK00008B/2/P

9 780387 958620